biological data mining and its applications in healthcare

SCIENCE, ENGINEERING, AND BIOLOGY INFORMATICS

Series Editor: Jason T. L. Wang
(New Jersey Institute of Technology, USA)

Published:

Vol. 1: Advanced Analysis of Gene Expression Microarray Data
(*Aidong Zhang*)

Vol. 2: Life Science Data Mining
(*Stephen T. C. Wong & Chung-Sheng Li*)

Vol. 3: Analysis of Biological Data: A Soft Computing Approach
(*Sanghamitra Bandyopadhyay, Ujjwal Maulik & Jason T. L. Wang*)

Vol. 4: Machine Learning Approaches to Bioinformatics
(*Zheng Rong Yang*)

Vol. 5: Biodata Mining and Visualization: Novel Approaches
(*Ilkka Havukkala*)

Vol. 6: Database Technology for Life Sciences and Medicine
(*Claudia Plant & Christian Böhm*)

Vol. 7: Advances in Genomic Sequence Analysis and Pattern Discovery
(*Laura Elnitski, Helen Piontkivska & Lonnie R. Welch*)

Vol. 8: Biological Data Mining and Its Applications in Healthcare
(*Xiaoli Li, See-Kiong Ng & Jason T. L. Wang*)

biological data mining and its applications in healthcare

editors

Xiaoli Li

*A*STAR, Singapore & Nanyang Technological University, Singapore*

See-Kiong Ng

*A*STAR, Singapore*

Jason T L Wang

New Jersey Institute of Technology, USA

NEW JERSEY • LONDON • SINGAPORE • BEIJING • SHANGHAI • HONG KONG • TAIPEI • CHENNAI

Published by

World Scientific Publishing Co. Pte. Ltd.
5 Toh Tuck Link, Singapore 596224
USA office: 27 Warren Street, Suite 401-402, Hackensack, NJ 07601
UK office: 57 Shelton Street, Covent Garden, London WC2H 9HE

Library of Congress Cataloging-in-Publication Data
Biological data mining and its applications in healthcare / [edited by] Xiaoli Li (A*STAR, Singapore & Nanyang Technological University, Singapore), See-Kiong Ng (A*STAR, Singapore), & Jason T.L. Wang (New Jersey Institute of Technology, USA).
pages cm
Includes bibliographical references and index.
ISBN 978-9814551007 (hardcover : alk. paper)
1. Medical informatics. 2. Bioinformatics. 3. Data mining. I. Li, Xiao-Li, 1969– editor of compilation. II. Ng, See-Kiong, editor of compilation. III. Wang, Jason T. L., editor of compilation.
R859.7.D35B56 2013
610.285--dc23

2013037382

British Library Cataloguing-in-Publication Data
A catalogue record for this book is available from the British Library.

Printed in Singapore

Preface

1. Introduction

Biologists and clinicians are stepping up their efforts in unraveling the biological processes that underlie disease pathways in the clinical contexts. This has resulted in a flood of biological and clinical data ranging from genomic sequences to DNA microarrays, protein and small molecule structures, biomolecular interactions, Gene Ontology, disease pathways, biomedical images, and electronic health records. We are in an unprecedented scenario where our capability to generate biomedical data has greatly surpassed our ability to mine and analyze the data effectively. The focus of biomedical research has shifted from data generation to knowledge discovery, or more specifically, *biological data mining*, as we enter this new era of data-intensive scientific discovery. The emphasis is no longer on enabling the biologists and clinicians to generate more data rapidly, but on converting them into useful knowledge through data mining and acting upon it in a timely manner. In this scenario, effective biological data mining will be crucial for developing better understanding of intrinsic disease mechanisms to discover new drugs and developing informed clinical decision making and support that benefit the patients.

However, in order to translate the vast amount of biomedical data into useful insights for clinical and healthcare applications, there are data mining difficulties that have to be overcome such as handling noisy and incomplete data (e.g. notoriously noisy gene expression data and protein-protein interactions data, or PPIs, with high false positive and false negative rates), processing compute-intensive tasks (e.g. large-scale graph indexing, searching, and mining), integrating various data sources (e.g. linking genomic and proteomic data with clinical databases) and exploiting biomedical data with ethical and privacy protection. All these issues pose new challenges for data mining scientists working in the data-intensive post-genomic era.

The objective of this book is to disseminate inspiring research results and exemplary best practices of data mining approaches to cross-disciplinary researchers and practitioners from the data mining disciplines, the life sciences and healthcare domains. The book will cover the fundamentals of data mining techniques designed to tackle the data analysis challenges mentioned above, and demonstrate with real applications how data mining can enable biologists and healthcare scientists to make insightful observations and invaluable discoveries from their biomedical data.

Each chapter of this book will start with a section to introduce a specific class of data mining techniques, written in a tutorial style so that even non-computational readers such as biologists and healthcare researchers can appreciate them. This is then followed by a detailed case study on how to use the data mining techniques in a real-world biological or clinical application. In this way, we hope that the readers of this book can be inspired to apply the computational techniques to address their own research challenges.

It is our hope that researchers in the life sciences and healthcare will find *in silico* data mining tools useful, which can complement their work *in vitro* and *in vivo*. At the same time, we also hope that computer scientists, mathematicians, and statisticians will be attracted by the new data mining challenges that are important for knowledge discovery in biomedical research, and join their life sciences and healthcare colleagues in their impactful research to benefit humankind.

2. Data Mining and its Applications in Healthcare

Data mining has become a crucial research area for life sciences and healthcare applications. It focuses on the methodologies and processes for automatically or semi-automatically extracting useful patterns, insights or knowledge from large amount of data. Its importance has risen with the rapid growth of online information on the Web as well as databases used by industries (e.g. various biological or clinical databases), resulting in the increasingly urgent need to derive novel knowledge from the large amounts of data.

Data mining (or knowledge discovery) is itself a multi-discipline area, involving machine learning, statistics, artificial intelligence, databases, pattern recognition, and data visualization. Generally, data mining consists of an iterative sequence of the following steps [1]:

1) **data cleaning**: removing noise and inconsistent data from the original data;
2) **data integration**: combining multiple data sources consistently;
3) **data selection**: identifying and retrieving only the data that are relevant to the analysis task from the database;
4) **data transformation**: transforming and consolidating data into a format that is appropriate for mining by performing summary or aggregation operations;
5) **data mining for knowledge discovery**: applying intelligent machine learning methods to extract data patterns - this is the key process in knowledge discovery;
6) **pattern evaluation**: identifying interesting patterns that represent useful knowledge based on interestingness measures;
7) **knowledge presentation**: using intuitive visualization and effective knowledge representation techniques to present the discovered knowledge to the user.

Steps 1-4 are typically considered as data pre-processing steps for preparing the given data for the data mining task in Step 5 [2]. Steps 6 and 7 are often referred as decision support because they involve choosing interesting patterns/knowledge and presenting them to the user in a user-friendly manner for users to do further studies as well as decision making.

Step 5, which is the key process for knowledge discovery, typically involves one or more of the following tasks [3]:

- Association rule mining (Dependency modeling): This is invented and extensively studied by the data mining community. Its task is to detect relationships/associations between variables/features/items. One classic example is market basket analysis in which associations in supermarket customers' purchasing habits are mined from their sales transaction records ("market baskets") that can be translated useful insights for market campaigns. For example, the association rule {bread}=>{milk} can be mined in most supermarket data, indicating that if a customer has bought bread, he or she is likely to buy milk as well. With the association rules mined from the shoppers' transactional records, a supermarket/shop can automatically detect which products are frequently bought together and use this knowledge for marketing purposes, e.g., promotional bundle pricing or product placements. In the biomedical domain, association rule mining can be performed on gene expression data (or other medical data) to discover association rules where the antecedents are the biological features and their value ranges (cancer genes and corresponding gene expression values under different conditions; clinical test and corresponding readings/values) and the consequents are the class labels (cancer or non-cancer). The knowledge discovered can then be used for building a diagnostic system to assist doctors for decision making.

- Cluster analysis or clustering: This task's objective is to segment a set of objects into groups such that the objects within the same group are more similar to each other than compared to those in other groups. In machine learning, cluster analysis is a form of *unsupervised* learning, since there is no need for users or domain experts to provide training examples for clustering algorithms. One example use of cluster analysis in biomedical research is to segment gene expression data into groups where genes in each group share similar gene expression profiles, in order to discover genes that have the same biological functions.
- Classification: This task's goal is to assign the given input data into one of a known number of categories/classes. A classic example is spam filtering, in which the task is to classify a new email as a legitimate message or just spam. To build a classifier, a user must first collect a set of training examples that are labeled with the predefined known classes (e.g., known spam messages and different types of legitimate email messages). A machine learning algorithm is then applied to the training data to build a classification model

(classifier) that can be employed subsequently to assign the predefined classes to examples in a test set (for evaluation) or future instances (in practice). An application of classification in biomedical research is to predict the biological functions of novel proteins or genes. Here, proteins with known biological functions are first used as training examples to build a classification model which can be subsequently used to classify unknown proteins into one or more biological families with different functions. In this book, we also introduce the use of classification methods for predicting protein-protein interactions as well as drug-target interactions.

- Regression analysis: This task aims to find a mathematical function which models the data with the least error, where the focus is on the relationship between a dependent variable and one or more independent variables. Similar to classification, regression also requires training examples for building a regression function. In this case, each training example is associated with a numerical value instead of a class label as in the classification scenario. The difference between regression and classification is that regression handles numerical or continuous class attributes, whereas classification handles discrete or categorical class attributes. In this book, we have a chapter for trend analysis where the regression analysis is used as one of trend analysis techniques.
- Anomaly detection (Outlier detection): In this task, we attempt to detect data records/examples which do not conform to an expected or established normal behavior. The results could be interesting data records or erroneous records which require further investigation.

3. Organization of the Book

This book is organized into 4 major parts, covering the following topics: Sequence Analysis (Part I), Biological Network Mining (Part II), Classification, Trend Analysis and 3D Medical Images (Part III), and Text Mining and its Biomedical Applications (Part IV).

Part I: Sequence Analysis

Chapter 1, entitled "Mining the sequence databases for homology detection: application to recognition of functions of trypanosoma brucei brucei proteins and drug targets," utilizes sensitive profile-based sequence database search algorithms to recognize evolutionary related proteins when the amino acid sequence similarity is very low in the repertoire of functions and 3D structures of parasitic proteins in *Trypanosoma brucei brucei*, a causative agent of African sleeping sickness. The information of parasitic proteins in metabolic pathways and their homology to the targets of FDA-approved drugs are then combined for recognizing attractive drug targets.

While the genomes of many organisms have been sequenced, deciphering them is a herculean task due to the tremendous amount of data and also the complexity of the genetic regulation mechanism. While biochemical experiments can be used to detect the genes and their regulatory regions, these experiments are very expensive and time-consuming to conduct. Chapter 2, entitled "Identification of genes and their regulatory regions based on multiple physical and structural properties of a DNA sequence," addresses this by applying signal processing and machine learning techniques for identifying genes and their regulatory regions.

The advent of next generation sequencing has led to a further deluge of biological sequence data. Although pairwise statistical significance between sequences can accurately identify related sequences (homologs), it is computationally intensive when the datasets are huge. Chapter 3, entitled "Mining genomic sequence data for related sequences using pairwise statistical significance," resorts to high performance computing techniques for accelerating the computation.

Part II: Biological Network Mining

Biological networks describe how different biochemical entities interact with one another to perform vital functions in an organism. The ability to data mine such networks efficiently is important for biological knowledge discovery. Chapter 4, entitled "Indexing for similarity queries on biological networks," discusses three network indexing methods, namely feature based indexing, tree based indexing and reference based indexing, for efficiently accessing and querying these biological networks. It also provides an experimental comparison of applying the three methods for querying gene regulatory networks in terms of running time and accuracy.

Chapter 5, entitled "Theory and method of completion for a Boolean regulatory network using observed data," addresses the knowledge completion problem on Boolean networks, in which the given existing knowledge is modified to be consistent with observed data. It also detects topological changes in signaling pathways after cell state alteration, and presents an integer programming based method. The proposed method is applied to a data set of gene expression profiles of colorectal cancer downloaded from the Gene Expression Omnibus and the signaling pathway data of colorectal cancer downloaded from the KEGG database.

For biological graph mining, the problem of discovering interesting subgraph patterns in a database of graphs or a single graph is an important one, as the set of interesting subgraphs has immediate applications in graph clustering, graph classification, and graph indexing. Chapter 6, entitled "Graph mining and applications," studies the frequent subgraph pattern mining problem and the problem of mining a summarized set of frequent patterns. As case studies of biological graph mining techniques, we describe how to apply frequent subgraphs in chemical compound classification and mining family-specific protein structural motifs.

Chapter 7, entitled "On the integration of prior knowledge in the inference of regulatory networks," studies the integration of prior knowledge and genomic data for the inference of regulatory networks. In this chapter, we show how to apply existing tools to retrieve prior knowledge from different sources such as PubMed and structured biological databases. Then, we use network inference methods to combine genomic data and prior knowledge. A case study based on two publicly available cancer data sets demonstrates the usefulness of prior knowledge for network inference.

Part III: Classification, Trend Analysis and 3D Medical Images

In Chapter 8, entitled "Classification and its application to drug-target prediction," we discuss how to apply classification methods to a critical problem in drug discovery pipeline – drug-target prediction. The predicted drug-target interactions may lead to the discovery of new drugs that interact with desired protein targets. They are also useful for understanding the possible causes of side effects of existing drugs that could interact with undesired protein targets.

Protein interaction research is useful for predicting proteins' cellular functions, providing insights into disease mechanisms, and developing new drugs to prevent the diseases. Given that the current human protein interaction map is far from complete, Chapter 9, entitled "Characterization and prediction of human protein-protein interactions," introduces how to apply classification methods to predict novel human protein-protein interactions to increase the coverage of the human interactome.

Chapter 10, entitled "Trend analysis," provides a survey on six commonly used techniques for trend analysis, including Age-period-cohort model, Joinpoint regression, Time series analysis, Cox-Stuart trend test, RUNS test, and Functional data analysis. Trend analysis can be defined as the practice of collecting data and attempting to spot a certain pattern, or trend in the data. Some successful applications of the trend analysis techniques are presented in the chapter as case studies.

In recent years, three dimensional (3D) medical imaging technique has rapidly advanced and poised to become a major improvement in patient care. From the 3D models of a patient, anatomical structures can be identified and extracted, and diagnosis and surgical simulation can be customized. Computational reconstruction of 3D models is critical for medical diagnosis and treatment, but the complex imaging processing requires considerable resources and advanced training. Chapter 11, entitled "Data acquisition and preprocessing on three dimensional medical images," introduces 3D image acquisition, segmentation and registration to effectively transform the unstructured image data into structured numerical data for further data mining tasks.

Part IV: Text Mining and its Biomedical Applications

Chapter 12, entitled "Text mining in biomedicine and healthcare," provides an overview of essential text-mining technologies that can be applied in biomedicine and healthcare. These include: entity recognition, entity linking, relation extraction, and co-reference resolution. For demonstration, a case study of a biomedical text mining database system that automatically recognizes and collects cardiovascular disease related genes is presented in this chapter.

Chapter 13, entitled "Learning to rank biomedical documents with only positive and unlabeled examples: a case study," applies positive and unlabelled learning algorithms to an RNA-protein binding dataset collected from PubMed. Different from traditional machine learning, PU learning studies how to learn a model with positive and unlabelled data as they are readily available in many real-world applications. The chapter systematically studies how feature selection and the proportion of positive examples in unlabelled data affect the performance of the PU learning methods.

Elucidation of protein interaction networks is essential for understanding biological processes and mechanisms. Significant information regarding protein interaction networks are available in the literature as free-text. However, such information is difficult to retrieve and synthesize. Chapter 14, entitled "Automated mining of disease-specific protein interaction networks based on biomedical literature," describes machine learning techniques and web-based tools for protein interaction extraction. This chapter also provides two case studies to illustrate the use of automated text-mining in biomedical research.

4. Conclusions

Biological data mining is becoming increasingly integrated into the entire pipeline of biological and medical discovery process. There is now an ever-pressing need for data mining researchers to collaborate with the biologists and clinical scientists. Data mining researchers have now the opportunity to contribute to the development of the life and clinical sciences by creating novel computational techniques for discovering useful knowledge from large-scale real-world biomedical data. In fact, there are abundant opportunities for data mining researchers to cross over from the computation domain into the biomedical domain to contribute to the meaningful scientific pursuit with the biologists and healthcare scientists, as shown by the various chapters in this book.

The ultimate success of the biological data mining for healthcare applications will therefore depend on parallel improvements both in the biological and clinical experimental techniques from the biologists and clinicians to provide rich and clean biological/clinical datasets for data mining community, and in the advanced computing techniques from the computer scientists, mathematicians, and statisticians to provide efficient and effective ways to exploit the data for knowledge discovery. The ultimate goal is to enable biologists and clinicians to better understand the biological data mining techniques and to be able to apply these techniques in their journeys to understand life processes and build better healthcare applications to benefit all of us.

We are grateful to the authors who contributed to the exciting and important research topics of developing biological data mining approaches for applications in biology and healthcare. Our heartfelt thanks also go to our book reviewers who have provided very useful feedback and comments, and to the publishing team at World Scientific for providing invaluable contributions and guidance throughout the whole process from inception of the initial idea to the final publication of this book.

Xiao-Li Li, See-Kiong Ng and Jason T.L. Wang

References

1. J. Han and M. Kamber, *Data Mining: Concepts and Techniques*, Morgan Kaufmann, 2005.
2. U. S. Fayyad, P. S. Gregory and S. Padhraic, "From Data Mining to Knowledge Discovery in Databases," *American Association for Artificial Intelligence*, 1996.
3. X.-L. Li and S.-K. Ng, *Computational Knowledge Discovery for Bioinformatics Research*, USA: IGI Global, 2012.

Contents

1. Sequence Analysis

Chapter 1

Mining the Sequence Databases for Homology Detection: Application to Recognition of Functions of *Trypanosoma brucei brucei* Proteins and Drug Targets

G. Ramakrishnan[1, 2], V.S. Gowri[2, ‡], R. Mudgal[1, 2], N.R. Chandra[3] and N. Srinivasan[2]

[1]*Indian Institute of Science Mathematics Initiative,*
[2]*Molecular Biophysics Unit,*
[3]*Department of Biochemistry, Indian Institute of Science, Bangalore-560012, India.*
[‡]*Present Address: School of Life Sciences, Jawaharlal Nehru University, New Delhi- 110067, India*

With the amount of data deluge as a result of high-throughput sequencing techniques and structural genomics initiatives, there comes a need to leverage the large-scale data. Consequently, the role of computational methods to characterize genes and proteins solely from their sequence information becomes increasingly important. Over the past decade, development of sensitive profile-based sequence database search algorithms has improved the quality of structural and functional inferences from protein sequence. This chapter highlights the use of such sensitive approaches in recognition of evolutionary related proteins when the amino acid sequence similarity is very low. We further demonstrate the use of sequence database mining based remote homology detection methods in exploring the repertoire of functions and three dimensional structures of parasitic proteins in *Trypanosoma brucei brucei*, causative agent of African sleeping sickness. With an emphasis on various metabolic pathways, sequence-function and structure-function relationships are investigated. Integrating the information of parasitic proteins in metabolic pathways along with their homology to targets of FDA-approved drugs, attractive drug targets have been proposed.

1. Introduction

Over 17 million protein sequences have been deposited in the public databases, and this number has been growing rapidly (http://ncbi.nlm.nih.gov/RefSeq). On the other hand, over 80,000 protein structures have been experimentally determined so far. With the rapidly growing disparity in the data deluge, the role of computational methods to characterize the function of the proteins from their sequence becomes increasingly important.

The intimate relationship between protein structure and its function has led to the view that a reliable prediction of the structure of a protein from its sequence could give useful insights on its function. Many structure-based approaches for function prediction today do provide reliable models for a substantial fraction of the protein space[1]. The widening gap between deluge of sequences and the experimental characterization of the respective proteins necessitates the utility of sequence-based as well as structure-based sensitive remote homology search techniques in identifying evolutionary relationships.

This chapter starts with description of remote homology detection methods for proteins with a focus on recognition of evolutionarily related proteins when the amino acid sequence similarity is very low. In particular, the power of profile-based sequence database search techniques and techniques involving matching of Hidden Markov Models (HMM) of protein domain families will be outlined. We will provide description of these extremely sensitive sequence search techniques in simple language with the tone of tutorial.

We shall then demonstrate the use of such techniques on the proteins encoded in the genome of *Trypanosoma brucei brucei* (one of the subspecies of *Trypanosoma brucei*) which causes African sleeping sickness. African trypanosomes are parasitic protozoa that belong to the class Kinetoplastida. These protozoan parasites are important mainly because of their energy metabolism. The metabolic enzymes of trypanosomes are very different from the host enzymes as these are localized in a specialized organelle called 'glycosomes'. The life cycle of *Trypanosoma brucei* involves the insect host which is the tsetse fly and the mammalian host such as human, cattle and other life forms, depending on the type of subspecies infecting the host.

By application of sequence database mining based remote homology detection methods, we will recognize the repertoire of functions and three dimensional structures of proteins of the parasite, followed by integrating this dataset with information on various metabolic pathways. We will then combine the information of parasitic proteins in metabolic pathways with their homology to the targets of FDA-approved drugs, thereby recognizing attractive drug targets.

2. Remote homology driven approaches for protein function annotation

Database similarity searches have become a mainstay of bioinformatics, where protein homology detection plays a pivotal role in understanding the evolution of protein structures, functions and interactions. Many of the developments in protein bioinformatics can be traced back to an initial step of homology detection.

Studies on evolution are largely influenced by protein homology detection and two proteins are said to be homologous if there exists a protein, an ancestor, from which these two proteins have evolved. Homology in a literal sense means descent from a common ancestor. To determine the likeliness of two proteins being evolved from a common ancestor, calculations according to the model of evolution could be assessed, which when high, can very well support the likelihood within the framework of the evolutionary model.

When a protein sequence is found to be homologous to a protein of known function, then it raises the possibility of both sharing functional features. This follows from the fact that functional residues are usually conserved during the course of evolution, and hence evolutionarily related proteins show high functional similarity.

Efforts have been made for decades to explore closely related homologues for protein function annotation, and detecting remote (distantly related) homologues in order to explore protein sequence/ structure space. The progress made in remote homology detection will be described in the following subsections, highlighting the use of extremely sensitive sequence search techniques. In the interest of the scope of this chapter, the description of each technique is limited to the basic understanding of principles of the algorithm employed. Readers are encouraged to refer to the original publications for details in mathematical basis of the techniques.

2.1. *Sequence-based approaches for remote homology detection*

Identification of well-characterized homologues of protein sequences is usually identified by matching pairs of sequence and the most widely used tool for sequence comparison and database searching is BLAST (Basic Local Alignment Search Tool)[2]. Chances of reliable detection of evolutionary relationships become smaller when the sequence identities of the related proteins go below 30%[3]. To improve the effectiveness of remote homology detection, sensitive search procedures based on the use of profiles such as Hidden Markov Models (HMMs) and Position Specific Scoring Matrices (PSSMs) were developed. Description of such sensitive methods followed by the assessment of significant sequence alignments are presented in the following subsections.

2.1.1. *Iterated searches using PSI-BLAST*

Position-Specific Iterated (PSI)-BLAST is a protein sequence profile search method that is far more capable of detecting remote homologues[4] than single query alone, as in BLASTp. The power of the profile methods is enhanced through iterative search procedure. The procedure[5] involves construction of multiple sequence alignment (MSA), followed by a PSSM profile from the statistically significant hits found for the query in an initial BLAST search. Such a position-specific matrix can be thought of as a consensus sequence used to detect more distantly related proteins not identified by BLAST alone. The original query sequence serves as the template for both the MSA and the PSSM, whose lengths are identical to the query. The profile is constructed based on substitutions occurring in the homologues of the query which represents a more accurate model. This profile is re-searched against the database which drastically improves detection of remote homologues[4]. The procedure is iterated as often as desired or until convergence when no new statistically significant hits are found.

While PSI-BLAST is quite powerful in general, many developments have been made to improve the algorithm to overcome the limitations posed by the heuristics employed for speed of execution and sensitivity as well. The recent developments include use of compositionally biased profiles[6], multiple profiles[7-9] from alignments[10], intermediate sequence search approach[11,12] and use of sequence context-specific profiles[13].

Query dependence of the profiles generated in PSI-BLAST can be circumvented by the use of multiple profiles developed by our group, which forms the focus of the next subsection.

2.1.2. *Multi-profiles approach to improve sensitivity*

Before taking a look at the applicability of the algorithm to protein domain families, for structure/function annotation to a protein sequence, we shall brief on protein domains, domain families and domain family databases.

Domains are distinct functional and/or structural units of a protein, responsible for a function/interaction, contributing to the overall role of a protein. Typically there are domains of four types:

- Structural domains: Some polypeptide chains fold into two or more compact regions that may be connected by a flexible segment of polypeptide chain. These compact globular units are structural domains.
- Functional domains: These are associated with a specific biochemical function and could comprise of more than one compact unit.
- Folding domains: These are modules in a multidomain protein which are capable of folding into a native-like form, independent of other domains in the protein.
- Sequence domains: These correspond to the existence of homologues of a given protein either as a part of a larger protein or as single domain.

Based on the similarities between protein domains, derived considering structure, sequence, and evolutionary information, domains are grouped into domain families often organized into hierarchies. SCOP[14,15] (Structural Classification of Proteins) database classifies protein domains based on structural similarity and clusters them in a hierarchical manner, beginning from the top as "Class", "Fold", "Superfamily" and "Family". Classification scheme dependent on sequence information alone is comprehended by Pfam[16,17] database, which aims in providing comprehensive collection of sequence domains, defined based on their biochemical functions and amino acid sequences. Both the databases form the main resources in annotating proteins of *T. brucei* which will be elaborated in the next section.

As described in the previous section, use of "dynamic" PSSMs, as in PSI-BLAST, is known to be extremely effective in the detection of distant homologues of a protein in sequence database. Complementary to PSI-BLAST, a reverse approach that enables rapid comparison of a query sequence against a library of PSSMs known as RPS-BLAST[18] (Reverse PSI-BLAST), similar to IMPALA[19] (Integrating Matrix Profiles And Local Alignments), has also been shown to be effective in facilitating functional annotation of a sequence in a sensitive manner. Such an approach forms a basis for the development of database of multiple profiles, i.e. MulPSSM.

MulPSSM[7-9] (http://mulpssm.mbu.iisc.ernet.in), developed by our group earlier, is a database of PSSMs for a large number of sequence and structural families of protein domains with multiple PSSMs for every domain family. An RPS-BLAST based use of such multiple PSSM profiles corresponding to an alignment, with multiple sequences in a domain family used as reference, improves the sensitivity of the remote homologue detection and function annotation dramatically[7-9]. Current release of MulPSSM comprises of 385,258 profiles for 13,672 sequence based domain families obtained from Pfam (version 26.0) and 14,235 profiles corresponding to 3,856 structural families in PALI[10] database based on SCOP (version 1.75).The number of domain assignments done with the help of multi-profiles approach is reportedly higher than that domains identified using single profile approach[7-9].

The following subsection shall highlight another development involving the employment of intermediately related sequences in detecting remote relationships between proteins.

2.1.3. *Cascade PSI-BLAST*

When sequences diverge to the extent that they retain little "memory" of each other i.e. beyond the point of being recognized as homologues by simple direct comparison, they can be related through a third sequence that is suitably intermediate between them[20-22]. Cascade PSI-BLAST[11] rigorously employs intermediate sequences in detecting remote relationships between proteins. In this approach, relationships are detected using PSI-BLAST which involves multiple rounds of iteration. An initial set of homologues are identified for a protein in a "first generation" search by querying a database. A "second generation" search in the database is propagated, using each of the homologues identified in the previous generation as queries which facilitates the recognition of homologues not detected earlier. This non-directed search process can be viewed as iteration of iterations which is continued until no new hits are detectable. An assessment of this cascaded approach on diverse folds showed an improved performance by detecting 15% more relationships within a family and 35% more relationships within a superfamily. Such a propagated search would aid in distant homologue detection effectively.

An application of the ideology described above was employed in generation of "protein-like" artificial sequences that could serve to bridge gaps in protein sequence space[12]. These designed sequences function as intermediates in linking clusters of distantly related proteins which facilitate detection of distant and non-obvious similarities.

Apart from the use of PSSMs for effective identification of remote homologues, use of hidden Markov models have also been proved to be effective, and their importance in terms of speed and sensitivity forms the basis of further subsections.

2.1.4. *Hidden Markov Models*

To make the most of the sequence data, maximizing the power of computational sequence comparison tools to detect remote homologues would facilitate in understanding clues regarding structure, function and evolution. Hidden Markov Models[23,24] or HMMs, have been shown to give better sensitivity and alignment accuracy than PSSM-based methods[25]. Probabilistic modeling of gaps, unlike PSSMs and the estimation of conservation of amino acids provide a fine-tuned framework to detect homology. This enables construction of sophisticated profiles, thus contributing significantly to the improved performance of the HMMs. In the past, Eddy and coworkers, who have been developing and maintaining the HMMER suite of programs, (http://hmmer.janelia.org) have made significant improvements towards the heuristics in the acceleration pipeline implemented by HMMER, thus enabling HMMER3 to be 100-fold faster and significantly more powerful in performance. Such advances ensure effectiveness in detection of homologues for protein annotation efforts.

2.1.5. *Profile-profile matching algorithms*

The sensitivity of sequence-profile methods and their developments as described in previous sections has been fairly efficient in identifying distant homologues. An increase in sensitivity of remote homology detection is achieved, if instead of alignment between sequence and profile, two profiles are aligned. Since the HMM profiles use more information per MSA column than typical PSSM based profiles, alignment of HMM profiles is a bit difficult exercise. The earliest attempt was made in COACH[26] method, which aligns an MSA with a profile.

Development of HHsearch[27] method facilitated the development of more sophisticated methods for profile-profile matches. The principle behind this method was to perceive the alignment between two profiles as an alignment between states of HMMs, co-emitting an amino-acid sequence. The match between two columns of the profiles is calculated on the basis of log of the sum of odds ratio (ratio of the probability of amino acid to its random probability) for each of the 20 amino acid types. Additionally, the column score also takes into account the confidence in a given predicted secondary structure state. If the confidence in both the profiles are high and equal as well, then the match score is enhanced by a positive score, which is otherwise penalized for misalignment. This ensures confidence in the correct alignment, wherever the secondary structure is predicted with high accuracy. HHsearch has been employed in the algorithm of HHPred[27] which has been consistent in being amongst top-ranked servers for structure prediction[1,28,29].

A recent development in HMM-HMM match with an improved sensitivity and alignment accuracy is AlignHUSH[30] (http://alignhush.mbu.iisc.ernet.in/). This algorithm uses residue conservation, structural information in the form of predicted secondary structure probabilities, taking into account the hydrophobicity of amino acids of not only the columns being aligned, but also adjacent columns, thus enabling it to differentiate HMMs with similar secondary structure but different topology.

Employing this approach onto sequence and structural domain family databases (Pfam and SCOP respectively), two kinds of relationships are derived:

- Pfam-SCOP: Relating Pfam domain families to family with known structural information.
- PNSFs: Identification of related Pfam domain families forming potentially new superfamilies (PNSFs).

SUPFAM database (http://supfam.mbu.iisc.ernet.in/) developed, that comprehends the above derived relationships[31,32], has been of profound use for annotation of uncharacterized protein families (UPFs) or domain families of yet unknown structure/function (DUFs in Pfam database).

Another very recent development, improvising on accuracy and speed of profile-profile alignment is HHblits[33]. This approach is an HMM-HMM-based lightning-fast iterative sequence search which extends HHsearch to enable the same. The protocol involves the conversion of query (or input MSA) to an HMM, which is performed by adding pseudocounts of amino acids that are physicochemically similar to amino acids in the query, depending on local sequence context[13] (13 positions around each residue), which improves the alignment quality and sensitivity considerably. HHblits then searches, the HMM database generated by clustering large sequence database and adds sequences from HMMs that are statistically significant to the query MSA, from which HMM for next iteration is built. Instead of implementing insertion and deletion probabilities of the HMMs, 20 amino acid probabilities in each HMM column are discretized into an alphabet of 219 letters. The score of each query HMM column is then calculated with each of the 219 letters, which results in 219-row extended sequence profile. Such an extended query sequence profile is aligned to extended database of sequences with the help of prefilters. Statistically significant hits that have passed the prefilter are realigned with a local maximum accuracy algorithm. Exploiting mathematical advantages of probabilistic modeling, this method takes its credit in a better alignment quality, speed of execution not compromising on sensitivity[33].

Table 1 summarizes the progress made in development of tools from sequence alignment to the use of sensitive profile-based approaches for effective remote homologue detection, discussed so far. Such advancements in detection of remote homologues facilitate the exploration of unknown repertoire in protein sequence space.

2.2. *Assessment of significant sequence alignments*

To interpret biological relevance of a finding it is important to know if the hit constitutes evidence for homology and the likeliness that it is expected by chance. Measures such as score, bit-score, p-value and e-value aid in assessing the statistical significance of a sequence alignments[2]. A score *S*, is a numerical value that describes overall quality of the alignment, which depends on scoring systems used (substitution matrix, gap penalty). Higher is the score, higher the similarity, and better the alignment.

Table 1. Summary of the techniques discussed.

Technique	Web address	Algorithm	Limitations	References
BLAST	http://blast.ncbi.nlm.nih.gov	Determines statistically significant regions of local similarity	Reliable detection of homologues becomes skewed at very low sequence identities (<30%)	[2, 3]
PSI-BLAST	http://blast.ncbi.nlm.nih.gov	More effective than BLAST. Employs iterative searches with the help of PSSMs.	Questionable sensitivity for increasing number of sequences in the PSSM.	[4, 34]
Cascade PSI-BLAST	http://cascade.mbu.iisc.ernet.in	Rigorously employs intermediate sequences to detect remote homologues.	Heuristics employed in speed.	[11]
MulPSSM	http://mulpssm.mbu.iisc.ernet.in	RPS-BLAST, a reverse approach for rapid comparison of query against a library of PSSMs.	Speed. Increase in number of multiple profiles per protein family improvises sensitivity of the approach, but compromises on speed of execution.	[7-9]
HMMER	http://hmmer.janelia.org	With probabilistic modeling of gaps, HMMs serve a fine tuned framework to detect homology.	An option for profile-profile alignment for enhanced sensitivity.	[23-25]
HMM-HMM match: HHPred, AlignHUSH, HHblits	http://toolkit.tuebingen.mpg.de/hhpred http://alignhush.mbu.iisc.ernet.in/ http://toolkit.tuebingen.mpg.de/hhblits	Determining statistical significance of the alignment between states of HMMs taking structural information into account has proven to ensure confidence in accuracy (HHPred). This is further enhanced with the incorporation of hydrophobicity information of the aligned columns and their adjacent columns (AlignHUSH). Employing an iterative HMM-HMM search with the help of context-specific profiles, the improvisation has proven to be significantly more powerful in performance (HHblits).		[13, 27, 30, 33]

Bit score *S'*, is a normalized score that allows to estimate the magnitude of the search space to be searched through, to get a score better than *S*, by chance. *S'* is given by,

$$S' = \frac{\lambda s - lnk}{ln2} \tag{1}$$

K and λ are statistical parameters that can be thought of as natural scales for the search space size and the scoring system respectively. P-value is a "dissimilarity score" associated to score *S* and it is the probability to obtain a score atleast equal to *S* by chance. E-value (Expectation value) associated to score *S*, gives a measure on the number of hits, with a score equivalent to or better than *S*, that are expected to occur in a database by chance. Lower the E-value, the more significant the score is. E-value and P-value are inverses of one another in terms of statistical measures. E-value is given by,

$$E = Kmne^{-\lambda S} \tag{2}$$

Where *m* is the length of database, *n* is the length of query sequence, *K* and λ are statistical parameters, *S* is the score. E-value and bit-scores as given in equations (1) and (2) are the widely used measures in assessing the significance of a hit to the query protein sequence.

Sensitive search approaches discussed in previous subsections may be very powerful in detecting evolutionarily distant homologues, but have their own error-rates in distinguishing a noise from a true hit. Thus, idealizing one approach for the purpose of protein function annotation is unjustified. By employing multiple sensitive sequence search approaches, wherein more than one approach identifies a true hit for the query, the possibilities of occurrence of false positives populating results can be reduced.

The next section would demonstrate the use of such sensitive search techniques in exploring genome-wide structural and functional repertoire of proteins in encoded in the genome of *Trypanosoma brucei.*

3. Trypanosoma brucei: A case study

The protozoan flagellate parasite *Trypanosoma brucei* is a causative agent of human African trypanosomiasis (HAT), also known as sleeping sickness, which primarily affects the poorest rural populations in some of the least developed countries in Central Africa (http://www.who.int/trypanosomiasis_african/parasite/en/). It remains a major threat, with 300,000 to 500,000 cases per year, and is invariably fatal if left untreated. The current treatments are inadequate, with the four available drugs (pentamidine, suramin, melarsoprol, eflornithine) being unsatisfactory because of low efficacy, severe side effects, and difficulty in administration[35,36]. Moreover, with the absence of prophylactic chemotherapy and little or no prospect of vaccine, there comes a need for new therapeutics for the treatment of HAT.

Transmission of the trypanosome is aided by an insect vector, the tsetse fly. The disease progresses from a bite from an infected tsetse fly, wherein the trypanosome makes a series of transitions to complete its life cycle. The differentiations in its life cycle, in different environments pertaining to insect and mammalian host, produce characteristic cell types defined by their properties to overcome the challenges faced by the parasite in terms of invading the host immunity, and environmental condition depicted by different tissues[37,38].

During a blood meal, the infective trypanosome forms, known as metacyclic forms, present in the salivary glands of the tsetse fly are injected into the mammalian host. The parasite in the human host then transforms into a long, slender form in the blood which actively divides and colonize the blood. These trypanosomes have a suppressed mitochondrion and rely upon glycolytic pathways localized in a specialized organelle called glycosomes[39]. As the infection progresses, the long, slender forms subsequently differentiate and become short, stumpy forms which cannot divide any further. The short stumpy forms, on uptake by the fly in a blood meal are competent to form the procyclic forms. These procyclic forms then proliferate in the mid-gut into metacyclic forms which are then transferred back to mammals[39,40]. Understanding of the molecular machinery of antigenic variation, cell-type differentiation, the localization of the glycolytic enzymes, metabolic potential and its versatility in the parasites, accelerated after the complete genome sequence of *Trypanosoma brucei* was released[41]. Yet, almost 50% of the genes in the parasite have no known function[17,41] and hence many biochemical pathways and structural functions await discovery.

The following subsections shall elaborate on the structural and functional characterization of the proteins encoded in the parasitic genome, using sensitive profile-based approaches, which would be used as a guiding tool in prioritizing targets for chemotherapeutic interventions.

3.1. *Overview on structural and functional domain assignments in T. brucei proteome*

The 11 megabase-sized chromosomes of *T. brucei* genome encode for 8747 gene products obtained from ENSEMBL database (http://protists.ensembl.org/Trypanosoma_brucei/Info/Index). Based on the employment of sensitive sequence-based search techniques described in the previous section, 64.2% of the gene products could be associated with a structural/functional domain family along with a further 4.3%, which could be associated with a protein of known structure. In comparison to the available domain annotations for the trypanosome proteome[17], the current study reports a relatively higher proportion of annotated gene products. Table 2 summarizes the extent of domain assignments made. The functional and structural domain family assignments indicate probable functional roles of the assigned gene products. Of the total domain families (2300) assigned to the gene products, about 65% are associated with known structural information.

Table 2. Domain annotation of the gene products encoded in genome of *T. brucei*

Number of gene products in *T. brucei*	8747
Sequence coverage of the proteome by domains	64.2% (5614)
Number of domain assignments	8160
Number of domain families	2300
Fold assignments	374

Since structural domain assignments provide valuable insights, such as molecular details of the function of a protein, an attempt has been made in order to identify relationships, if any, between domain families of unknown structure with known structural families. SUPFAM database developed by our group, was used in order to enhance the structural information coverage on the parasitic proteome, apart from the use of sensitive profile-based search against the domain families of known structure (SCOP) for effective detection of remotely related proteins. Fig-1 gives the percentage distribution of the domain families with respect to their structural information. About 4% of the domain families with no known structure could be associated with a known structural family as shown.

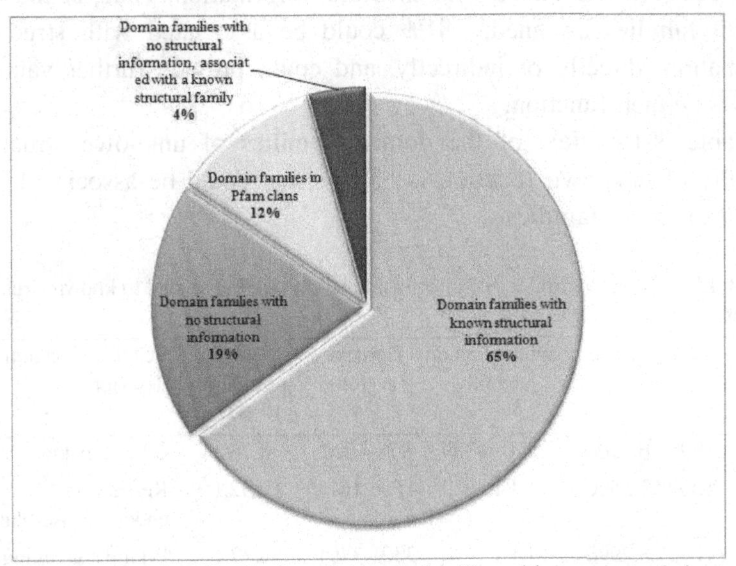

Fig. 1. Pie chart showing distribution of domain families with respect to their structural information.

Of the 2300 domain families assigned to 5614 gene products, 1500 domain families are directly associated with structural information which comprise 65% as shown in Fig-1. About 12% (260) of the Pfam domain families could be grouped into Pfam clans[42], which provide indirect structural information for the domain families of unknown structure. Pfam defines a clan as a collection of families that have arisen from a single evolutionary origin. Evidence of their evolutionary relationship can be in the form of similarity in tertiary structures, or from common sequence motifs, or similarity in their profile-HMMs, guided by the fact that a sequence significantly matches two profile-HMMs in the same region.

Of the remaining 23% of the domain families with no structural information, 4% (99) could be associated with known structural family with the help of SUPFAM database. Pfam database, a collection of protein domain families, also comprises of DUFs or Domains of Unknown Function, which are large set of functionally uncharacterized protein families[43]. Numerous attempts in the past[44-48] have been successful in identifying functions for DUFs. In *T. brucei* 14 of 160 such DUFs could be associated with structural information. Thus, of the total domain families assigned, ~81% could be associated with structural information, directly or indirectly, and could provide further valuable insights on their function.

Table 3 lists few of the domain families of unknown structure, typically of unknown function i.e. DUFs that could be associated with known structural families.

Table 3. List of domain families of unknown function (DUFs) related to known structural families

Sr. No.	Protein code	Pfam domain Family	Domain region	SCOP superfamily ID	SCOP superfamily description
1.	Tb927.6.3500	DUF4339	978-1022	d.76.1	GYF domain
2.	Tb927.7.5590	DUF3883	1506-1616	c.52.1	Restriction endonuclease-like
3.	Tb927.8.7990	DUF3638	3301-3491	c.37.1	P-loop containing nucleoside triphosphate hydrolase

Structure adopted by each of these DUFs was predicted with the use of fold recognition algorithm employed by PHYRE2[49] and compared with the associated structural superfamily to check for consistency in the predictions made. One of the instances is discussed further.

Fig-2a shows the predicted fold of the domain family DUF3883. The fold predicted for many members within DUF3883 family was restriction

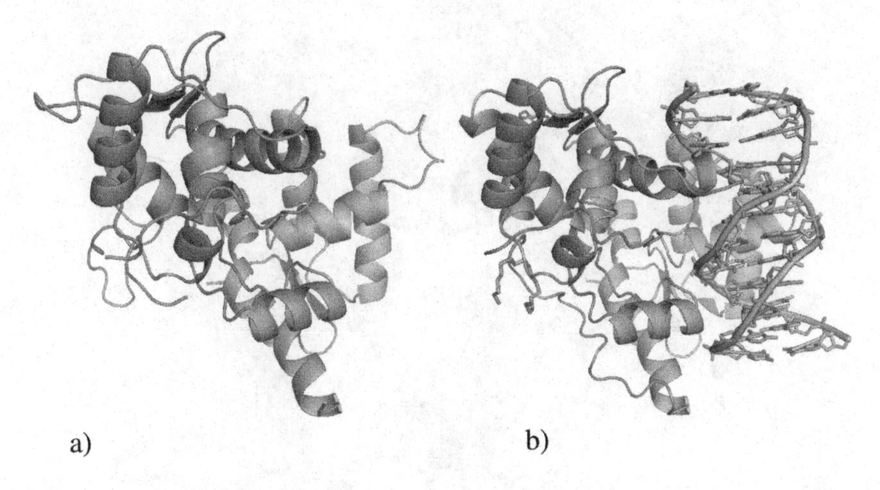

Fig. 2. a) Predicted fold of DUF3883 comprising of two winged-helix subdomains, b) one of the high confidence template (PDB ID: 2VLA) (protein of the fold: restriction endonuclease-like) using which the query was modeled, is a restriction endonuclease enzyme complexed with cognate DNA and PG6[a]. Images are rendered using PyMOL (http://pymol.org/).

endonuclease-like with >90% confidence which is consistent with our association made. Based on the benchmark studies[49,50], given a high confidence match(>90%) predicted by PHYRE2, the overall fold is likely to be correct.

The major groove contacts of amino acid residues located on both the helix-turn-helix motifs and the N-terminal arm of the restriction endonuclease i.e. the template, are involved in recognition of DNA sequences[51] which are in good consensus with the model (DUF3883 family) as shown in Fig-3 (highlighted in magenta). Assessment of function annotation transfer between DUF3883 and restriction endonuclease-like family was enabled with the help of structural alignment and putative structure/function annotation of DUF3883 family could thus be achieved.

[a] PG6: 1-(2-methoxy-ethoxy)-2-{2-[2-(2-methoxy-ethoxy)-ethoxy]-ethoxy}-ethane

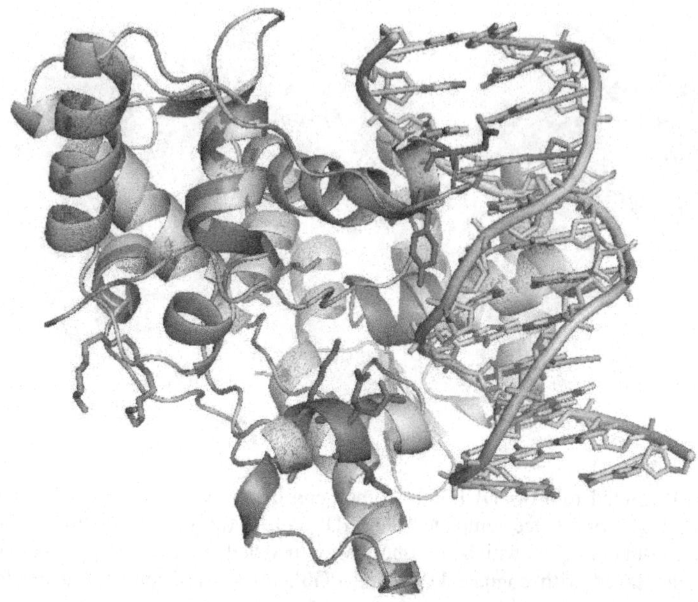

Fig. 3. Structural superimposition of the model (DUF3883) (in cyan) over its template (in yellow) RMSD=0.001. Residues obtained in good consensus (based on our analysis) between the DUF3883 family and restriction endonuclease-like family, are highlighted in magenta, which facilitate DNA binding.

Similar analysis for 99 such cases, as stated above, aided in extending structural and functional information through manual assessment for the families of unknown structure/function.

3.2. *Fold assignments*

Based on the observation that the number of folds in nature are limited and many different remotely homologous proteins adopt similar structures, an attempt was made to associate the rest of the gene products (3133), which could not be assigned any known structural/functional domain family, to a protein of known structure with the help of PHYRE2. This approach essentially predicts the fold acquired by the secondary structure elements of the gene product and transfer of a function using fold recognition algorithms is not always the case.

A total of 374 of 3133 gene products could be related to proteins of known structures with a confidence rate of >95%. Few of such associations are listed in Table 4. Given a high confidence match (>90%) between the query and the template, as stated previously, the overall fold is most certainly correct and the central core of the model tends to be accurate even at low sequence identities[49,50] i.e. <20%.

Table 4. Predicted folds of the gene products with no domain assignment, with >95% confidence and <20% sequence identity.

Sr. No.	Gene ID	Confidence	Sequence Identity	Query coverage	Template/Fold description
1.	Tb927.4.2770	98.6%	15%	251/341	Alkaline phosphatase-like
2.	Tb10.70.2350	97.1%	12%	370/472	Alpha-alpha superhelix
3.	Tb10.v4.0017	98.6%	10%	247/305	Membrane protein. Chain B of variant surface glycoprotein of *T.brucei*

The remaining 2759 gene products which could not be associated with a domain family or a protein of known structure could be subjected to remote homologue search against a non-redundant database with the help of PSI-BLAST or jackhammer (HMMER3). Inability to assign a domain for a gene product may be either because of the incomplete information on domain boundaries of the protein or it is too diverged to be detected by the algorithms employed. This exercise would result in the identification of orphans i.e. gene products specific to *T. brucei*, identification of conserved hypotheticals and identification of functional homologues that would require manual assessment in transfer of functions.

3.3. *Metabolic proteins in Trypanosoma brucei*

Understanding the adaptive changes brought about in the metabolic pathways of the parasite will underpin new drug target discovery. Metabolic compartmentalization and specialization of membrane bound organelles have long been under the light of drug target discovery. The parasite is capable of oxidizing fatty acids via β-oxidation in two organelles: glycosomes and mitochondria, which have received considerable attention in the recent past[37]. Import of nuclear encoded proteins to glycosomes has shown to bear similarities to peroxisomes in terms of signaling.

With its adaptive metabolic system, the parasite easily survives low oxygen tension within the host. Cyanide-insensitive trypanosome alternative oxidase (TAO), a terminal oxidase of respiratory chain has been recognized as an attractive drug target for chemotherapy[52] mainly because of its uniqueness with respect to host and its essentiality for the survival of the organism.

Other metabolic pathways such as, glycosylphosphatidylinositol anchor biosynthesis, essential for immune evasion, invasion and attachment to host cells; amino acid metabolism, trypanothione metabolism and, purine salvage and pyrimidine synthesis have been reported to be adapted for pathogenesis and survival within the host[41]. Exploration on structural and functional features of metabolic proteins will be discussed in the following subsections followed by the use of targets of FDA-approved drugs in identification of putative drug targets.

3.3.1. *Domain composition of metabolic proteins*

Trypanocyc[53], a metabolic pathway database, provides a comprehensive analysis of metabolic network in the organism linking biochemical data to reference genome, along with cross-species comparisons with other kinetoplastids. Information on metabolic pathways in *T. brucei* was retrieved from this database. Of 8747 protein coding genes in *T. brucei*, 307 genes are responsible for encoding metabolic proteins for 230 metabolic pathways. An enhanced structural and functional characterization of the metabolic proteins was achieved with the help of sensitive approaches as detailed in the previous subsection, and about 480 domain families could be associated with the metabolic proteins. The distribution of highly populated domain families in metabolic proteins of the parasite is shown in the form of bar graph in Fig-4.

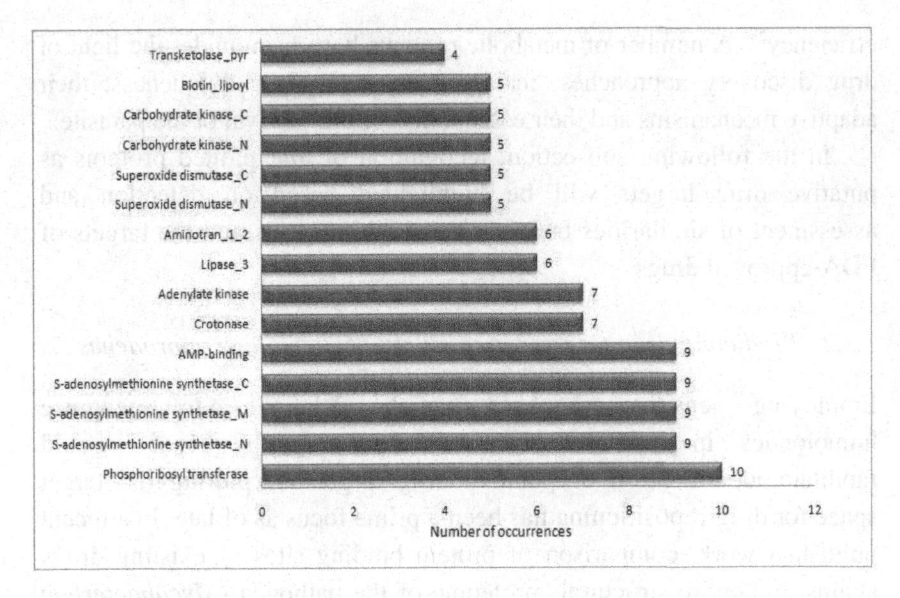

Fig. 4. Graphical representation of occurrences of 15 most populated domain families in metabolic proteins.

Experimental work on characterization of metabolic proteins has crystallized the essentiality of the enzymes in the parasite. For instance, as shown in Fig-4, phosphoribosyl transferase, which forms the most highly occurring domain family in the metabolic repertoire, is essential for the survival of the parasite as the parasite lacks enzymes for de novo purine nucleotide synthesis[54]. Moreover, hypoxanthine phosphoribosyl transferase (HPRT) has been exploited as a potential drug target for enzyme structure-based drug design in *T. cruzi*[55].

S-adenosylmethionine synthetases identified to be different from those of mammalian isoforms[56] have been indicated to be critical drug targets linking inhibition of polyamine synthesis to disruption of AdoMet metabolism.

Other domain families have also been explored for their role in pathogenesis and survival. Superoxide dismutase coding gene is reportedly essential for the parasite to survive drug-generated superoxide[57]. The unusually large adenylate kinase gene family in the flagellated parasite, in order to target adenylate kinases to compartmentalized metabolic organelles such as glycosomes, explains the metabolic organization and efficiency[58]. A number of metabolic proteins have been under the light of drug discovery approaches, mainly because of their uniqueness, their adaptive mechanisms and their essentiality for the survival of the parasite.

In the following subsection, recognition of unexploited proteins as putative drug targets will be highlighted based on detection and assessment of similarities between metabolic proteins and the targets of FDA-approved drugs.

3.3.2. *Predicting drug targets based on remote homology approaches*

Employing sensitive sequence search techniques to recognize homologues, in a dataset of protein drug target sequences, would facilitate identification of putative drug targets. Exploring the target space for drug-repositioning has been a prime focus as of late. In a recent published work, comparison of protein binding sites of existing drugs against the entire structural proteome of the pathogen (*Mycobacterium tuberculosis*) aided exploration of unexploited proteins that could serve as attractive drug targets.

DrugBank[36] is a database that encompasses bioinformatics and cheminformatics resources, to combine detailed drug data with comprehensive drug target information. Data in the current release of DrugBank database is summarized in the Table 5.

In the current analysis, the drug targets of the approved drugs, having an effect (toxicity/targets) on humans were discarded from the dataset. Such an exclusion was done in order to ensure that the binding sites of the putative drug targets identified in the parasite are not homologous to human (host) proteins.

Table 5. List of number of entries in each category of drugs in DrugBank

Sr. No.	Drug type	Number of entries
1.	FDA-approved small molecules	1447
2.	FDA-approved protein/peptide drugs	131
3.	Nutraceuticals	85
4.	Experimental	5080

Upon the application of sequence comparison methods, 32 metabolic proteins of the parasite could be identified to be homologous to 29 protein drug target sequences which correspond to 34 FDA-approved drugs. Table 6 lists a few of such homologues identified.

From the results obtained, cases were identified where more than one drug target was found to be homologous to a single gene product. To obtain a better visualization, metabolic proteins and their homology to the drug targets were rendered in a network fashion. In terms of graph theory, two metabolic proteins considered as "nodes" were connected with an "edge" if they shared homology to the same drug target. Thus a network was constructed using the metabolic proteins as nodes and homologous drug targets as edges. The edges were weighted in a manner that the information on number of drug targets being homologous to the proteins was taken into account.

Table 6. List of genes in *T. brucei* and the identified drug target homologues

Sr. No.	Gene ID	Drug target identified as homologue	Source of the drug target	Corresponding Drug
1.	Tb10.6k15.3140	Dihydropteroate synthase Folylpolyglutamate synthase	E.coli	DB01015: Sulfamethoxazole
2.	Tb11.03.0090	Large structural protein RNA-directed RNA polymerase catalytic subunut	Human parainfluenza 2 vius Influenza A virus	DB00811: Ribavirin
3.	Tb11.01.8470	Fumarate reductase	E.coli	DB00730: Thiabendazole

More the number of drug targets, homologous to two proteins, the thicker would be the edge between them. Fig-5 shows clusters of connected components. 23 of 32 metabolic proteins identified, share at least one drug target homologue.

One of the connected components, a star-shaped entity, comprises of five metabolic proteins. Upon closer inspection, these four of the five gene products were recognized to be dihydrolipoyl dehydrogenases and one of them, a succinate dehydrogenase. Both dihydrolipoyl dehydrogenases and succinate dehydrogenases have been regarded as attractive drug targets in other species[60] such as *Mycobacteria* and *Plasmodium.*

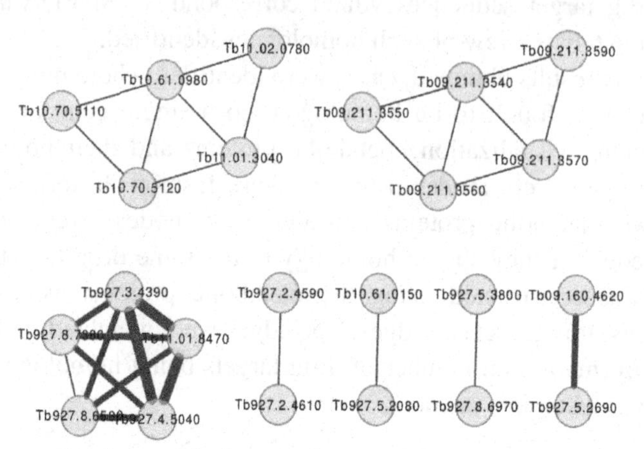

Fig. 5. Figure showing small clusters of connected components comprising of 23 metabolic proteins which are connected based on their shared homology to drug targets. Network visualization was done using cytoscape (http://www.cytoscape.org/).

Table 7 summarizes the domain assignments of the 5 gene products along with the information on the drug(s) that have a possibility of being re-purposed for the parasite. Each drug corresponds to the protein drug target homologous to the parasitic protein.

Table 7. Domain assignments for the gene products identified to be putative drug targets

Sr. No.	Gene ID	Domain family	Domain region	Drugs
1.	Tb11.01.8470	Pyridine nucleotide-disulphide oxidoreductase Dimerization domain	13-329 358-467	Nitrofurazone Azelaic Acid Thiabendazole
2.	Tb927.3.4390	Pyridine nucleotide-disulphide oxidoreductase Dimerization domain	14-329 362-472	Nitrofurazone Azelaic Acid Thiabendazole
3.	Tb927.4.5040	Pyridine nucleotide-disulphide oxidoreductase Dimerization domain	50-380 414-524	Nitrofurazone Azelaic Acid Thiabendazole
4.	Tb927.8.7380	Pyridine nucleotide-disulphide oxidoreductase Dimerization domain	28-361 391-501	Nitrofurazone Azelaic Acid
5.	Tb927.8.6580	FAD binding domain Fumarate reductase C-terminal	25-419 475-609	Azelaic Acid Thiabendazole

Nitrofurazone, an anti-infective agent against gram-positive and gram-negative bacteria, has been already exploited for its use as a trypanocidal agent[61]. Being administered orally, it is used in the treatment of trypanosomiasis.

Azelaic acid is a natural substance found in wheat, rye and barley, and also produced by yeast that thrives on human skin. It is a saturated dicarboxylic acid, which is used as an antineoplastic and dermatologic agent, known to cure varied skin conditions. Though being effective against various aerobic and anaerobic microorganisms, likeliness of this small molecule being re-purposed for the proteins in *T. brucei*, needs further elucidations on structure of the parasitic protein and its binding sites.

Thiabendazole is a fungicide and a parasiticide used against variety of nematodes. Mechanism of action of the drug remains unknown, but is speculated to act on the helminth enzyme fumarate reductase. As conceived from the table, four of five gene products, comprising of oxidoreductase domains, are homologous to a drug target on which Thiabendazole acts upon. These could be taken up further for prioritization. Structural and binding site elucidations are essential before it is regarded as a potential drug target.

Thus, the identification of unexploited proteins coupled with structural elucidations form a basis of recognition of attractive drug targets. Moreover, information on gene expression and gene localization could provide further insights in prioritizing the predicted drug targets.

4. Conclusions

The molecular basis of life involves a complex interplay of various biological molecules with proteins acting in concert. Understanding its functional design, thereby exploring the physiological processes and differences at the molecular level, has been a major endeavor for molecular biologists. At molecular level, the basis of usage of sequence information to address complex issues forms no exception. With an exponential increase in sequence data compared to the relatively slow paced experimental characterization of the proteins, an exponential expansion on the computational approaches has been achieved. From pairwise sequence alignment tools to the use of highly sensitive profile-based searches, the significant advancements made in speed, sensitivity and accuracy for remote homologue detection have largely influenced the exploration of protein sequence/structure space. This chapter highlights the progress made in remote homologue detection approaches followed by demonstrating the use of such techniques in exploring structural and functional repertoire of proteins encoded in the genome of *T. brucei.* Such an approach resulted in annotation of about 68% of the gene products which is substantially higher in proportion (15-18%) as compared to currently available annotations.[17] Later part of the analysis presented in the chapter elucidates the recognition of unexploited metabolic proteins as putative drug targets.

After reductionism, which has formed the basis of most of the analytical theories, molecular biology is now experiencing a paradigm shift towards holistic view of biological processes, and large-scale data integration of different types has thus become of fundamental importance. Being affected by data deluge, a data-intensive science is emerging which would eventually bring about improvisation in the aspects of medicine and the environment.

Acknowledgments

This research was supported by Mathematical Biology programme, Department of Science and Technology and by the Department of Biotechnology, Government of India.

References

1. Moult J., Fidelis K., Kryshtafovych A. and Tramontano A. *Proteins,* 79 (2011).
2. Altschul S.F., Gish W., Miller W., Myers E.W. and Lipman D.J. *J.Mol.Biol.,* 15 (1990).
3. Park J., Karplus K., Barrett C., Hughey R., Haussler D., Hubbard T. and Chothia C. *J.Mol.Biol.,* 284 (1998).
4. Altschul S.F., Madden T.L., Schaffer A.A., Zhang J., Zhang Z., Miller W. and Lipman D.J. *Nucleic Acids Res.,* 25 (1997).
5. National Center for Biotechnology Information. NCBI PSI-BLAST. http://www.ncbi.nlm.nih.gov/BLAST/tutorial/Altschul-2.html (Accessed 15 Sep 2012).

6. Ye J., McGinnis S. and Madden T.L. *Nucleic Acids Res.,* 34:W (2006).
7. Anand B., Gowri V.S. and Srinivasan N. *Bioinformatics,* 21 (2005).
8. Gowri V.S., Krishnadev O., Swamy C.S. and Srinivasan N. *Nucleic Acids Res.,* 30 (2006).
9. Gowri V.S., Tina K.G., Krishnadev O. and Srinivasan N. *Protein,* 67 (2007).
10. Balaji S., Sujatha S., Kumar S.S.C. and Srinivasan N. *Nucleic Acids Res.,* 29 (2001).
11. Sandhya S., Chakrabarti S., Abhinandan K.R., Sowdhamini R. and Srinivasan N. *J.Biomol.Struct.Dyn.,* 23 (2005).
12. Sandhya S., Mudgal R., Jayadev C., Abhinandan K.R., Sowdhamini R. and Srinivasan N. *Mol.Biosystems,* 8 (2012).
13. Biegert A. and Soding J. *Proc.Natl.Acad.U.S.A.,* 106 (2009).
14. Murzin A.G., Brenner S.E., Hubbard T.J.P. and Chothia C. *J.Mol.Biol.,* 247 (1995).
15. Andreeva A., Howorth D., ChandoniaJ.-M, Brenner S.E., Hubbard T.J.P., Chothia C. and Murzin A.G., *Nucleic Acids Res.,* 32:D (2008).
16. Sonnhammer E.L., Eddy S.R. and Durbin R. *Proteins,* 28 (1997).
17. Punta M., Coggill P.C., Eberhardt R.Y., Mistry J., Tate J., Boursnell C., Pang N., Forslund K., Ceric G., Clements J., Heger A., Holm L., Sonnhammer E.L., Eddy S.R., Bateman A. and Finn R.D. *Nucleic Acids Res.,* 40:D (2012).
18. Marchler-Bauer A, Panchenko A.R., Shoemaker B.A., Thiessen P.A., Geer L.Y. and Bryant S.H. *Nucleic Acids Res.,* 30 (2002).
19. Schaffer A.A., Wolf Y.I., Ponting C.P., Koonin E.V., Aravind L. and Altshul S.F. *Bioinformatics,* 15 (1999).
20. Park J., Teichmann S.A., Hubbard T. and Chothia C. *J.Mol.Biol.,* 273 (1997).
21. Li W., Pio F., Pawlowski K. and Godzik A. *Bioinformatics,* 16 (2000).
22. Sandhya S., Kishore S., Sowdhamini R. and Srinivasan N. *FEBS Lett.,* 552 (2003).
23. Eddy S.R. *Bioinformatics,*14 (1998).
24. Krogh A., Brown M., Mian L.S., Sjolander K. and Haussler D. *J.Mol.Biol.,* 235 (1994).
25. Madera M. and Gough J. *Nucleic Acids Res.,* 30 (2002).
26. Edgar R.C. and Sjolander K. *Bioinformatics,* 20 (2004).
27. Soding J. *Bioinformatics,* 21 (2005).
28. Moult J., Fidelis K., Kryshtafovych A., Rost B. and Tramontano A. *Proteins,* **S9**, 77 (2009).
29. Moult J., Fidelis K., Kryshtafovych A., Rost B., Hubbard T. and Tramontano A. *Proteins,* **S8**, 69 (2009).
30. Krishnadev O. and Srinivasan N. *BMC Bioinformatics,* 12 (2011).
31. Pandit S.B., Gosar D., Abhiman S., Sujatha S., Dixit S.S., Mhatre N.S., Sowdhamini R. and Srinivasan N. *Nucleic Acids Res.,* 30 (2002).
32. Pandit S.B., Bhadhra N., Gowri V.S., Balaji S., Anand B. and Srinivasan N. *BMC Bioinformatics,* 5 (2004).
33. Remmert M., Biegert A., Hauser A. and Soding J. *Nat.Methods.,* 9 (2012).

34. Bhagwat M. and Aravind L. *PSI-BLAST tutorial.* In: Bergman N.H., Ed., *Comparative genomics,* Humana Press, Vol. 1 & 2 (Totowa, New Jersey, 2007).
35. Zhao Y., Wang Q., Meng Q., Ding D., Yang H., Gao G., Li D., Zhu W. and Zhou H.. *Bioorg. Med. Chem.,* 20 (2012).
36. Knox C, Law V., Jewison T., Liu P., Ly S., Frolkis A., Pon A., Banco K., Mak C., Neveu V., Djoumbou Y., Eisner R., Guo A.C. and Wishart D.S. *Nucleic Acids Res.,* 39:D (2011).
37. Gull K. *Curr. Pharm. Des.* 8 (2002).
38. Welburn S.C. and Maudlin I. *Parasitol. Today,* 15 (1999).
39. Opperdoes F.R., Baudhin P., Coppens I., Roe C.D., Edwards S.W., Weijers P.J., and Misset O. *J. Cell Bio.,* 98 (1984).
40. Matthews K.R. *Parasitol. Today,* 15 (1999).
41. Berriman M., Ghedin E., Hertz-Fowler C., Blandin G., Renauld H., Bartholomeu D.C., Lennard N.J., Caler E., Hamlin N.E., Haas B., Bohme U., Hannick L., Aslett M.A., Shallom J., Marcello L., Hou L., Wickstead B., Alsmark. U.C.M., Arrowsmith C., Atkin R.J., Barron A.J., Bringaud F., Brooks K., Carrington M., Cherecvach I., Chilingworth T-J., Churcher C., Clark L.N., Corton C.H., Cronin A., Davies R.M., Doggett J., Djikeng A., Feldblyum T., Field M.C., Fraser A., Goodhead I., Hance Z., Harper D., Harris B.R., Hauser H., Hostetler J., Ivens A., Jagels K., Johnson D., Johnson J., Jones K., Kerhornou A.X., Koo H., Larke N., Landfear S., Larkin C., Leech V., Line A., Lord A., MacLeod A., Mooney P.J., Moule S., Martin D.M.A., Morgan G.W., Mungall K., Norbertczak H., Ormond D., Pai G., Peacock C.S., Peterson J., Quail M.A., Rabbinowitsch E., Rajendream M-A, Reitter C., Salzberg S.L., Sanders M., Schobel S., Sharp S., Simmonds M., Simpson A.J., Tallon L., Michael T., Tait A., Tivey A.R., Aken S.V., Walker D., Wanless D., Wang s., White B., White O., Whitehead S., Woodward J., Wortman J., Adams M.D., Embley T.M., Gull K., Ullu E., Barry J.D., Fairlamb A.H., Opperdoes F.R., Barrell B.G., Donelson J.E., Hall N., Fraser C.M., Melville S.E. and El-Sayed N.M. *Science,* 309 (2005).
42. Finn R.D., Mistry J., Schuster-Bockler B., Griffith-Jones S., Hollich V., Lassman T., Moxon S., Marshall M., Khanna A., Durbin R., Eddy S.R., Sonnhammer E.L. and Bateman A. *Nucleic Acids Res.,* 34 (2005).
43. Bateman A., Coggill P. and Finn R.D. Acta Crystallogr.Sect.F Struct.Biol.Cryst.Commun., 66 (2010).
44. Dlakic M. *Bioinformatics,* 22 (2006).
45. Martzen M.R., McCraith S.M., Spinelli S.L., Torres F.M., Fields S., Grayhack E.J. and Phizicky E.M. *Science,* 286 (1999).
46. Karras G.I., Kustatscher G., Buhecha H.R., Allen M.D., Pugieux C., Sait F., Bycroft M. and Ladurner A.G. *EMBO J.,* 24 (2005).
47. Schulze-Gahmen U., Pelaschier J., Yokota H., Kim R. and Kim S-H. *Proteins,* 50 (2003).

48. Krishna S.S., Tautz L., Xu Q., McMullan D., Miller M.D., Abdubek P., Ambing E., Astakhova T., Axelrod H.L., Carlton D., Chiu H-J., Clayton T., DiDonato M., Duan L., Elsliger M-A., Grzechnik J.H., Hampton E., Han G.W., Haugen J., Jaroszewski L., Jin K.K., Klock H.E., Knuth M.W., Koesema E., Morse A.T., Mustelin T., Nigoghossian E., Oomamachen S., Reyes R., Rife C.L., van den Bedem H., Weekes D., White A., Hodgson K.O., Wooley J., Deacon A.M., Godzik A., Lesley S.A. and Wilson I.A. *Proteins,* 69 (2007).
49. Bennett-Lovsey R.M., Herbert A.D., Sternberg M.J. and Kelly L.A. *Proteins*, 70 (2008).
50. Kelly L.A. and Sternberg M.J. *Nat. Protoc.*, 4 (2009).
51. Sukackaite R., Grazulis S., Bochtler M. and Siksnys V. *J.Mol.Biol.,* 378, (2008).
52. Nihei C., Fukai Y. and Kita K. *Biochim.Biophysic.Acta.*, 1587 (2002).
53. Chukualim B., Peters N., Hertz-Fowler C. and Berriman M. *BMC Bioinformatics,* **P5**, 9 (2008).
54. Fijolek A. *Salvage and de novo synthesis of nucleotides in Trypanosoma brucei and mammalian cells.* PhD Thesis. Medical Biochemistry and Biophysics, Umeå University, Umeå. (2008).
55. Eakin A.E., Guerra A., Focia P.J., Torres-Martinez J. and Craig S.P. *Antimicrob.Agents Chemother.,* 41 (1997).
56. Yarlett N., Garofalo J., Goldberg B., Ciminelli M.A., Ruggiero V., Sufrin J.R. and Bacchi C.J. *Biochim.Biophys.Acta.,* 1181 (1993).
57. Prathalingam S.R., Wilkinson S.R., Horn D. and Kelly J.M. *Antimicrob.Agents Chemother.,* 51 (2007).
58. Ginger M.L., Ngazoa E.S., Pereira C.A., Pullen T.J., Kabiri M., Becker K., Gull K. and Steverding D. *J.Biol.Chem.,* 280 (2005).
59. Kinnings S.L., Xie L., Fung K.H., Jackson R.M., Xie L. and Bourne P.E. *PloS Comp.Biol.,* 6 (2010).
60. Aguero F., Al-Lazikani B., Aslett M., Berriman M., Buckner F.S., Campbell R.K., Carmona S., Carruthers I.M., Edith Chan A.W., Chen F., Crowther G.J., Doyle M.A., Hertz-Fowler C., Hopkins A.L., Gregg M., Nwaka S., Overington J.P., Pain A., Paolin G.V., Pieper U., Ralph S.A., Riechers A., Roos D.S., Sali A., Shanmugam D., Suzuki T., Van Voorhis W.C. and Verlinde C.L.M.J. *Nat. Rev. Drug. Disc.,* 7 (2009)
61. Baker J.R. Br.J.Pharmacol.Chemother., 14 (1959).

Chapter 2

Identification of Genes and their Regulatory Regions Based on Multiple Physical and Structural Properties of a DNA Sequence

Xi Yang, Nancy Yu Song and Hong Yan

Department of Electronic Engineering
City University of Hong Kong, Kowloon, Hong Kong

Genomic sequences contain many functional regions, such as exons and promoters. Identification of these regions is a very important task in molecular biology. It is costly and time consuming to carry out such identification based on biological experiments. Computational methods to identify the functional regions provide a good alternative. In this chapter, we briefly review the current methods of exon and promoter prediction. Then we present our work on the identification of short human exons based on autoregressive (AR) models and multi-feature spectral analysis, and eukaryotic promoter predictions based on both sequence and structural properties using Isomap and support vector machines. Methods for the recognition of disease-related genes and regulatory regions are also introduced. Our computational methods for functional region identification have been tested on many genomic datasets with good results.

1. Introduction

Over the past decade, the genomes of many organisms have been sequenced. However, until now the genomic data are far from fully understood due to the tremendous amount of data and also the complexity of the genetic regulation mechanism. Using biochemical experiments to detect the genes and their regulatory regions is undoubtedly the most accurate method since they provide first-hand information for the understanding of gene expression, but these experiments are very expensive and time-consuming to conduct. For most species, it is almost impossible to completely annotate their genome through experimentation only. Hence there exists a strong demand for fast and accurate computer tools to perform large-scale genomic sequence analysis. The genome of eukaryotic organisms can be primarily divided into gene regions and intergenic regions. A gene region refers to the transcribed region while the intergenic region refers to the region between two transcribed regions [1]. A gene is further divided into exons and introns [2]. Introns are removed by *cis*-splicing when the splicing occurs in one precursor RNA molecule or by trans-splicing when two or more precursor RNA molecules are cleaved and ligated, leaving only exons represented in the mature form of RNA molecules that are translated into amino acids [3]. Strictly speaking, genes include both translated (protein-coding) and untranslated regions (UTRs). UTRs play an important role in translational regulation. However, very few gene predictors predict UTRs except a gene-prediction system named N-SCAN which predicts 5' UTRs as part of an integrated gene prediction process [4]. Computational gene prediction methods nowadays still focus on the prediction of translated regions. Gene recognition includes determining transcription start sites (TSSs) and distinguishing exons from introns. On the other hand, transcriptional regulatory elements, such as promoters, enhancers, silencers, insulators and locus control regions, are distributed across the large genome, mainly in the intergenic regions [5]. Identification of regulatory regions is very important for the understanding of transcriptional regulatory mechanisms and genomic functional annotation. More importantly, it is revealing for the pathogenesis to discriminate the exons from the flanking intronic regions in a gene and the transcriptional regulatory elements from the non-functional regions. For example, there are more than 6000 known single-gene disorders that cause diseases by a mutation in one gene [6], and also many polygenic disorders that are caused by several genes [7]. Genetic phenotypes for the gene disorders include Alzheimer disease, cancers, heart disease, etc [8]. Promoter mutations that increase amyloid precursor-protein expression are found to be correlated with Alzheimer disease [9], and similarly, a promoter mutation in the erythroid-specific 5-aminolevulinate synthase ($ALAS_2$) gene may cause X-linked sideroblastic anemia [10].

The information obtained from the sequencing and biological experiments can be used as training data to construct mathematical models for genes and regulatory elements. The most commonly used algorithms include the hidden Markov model [11-15], dynamic programming [16,17], neural network [18-20], discriminant analysis [21], Fourier analysis [22] and multivariate entropy distance method [23,24]. For gene prediction, most methods can achieve good discrimination between long coding and non-coding regions, but they all have their own limitations and may generate different results for the same genome. How to identify the exact boundaries between exons and introns, and how to decrease the chances of gene fusion (when two or multiple genes are too close) and gene fragmentation (when the intron between two exons is too large) remain difficult problems for most gene prediction techniques. For promoter prediction, many methods are only workable for specific groups of promoters. Because eukaryotic promoters are highly diverse, no ubiquitous sequence patterns applicable to all eukaryotic promoters have yet been found.

Recently it has been argued that besides carrying the sequence composition information, the linear DNA molecule also has very distinct physicochemical properties that determine to a large extent its topological structure and binding affinity to RNA polymerase and other protein factors during the formation of transcriptional complexes [25,26]. Over the past twenty years, abundant experiments have been carried out to measure the physicochemical properties of DNA molecules under different conditions, based on which people have summarized a set of empirical physicochemical parameters for various short DNA segments. These parameters have been taken into account in recent studies for the description of coding, non-coding and regulatory regions in the genome of specific species, including yeast, Arabidopsis, rice, Plasmodium falciparum, mouse, human, etc [27-29].

In this chapter, we will briefly review the current methods of gene and promoter prediction. Then we will introduce in more detail our work on the classification of short human exons and introns based on statistical features, studies on spectral properties of short genes using the wavelet subspace Hilbert-Huang transform, short exon detection based on AR models and multifeature spectral analysis, and eukaryotic promoter prediction based on both sequence and structural properties.

2. Gene prediction methods

2.1. *Background*

Genes are small DNA fragments which are scattered on the whole genome. Identifying genes on the genome sequence can be an extremely challenging task because only a very small portion of genome sequence data contains genes. Ever since biology entered the genomic era, sequencing costs have been falling so rapidly that nowadays a single laboratory can sequence large, even human-sized, genomes. However, on the other hand, the difficulty of genome annotation is increasing. This is caused by several factors. First, the current genome assemblies on the second-generation sequencing platforms rarely attain the level of contiguity of the classic shot gun assemblies. Second, there is a lack of pre-existing gene models for today's genomes which are available in the first generation of genome projects. This poses great challenges for genome annotation, especially gene finding. Third, updating and merging annotation data sets is very complex because different groups use different annotation procedures. Finally, today's genome annotation projects are usually smaller-scale. Not many bioinformatics experts work on the same genomes as those in the first generation genome assemblies. As a result, genome annotation remains a great challenge [28]. In the early days, most of the genome annotation work was based on cDNA sequence data from biology experiments. Large-scale cDNA sequencing projects provide a rich database for further gene prediction. Nevertheless, even if all human genes were determined by experiments one day, there will still exist a strong demand for computational and bioinformatics tools in order to understand the nature of genes. Gene structure prediction is a huge intellectual and practical challenge. Eukaryotic genes are usually comprised of blocks of exons and introns. Exons can be generally classified into four groups: 5' exons, internal exons, 3' exons and intronless exons. The definition of the term 'exon' in many texts becomes so confused that it is used interchangeably with the term 'coding sequence' (CDS). Almost all gene prediction papers focus on predicting simply the coding regions on exons which ignores the untranslated regions (UTRs) [29].

Some approaches based on probabilistic models for gene finding require some pre-existing genomic information for a species obtained from experiments. The pre-existing data is used to train a probabilistic model such as the hidden Markov Model. After the model is built, genome annotation can be carried out without any previous experiment information of the genomic sequence to be annotated. Software implemented based on these approaches include GENESCAN [30] and TWINSCAN [31].

Some other approaches are based on digital signal processing (DSP) algorithms. The property that the coding sequence contains 3-periodicity has been known for some time [32]. This characteristic can be utilized to predict the CDS and distinguish intronic and exonic regions. The term 'exonic region' here only refers to CDS without taking into account UTRs. Main DSP methods include the discrete Fourier transform (DFT), digital filters, wavelet transform and parametric spectral analysis. All these algorithms are better able to distinguish the regions with and without the 3-periodicity. The Fourier transform was firstly used for exon identification [33]. However, the spectrum obtained by the Fourier transform contains windowing artifacts and spurious spectral peaks. Akhtar *et al.* proposed an optimized period-3 method called a paired and weighted spectral rotation (PWSR) measure which takes into account both computational complexity and the relative accuracy of gene prediction [34]. Vaidyanathan and Yoon proposed a method which deploys an antinotch digital filter to find the signal energy at the $2\pi/3$ frequency [35]. Entropy measures are also employed in exon detection. A complexity measure based on the entropic segmentation of DNA sequences into homogeneous domains is defined by Román-Roldán *et al.*[36] Yan and Pham proposed an autoregressive (AR) model-based sequence analysis method to estimate the power spectral density [37]. The AR model-based analysis is able to produce stronger power spectral density peaks and weaker artifacts than the DFT. Choong and Yan further proposed multi-scale parametric spectral analysis for exon detection based on the AR model [38]. This method is proven to be better than the DFT and previous AR model based methods because more accurate prediction is achieved as shown in [38]. Jiang and Yan also used wavelet subspace Hilbert-Huang transform to identify exon regions [39].

Tina and Tessamma, proposed to de-noise the signals in the coding regions using the discrete wavelet transform [40]. Song and Yan proposed to convert the symbolic DNA sequences into numerical ones based on DNA structural features before applying the AR model to find exons [41]. Zhang and Yan proposed an improved exon prediction method based on empirical mode decomposition (EMD) and the Fourier transform [42].

The methods based on probabilistic models generally perform more accurately than those based on signal processing. However, they need a large amount of experimental data to train the model. It is very difficult to predict the genes of a new species based on the probabilistic models because the gene database obtained by biological experiments may not exist. The methods based on signal processing can overcome this difficulty because no prior knowledge is needed for prediction.

Biological experiments for exon identification are laborious and costly to conduct. Computational methods for exon identification are much cheaper and more convenient. Computational methods predicting exonic and intronic regions depend on the original DNA sequences, while biological experiments identify these regions in a reverse way. Biological experiments obtain the final protein products first, then trace back to the original DNA sequences. Therefore, computational exon predictions provide a faster and easier way to identify disease-causing genes. The exon prediction results also give biologists more information about the possible alternative splicing of a gene sequence.

2.2. *Exon prediction based on the AR model and multifeature spectral analysis*

A main difficulty in employing DSP methods for exon prediction is to transform symbolic DNA sequences into numerical ones. The most straightforward way is to assign 1 to 'A', 2 to 'C', 3 to 'G' and 4 to 'T'. Another way is to use single base binary representation. For a DNA sequence $x[n]$, we can construct four indicator sequences as:

$$x_i[n] = \begin{cases} 1 & \text{if } x[n] = i \\ 0 & \text{otherwise} \end{cases} \quad (i \in \{A, C, G, T\}) \tag{1}$$

A better way is to use the double base (DB) curve representation [43]. There are four single nucleotide bases: A, C, G, T. The DB curve representation is defined as:

$$x_{b_1b_2}(n) = \sum_{i=1}^{n} s(i) \quad n = 1,2,\ldots N \tag{2}$$

where N is the length of the DNA sequence and the unit numeric value $s(n)$ is defined as:

$$s(n) = \begin{cases} +1 & \text{for base } \; b_1 \\ -1 & \text{for base } \; b_2 \\ 0 & \text{for other bases} \end{cases} \tag{3}$$

where $b_1, b_2 \in \{A, G, C, T\}$ and $b_1 \neq b_2$. Therefore the nucleotide bases can be classified into six double bases: AC, AG, AT, CG, CT and GT. The DB curve reflects the difference between two kinds of nucleotides along a DNA sequence. The DB curve representation is much more informative than the single base binary representation. The drawback is that the computation complexity increases because the number of signals to be processed increases from four to six.

Compared with doing the conversion based on subjectively assigned numbers, it is biologically more meaningful to do the conversion based on DNA structural properties. Figures 1(a) and (b) show the power spectral density (PSD) obtained for base pairs 6900-8100 of a DNA sequence with NCBI accession number Z20656. The actual exon positions are indicated by red rectangles. The shortest exon is only 27-bp long located at relative position 430. It is not difficult to see that there is no peak showing the existence of the 27-bp long exon in Figure 1(a) which is obtained from the indicator sequences while there is an obvious peak in the same position in Figure 1(b) which is obtained from the DNA propeller twist value. The result here shows that DNA structural properties can provide better results than simple numerical indicator sequences for the 1/3 frequency detection.

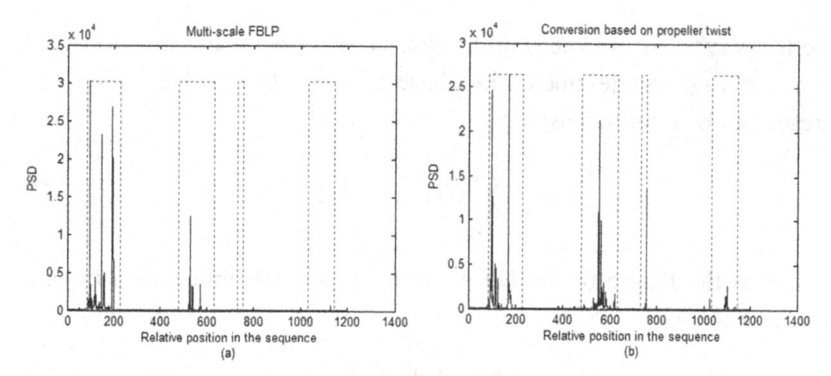

Fig. 1. (A) The PSD obtained from the multi-scale FBLP method [38] is applied to the indicator sequences. (B) The PSD obtained by applying the AR modeling method to the DNA propeller twist value. This diagram is adopted from Figure 3 in [41].

Some examples of the DNA properties are A-philicity, B-DNA twist, DNA bendability, DNA-bending stiffness, DNA denaturation, Duplex disrupt energy, Duplex free energy, GC trinucleotide content, nucleosome positioning, propeller twist, protein-DNA twist, protein-induced deformability, stacking energy, and Z-DNA stabilizing energy [44]. The detailed values of these properties can be found in the appendix of [45] and here we only list their minimum and maximum values (Table 1). Not every signal which was converted based on DNA structural properties performed well in gene prediction. An experiment in [41] showed that DNA-bending stiffness, disrupt energy, free energy and propeller twist performed better than the other DNA structural properties in exon finding for the human genome.

DNA sequences are not stationery from a DSP point of view. Therefore a moving window is employed in order to handle non-stationery data. The signal within a short piece of a sequence is assumed to be stationery. The size of the window shall be several times larger than the fundamental repeating unit, which in this case is three. For each DNA symbolic sequence, there may be several DNA numerical signals. It is necessary to apply the moving window to every numerical signal before applying DSP.

Table 1. DNA physical properties

Physical property	Min	Max
Stacking energy	-14.59 kcal	-3.82 kcal
Propeller twist	-18.66°	-8.11°
Nucleosome positioing	-36%	+45%
DNA bendability	-0.280	+0.194
A-philicity	0.13	1.04
Protein-induced deformability	1.6	12.1
Duplex disrupt energy	0.9 kcal	3.1 kcal
Duplex free energy	-2.1 kcal/mol	-0.9 kcal/mol
DNA denaturation	64.35 cal/mol	135.38 cal/mol
DNA-bending stiffness	20 nm	130 nm
B-DNA twist	30.6°	43.2°
Protein-DNA twist	31.5°	37.8°
Z-DNA stabilizing energy	5.9 kcal/mol	0.7 kcal/mol
GC trinucleotide content	0	3

According to the Heisenberg Uncertainty Principle, one cannot know what spectral components exist at what instance of time. What one can know is which frequencies exist at what intervals of time. In addition, the better the frequency resolution we have, the worse time resolution we get and vice versa. When we apply the principle to our problem, it becomes a trade-off between frequency resolution and position resolution. To keep the balance between frequency and position resolution, several different window sizes can be used in order to catch both short and long exon signals. Vertebrate genes consist of short exons separated by introns. The length of the intron is usually 10 or even 100 times longer than that of exons on average. This is especially true for human exons. The average length of human exons is 137 bp [46]. Therefore, the window sizes could be set to be within the range of 30 and 300 in order to obtain both satisfactory frequency and position information.

An autoregressive (AR) model is a spectral estimation technique. An AR model can overcome short signal problems, give a higher resolution and produce smaller artifacts for spectral estimation compared with DFT [35]. The details of the AR model are described below.

Let $\mathbf{S} = [y_1, y_2, y_3, \cdots y_t, \cdots y_n]$ be a stationary time series which follows an AR model of order *p*. The AR model in matrix form can be described as:

$$\mathbf{y} = \mathbf{Ya} + \boldsymbol{\varepsilon} \tag{4}$$

where **a** are the AR model coefficients and ε is a noise sequence which is assumed to be normally distributed, with zero mean and variance σ^2.

If we use the forward-backward linear prediction method, (4) can be written as:

$$\begin{bmatrix} y[p+1] \\ y[p+2] \\ \vdots \\ y[n] \\ y[1] \\ y[2] \\ \vdots \\ y[n-p] \end{bmatrix} = \begin{bmatrix} y[p] & y[p-1] & \cdots & y[1] \\ y[p+1] & y[p] & \cdots & y[2] \\ \vdots & \vdots & & \vdots \\ y[n-1] & y[n-2] & \cdots & y[n-p] \\ y[2] & y[3] & \cdots & y[p+1] \\ y[3] & y[4] & \cdots & y[p+2] \\ \vdots & \vdots & & \vdots \\ y[n-p+1] & y[n-p+2] & \cdots & y[n] \end{bmatrix} \times \begin{bmatrix} a_1 \\ a_2 \\ a_3 \\ \vdots \\ \vdots \\ \vdots \\ a_{p-1} \\ a_p \end{bmatrix} + \varepsilon_j \tag{5}$$

Equation (5) can be ill-conditioned or inconsistent in many applications. In these cases, we can use singular value decomposition (SVD) to overcome the problem. That is, matrix **Y** is decomposed into three matrices as follows:

$$\mathbf{Y}_{\mathbf{p \times [2 \times (n-p)]}} = \mathbf{U}_{\mathbf{p \times [2 \times (n-p)]}} \boldsymbol{\Lambda}_{\mathbf{[2 \times (n-p)] \times [2 \times (n-p)]}} \mathbf{V}^{\mathbf{T}}_{\mathbf{[2 \times (n-p)] \times [2 \times (n-p)]}} \tag{6}$$

where $\boldsymbol{\Lambda}$ is a diagonal matrix containing singular values:

$$\mathbf{\Lambda}_{[2\times(n-p)]\times[2\times(n-p)]} = \begin{bmatrix} \lambda_1 & 0 & 0 & 0 \\ 0 & \lambda_2 & 0 & 0 \\ \vdots & \vdots & \ddots & \vdots \\ 0 & 0 & 0 & \lambda_{2\times(n-p)} \end{bmatrix} = \text{diag}(\lambda_j) \qquad (7)$$

In order to reduce noise effect, we can rank singular values as:

$$\lambda_1 \leq \lambda_2 \leq \ldots \leq \lambda_{2\times(n-p)}$$

Then we replace small λ_j values with zero.

The AR coefficients can then be found from the following equation:

$$\mathbf{a} = \mathbf{V}_{[2\times(n-p)]\times[2\times(n-p)]}\mathbf{\Lambda}^{-1}{}_{[2\times(n-p)]\times[2\times(n-p)]}\mathbf{U}^{\mathbf{T}}{}_{p\times[2\times(n-p)]}\mathbf{y} \qquad (8)$$

where $\mathbf{\Lambda}^{-1}{}_{[2\times(n-p)]\times[2\times(n-p)]} = \text{diag}(1/\lambda_j)$. The prediction order p is chosen to be $N/2$ where N refers to window size. The reason for selecting this order is that Lang and McClellan recommended that the number of AR coefficients should be in the range of $N/3$ and $N/2$ for the best frequency estimation [47].

We apply singular value decompositions to **Y**, compute, rank the singular values and zero the small ones. Then we compute the noise-reduced **Y** by

$$\mathbf{Y} = \mathbf{U}\,\mathbf{\Lambda}\,\mathbf{V}^{\mathbf{T}} \qquad (9)$$

where $\mathbf{\Lambda}$ is a new diagonal matrix containing processed singular values.

Then, we average the values in each descending diagonal **Y** and put the averaged value back in their original positions. After that, we carry out singular value decomposition to compute the AR coefficients again according to Equations (6), (7) and (8). The noise reduction process can be iterated until the difference between the current **Y** is almost the same as the **Y** from the last iteration. However, more iterations means higher computation complexity. A threshold can be set for the difference value to keep the balance between noise reduction and computation complexity.

Finally, power spectral density (PSD) can be calculated based on the equation below:

$$P_{AR}(\omega) = \frac{\sigma^2}{\left|1 + \sum_{k=1}^{p} a_k \exp(-j\omega k)\right|^2} \tag{10}$$

where σ^2 is the variance of noise.

Usually there are several numerical signals obtained from a single DNA symbolic sequence. After applying the AR model to each numerical signal, the same number of PSD series can be obtained. The PSD series can be added together to achieve a better prediction performance as the noise inside the PSD series is further reduced. Exon prediction based on the AR model and multi-feature spectral analysis performs well for short human exon detection.

One common way to evaluate the performance of an exon prediction method is to use the Receiver Operating Characteristic (ROC) curve. Criteria such as true positive rate, true negative rate, false positive rate and false negative rate are all clearly shown on the ROC curve. The larger the area under the ROC curve, the better the performance of the algorithm is [41].

3. Regulatory region (promoter) prediction methods

3.1. *Background*

A promoter is a region of a genomic DNA sequence located near a gene and contains elements to regulate the transcription of the gene. The binding of these critical elements with transcription factors (TFs) serves as direct docking platforms for RNA polymerase II complex to initiate the gene transcription (Figure 2).

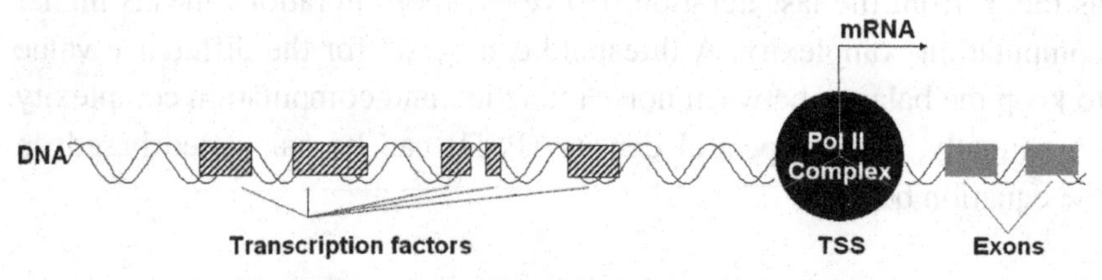

Fig. 2. The organization of the eukaryotic promoter

Computational methods of promoter prediction mostly rely on the conserved *cis*-acting sequence motifs, such as TATA-boxes, CpG islands and CAAT boxes. The techniques involved in promoter prediction mainly include support vector machine (SVM), artificial neural network (ANN), quadratic discriminant analysis (QDA), position weight matrix (PWM) and hidden Markov model (HMM). The performance of prediction can be evaluated by a series of standards, such as false positive rate, specificity, sensitivity, precision, recall, F-measure, correlation coefficient, etc.

Support vector machine (SVM) is the most widely used machine learning technique in promoter prediction. Based on the principle of structural risk minimization, SVM has an advantage in solving the problems of small sample size, high dimensionalities, non-linear classification and local minimal [48]. Given a set of training samples, each of which belongs to one of the two classes, SVM algorithms construct a non-probabilistic binary classifier that assigns the new samples into one class or the other. Firstly, the samples which are not linearly separable in the original space are mapped to high-dimensional feature space by a kernel function. The algorithms then calculate the optimal solutions that represent the hyperplanes in the high-dimensional feature space to separate these samples [49]. Gangal and Sharma applied SVM to build a human promoter prediction model named Prometheus [50]. The variation of DNA sequence content was seen as a dynamic process and the non-linear time series descriptors (Lyapunov component and Tsallis entropy) were used to depict features of DNA sequences. The model achieved an accuracy of more than 85% and successfully identified all twenty promoters experimentally verified on human chromosome 22. SVM-based promoter prediction methods are especially useful when the data points are not regularly distributed or have an unknown distribution. By choosing different thresholds that separate promoters from non-promoters, SVMs can provide some flexibility to decrease the influence of bias in the training samples. A major disadvantage of SVM-based methods is the lack of transparency of the results. The score of all promoter and non-promoter sequences cannot be represented as a parametric function of all the input features. Moreover, the quadratic programming (QP) optimization in the learning process, as a core step of SVMs, usually takes a long time.

An artificial neural network (ANN) combined with a genetic algorithm is an important tool in promoter prediction. The inputs to ANN are various features extracted from original DNA sequences. Choice of architecture for the network usually depends on which features are used and how they are encoded. The network for which the architecture is already determined is then trained based on the training set. The threshold on each neuron and the connection weights are continually adjusted during this learning process. The trained network is then tested on the test set. The most commonly used model evaluation method for promoter prediction is *k*-fold cross validation. Knudsen developed a vertebrate promoter prediction server Promoter2.0 that is based on four independent neural networks [18]. Each network consists of only one hidden and one output neuron, input neurons from a DNA sequence with modifiable length (each nucleotide type was encoded to a four-bit vector) and input neurons from other networks, were used to detect TATA-box, cap site, CCAAT-box and GC-box respectively and optimized by a genetic algorithm to discriminate between promoters and non-promoters. Arniker *et al.* proposed a neural network-based promoter classifier MultiNNProm [19]. It employed four different mapping functions to convert a DNA sequence into four numerical sequences, which were fed to four neural networks separately. Each of the four trained neural networks thus captured a particular promoter property embedded in a DNA sequence. Outputs from neural networks were then passed onto a probability builder function to calculate probabilities as to whether a tested sequence is a promoter or not. The four probability outputs were finally combined through an aggregation function with weights determined by a genetic algorithm and the final probability was used to distinguish promoters from non-promoters. Reese applied the time-delay neural network (TDNN) architecture in the promoter prediction tool for Drosophila melanogaster [20]. TDNN was originally designed for processing a speech sequence in a time series with local time shifts. This model is less influenced by the variable spacing between two features, the TATA-box and *Inr,* which are specified by the author, and can precisely identify them. ANN-based promoter prediction methods have a number of advantages. One is that the input and output features can be continuous (for example, score) or categorical (for example, logic 0 and 1 that correspond to non-promoter and promoter, respectively). Secondly, neural networks are capable of detecting and delivering possible interactions between input variables. Thirdly, neural network models can be developed using multiple architectures and training algorithms. ANN-based methods also have shortcomings. Just like most non-linear methods, the results lack transparency and have difficulty revealing the causal relationship between input features and the outcome. Over-fitting and local minimals are two additional main disadvantages of neural networks.

Quadratic discriminant analysis (QDA) is a classical multiple-variable statistical tool often used to analyze which variables are the best predictors of classifying samples into different groups. A QDA function [51] is

$$\delta_k(X) = -\frac{1}{2}\log|\Sigma_k| - \frac{1}{2}(X-\mu_k)^T \Sigma_k^{-1}(X-\mu_k) + \log \pi_k \quad (11)$$

where Σ_k, μ_k and π_k respectively represent a covariance matrix, a mean descriptive vector and prior probability for each class $k = 1, 2, \ldots, K$. Each sample is described by a r-dimensional vector X $(x_1, x_2, \ldots, x_r)$. In the case of promoter prediction, the number of classes is two, corresponding to promoter set I_1 and non-promoter set I_2. So the function can be expressed as

$$\delta = -\frac{1}{2}\log\frac{|\Sigma_1|}{|\Sigma_2|} - \frac{1}{2}(\xi_1 - \xi_2) + \log\frac{\pi_1}{\pi_2} \quad (12)$$

where $\xi_k = (X-\mu_k)^T \Sigma_k^{-1}(X-\mu_k)(k=1,2)$ stands for the Mahalanobis distance between X and μ_k. Davuluri *et al.* used three different quadratic discriminant functions (QDFs)-donor QDF, promoter QDF and first-exon QDF to characterize splice-donor sites, promoter regions and first exons, and incorporated them to construct a program named FirstEF to predict first exons and promoters of genes [21]. FirstEF was applied on a first-exon database and completed human chromosome 21 and 22 and showed a better performance than the commercial software PromoterInspector [52]. The separating surfaces between classes, which are represented by quadratic discriminant functions, can adopt more complex quadratic shapes. Consequently, the QDA-based promoter prediction model has greater freedom in matching the peculiarities of the data distribution. However, it may also lead to over-fitting of the data.

Position weight matrix (PWM) is a commonly used motif representation method. It captures the probability of observing the four nucleotides A, T, G and C at each specific position [53]. Firstly, a number of experimentally determined TF binding sites are collected and aligned. They then serve as the training set to construct the position weight matrices, which represent the common intrinsic features of the TF binding sites in this collection. By adopting certain optimization algorithms, a threshold value that works as a discrimination standard is determined. Then, for an input DNA sequence of specified length, the derived PWM is used to calculate the similarity of this new sequence with the TF binding sites in the training set. If the total score of the sequence is above a certain threshold, it is recognized as having at least one TF binding site (promoter). Since the PWM technique was introduced in promoter prediction in 1990 by Bucher [53], relative algorithms have been optimized continually and the position weight matrices for TATA-box, Initiator, CCAAT-box, GC-box in the three main resources of TF binding sites, namely Eukaryotic Promoter Database (EPD), JASPAR and TRANSFAC, have been updated several times [54-56]. Compared with motifs, PWMs can capture more quantitative information and thus can be a more accurate representation for specific TF binding sites. On the other hand, to produce a reliable PWM, a great number of TF binding sites are needed. This requirement cannot be easily satisfied. Another disadvantage of PWMs is that they can hardly reflect the dependence between the bases (when a change of one nucleotide in one position leads to a corresponding change in another position within the binding site) [57].

The Hidden Markov model (HMM) has been extensively used for sequence analysis. Before its application on promoter detection, this method had achieved some success in identifying genes or protein coding regions [11,12]. It was introduced first into eukaryotic promoter prediction by Audic and Claverie[13]. Firstly, the probability of each *k*-

mer being followed by A, T, G or C is computed by counting the occurrence of each (k+1)-mer in the data set, based on which a transition matrix T is established. For a DNA sequence W of length L, the probability of W being generated by a stochastic process based on the transition matrix T is as follows [12]

$$P(W|T) = p(s_0) \cdot p(n_k|s_0) \ldots p(n_{L-1}|s_{L-k-1}) \tag{13}$$

where s_i is the k-mer at the position i, n_k is the nucleotide at position k, $p(s_i)$ is the probability of the k-mer occurring at position s_i and $p(n_k|s_i)$ is the probability of nucleotide n_k following the k-mer s_i. Then for a given DNA sequence W, it will be classified as promoter or non-promoter according to the following probability [12]:

$$P(T_i|W) = \frac{P(W|T_i)P(T_i)}{\sum_{j=1}^{l} P(W|T_j)P(T_j)} \tag{14}$$

Where l = 2 corresponds to two transition matrices T_{prom} (T_1) and $T_{non-prom}$(T_2), and if $P(T_{prom}|W) > P(T_{non-prom}|W)$, W is a promoter, and if $P(T_{prom}|W) < P(T_{non-prom}|W)$, W is a non-promoter. Ohler *et al.* introduced the interpolated Markov chains (IMCs), which have been successfully applied into both speech recognition and eukaryotic promoter recognition [14]. The basic idea of the interpolation technique is to re-estimate the probability of subsequences shorter than k when the frequencies of a k-mer $s = n_1 \ldots n_k$ cannot be reliably estimated. The interpolation techniques include linear and rational interpolation, in which the new conditional probability $\hat{P}(n_k|n_1 \ldots n_{k-1})$ is defined as equation (15) and (16), respectively:

$$\hat{P}(n_k|n_1 \ldots n_{k-1}) = \rho_0 \frac{1}{M} + \rho_1 P(n_k) + \rho_2 P(n_k|n_{k-1}) + \cdots + \rho_k P(n_k|n_1 \ldots n_{k-1}) \tag{15}$$

$$\hat{P}(n_k|n') = \frac{\sum_{i=0}^{k} \rho_i \cdot g_i(n') \cdot P_i(n_k|n')}{\sum_{i=0}^{k} \rho_i \cdot g_i(n')} \tag{16}$$

In Equation (16), $g_i(n')$ is a function monotonically related to the reliability of the content $n' = n_1 \dots n_k$. The IMCs are beneficial in getting a balance between model context, the number of training samples and the absence of certain *k*-mers in the case of a larger context. Won *et al.* proposed an HMM architecture composed of interconnected blocks that represent transcription factor binding sites (TFBSs) and background regions of promoters respectively [15]. Four HMM block types, namely linear block, self-loop block, forward-jump blocks and zero blocks, were used for background description, and a type that represents forward and backward reading of a position-specific scoring matrix (PSSM block) was used to model TFBSs. Genetic algorithms were then used to optimize the HMM architecture that starts from a random one. The model was tested on artificial, human, mouse and muscle-specific promoter sets, and was found to have the ability to reconstruct the existing grammar of TFBSs, reveal the underlying regulation mechanism in the real promoter data and distinguish tissue-specific promoters from others. HMM-based promoter prediction methods have a sound statistical grounding and HMMs can be combined into larger HMMs, giving researchers a lot of freedom to manipulate the training and verification processes. Another excellent aspect of HMM-based methods is the transparency of the model, which can be easily read and made sense of. The disadvantages of HMM-based methods include the assumption of independence between states, which is usually untrue, over-fitting of the data and local maximums. Besides, HMM may also bring computational burdens because all the possible state transition paths are enumerated.

Besides the features of motifs with different lengths, DNA sequences also have very special structural and physical properties, which are the spatial and stereochemical basis for various regulatory mechanisms during the genetic process. It is generally thought that the DNA topology strongly influences transcription by immediately promoting the formation of specialized structures that are favored by polymerase, transcription factors or other auxiliary factors. Over the past twenty years, many experiments have been carried out to investigate the physical and structural aspects of the DNA molecules and different structural models have been recommended (Table 1). Instead of merely exploring the sequence content feature, more and more researchers have turned to using the physical and structural properties of DNA to predict promoters.

Ohler *et al.* introduced a joint model of promoter that combines these physical properties of DNA and sequence likelihoods [58]. In their model, a promoter is represented as a sequence of consecutive segments represented by joint likelihoods for DNA sequences and profiles of physical properties. Sequence likelihoods are modeled with interpolated Markov chains while their physical properties modeled with Gaussian distributions. The background uses two joint models for the coding and non-coding sequences, each consisting of a mixture of a sense and an anti-sense submodel. Using this model, they reduced the false positives obtained on the Drosophila test set by about 30%.

Florquin *et al.* plotted the profiles of physical properties for the promoters of Arabidopsis, rice, human and mouse and then performed clustering on these profiles [44]. By this method, they proved the existence of distinct types of core promoters and found that the structural profiles are much conserved within plants (Arabidopsis and rice) and animals (human and mouse) but greatly different between plants and animals. They indicated that the promoter regions can be separated from the non-promoter regions using these profiles.

Zeng *et al.* adopted a tetranucleotide model to calculate the rigidity profiles of human core promoters (-200 to +50 bp relative to the TSS) [26]. A single rigidity profile is rather noisy, so they smoothed each profile within a 100 bp window. Then the method of graph-based consensus clustering (GCC) was used to classify human promoters into different groups based on the similarity of their rigidity profiles. They found that promoters have a marked change around the TSS.

Next, we will introduce three promoter detection methods in detail. One is based on the DNA sequence content features and a cascade AdaBoost algorithm. The other one combines the motif features and a DNA structural property and makes use of a decision tree to construct a promoter recognition system. The third one is an SVM-based method, in which a collective structural profile is extracted from multiple DNA physical and structural properties by non-linear dimensionality reduction algorithm - isometric featuring mapping (Isomap) and used as discriminating features.

3.2. *Cascade AdaBoost algorithm*

A promoter identification method based on a cascade AdaBoost-based learning procedure is reviewed here [59]. The block diagram of this method is shown in Figure 3. In this approach, three different kinds of features including local distribution of pentamers, positional CpG island features and digitalized DNA sequence are extracted from DNA sequences and combined to build a high-dimensional input vector. Then a cascade AdaBoost algorithm is adopted to select a small subset of the most discriminating features and to train the classifiers. The performance of the model is evaluated by testing it on large-scale DNA sequences from different databases, such as EPD, DBTSS, GenBank and human chromosome 22. The result shows that the AdaBoost-based classifier outperforms some well-known ones, such as PromoterInspector [52], Dragon Promoter Finder (DPF) [60], and First Exon Finder (FirstEF) [21].

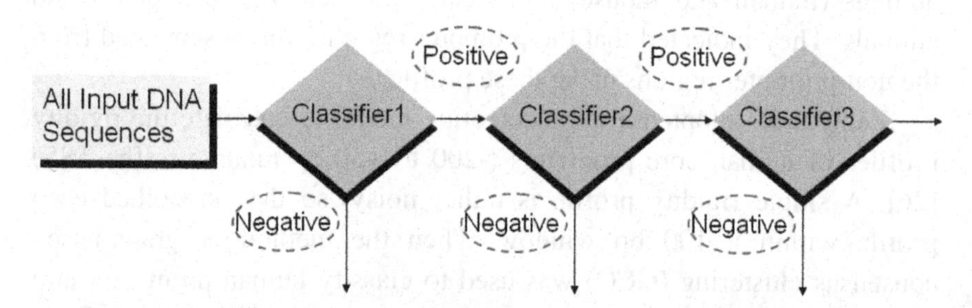

Fig. 3. Promoter identification based on a cascade AdaBoost algorithm (This diagram is adapted "adopted' from Figure 1 in [59])

3.3. *Hierarchical promoter prediction system based on signal, context and structural properties*

The choice of distinguishing features determines to a large extent the performance of promoter prediction. Here, three different kinds of features, namely signal, context and structural features of a given DNA sequence, are integrated as input features. By adopting decision trees, a hierarchical promoter prediction system is constructed [61]. The hierarchical system is tested on a database consisting of promoters, exons, introns and 3'UTR. Promoters are selected from regions of [−200, +50] relative to TSSs from 30,946 sequences in DBTSS as promoters, and 75,437 exons and 53,682 introns with 251 bp in length are selected from the Exon-Intron Database (EID) [62] and 80 538 3'UTR sequences with a length of 251 bp from the UTRdb database [63] as non-promoters. The performance of our hierarchical promoter prediction system is compared with KLC [64], FirstEF [21], McPromoter [58], Eponine [65] and ProStar [66] based on the 3-fold cross validation, and the result shows that the hierarchical system keeps the best balance between sensitivity (*Se*) and specificity (*Sp*) with the highest *Se* + *Sp*=1.7210 (Table 2).

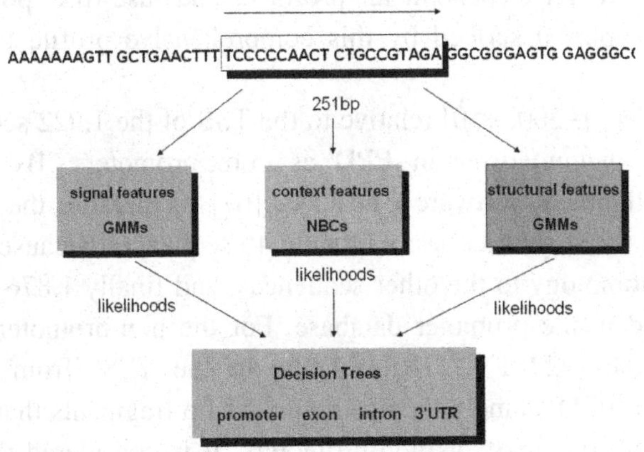

Fig. 4. The architecture of the hierarchical system that integrates signal, context and structure features for promoter recognition. Firstly, the system scans the entire human genome by a sliding window 251 bp in length, which moves over the sequence and extracts signal, context and sequence features within this window. Secondly, these features are fed into component classifiers such as GMMs and NBCs. Each component classifier output four conditional-class probabilities for categories of promoters, exons, introns and 3'UTRs. Finally, decision trees combine the twelve likelihoods to decide if this sequence belongs to either promoters or other genomic regions including exons, introns and 3'UTRs. (This diagram is adapted from Figure4 in [61]).

Table 2. Performance of different promoter predictior models based on the 3-fold cross validation

Methods	*Se*	*Sp*	*Se+Sp*
Hierarchical System	0.7917	0.9293	1.7210
KLC	0.5125	0.8618	1.3743
FirstEF	0.5457	0.8808	1.4265
McPromoter	0.4117	0.8671	1.2788
Eponine	0.2412	0.9663	1.2075
ProStar	0.3401	0.9636	1.3037

3.4. *Prediction of eukaryotic core promoters based on Isomap and support vector machine*

Instead of using PWMs for motifs or individual DNA physicochemical profiles as the descriptor to discriminate promoters and non-promoters, we apply the non-linear dimensionality reduction algorithm-Isomap to derive a comprehensive physicochemical profile from the fourteen original DNA physicochemical profiles, and use the positive and negative samples described by this comprehensive profile to train the SVM.

We select [−200, +50] relative to the TSS of the 1,922 sequences of *Drosophila melanogaster* in EPD as core promoters. By using the sequence alignment software ClustalX2 [67] to measure the homology between all the sequences, we eliminate 46 sequences because they have very high homology to the other sequences, and finally 1,876 sequences are included in the promoter database. For the non-promoter database, we select the [−2251, −2000] relative to the TSSs from the 1,922 sequences in EPD, namely the upstream 251 bp fragments that are 2,000 bp apart from the TSSs, as non-promoters. It is considered that critical structural response elements rarely reside in these regions. Thus they are competent for the role of non-promoters, and theoretically speaking, have different physicochemical or structural characteristics compared with core promoter regions. For the sake of an unbiased size of positive and negative dataset, we also eliminate 46 sequences that have the highest homology with others so that the non-promoter database consists of 1,876 sequences.

The DNA sequence of each promoter is converted into fourteen physicochemical profiles (Table 1). Therefore, each core promoter is depicted by a 251×14 matrix, and the non-linear dimensionality reduction algorithm-Isometric feature mapping (Isomap) is applied to extract a comprehensive physicochemical profile. Isomap is one of the most widely used algorithms in the manifold learning field. The advantage of manifold learning over linear dimensionality reduction techniques, such as principal component analysis (PCA) and singular value decomposition (SVD), lies in that it not only reflects relevance between dimensionalities but also makes an approximation to the true geometry of the data distribution in a high-dimensional space [68,69]. The basic strategy of manifold learning approaches is to extract, where one exists, the low-dimensional manifold upon which the points in high-dimensional space approximately reside [69]. Isomap has three steps:

(1) Build a neighborhood graph. This step determines which points are neighbors on the manifold *M*. If the distance $d_X(i, j)$ between two points *i*, *j* in the input space *X* satisfies the criteria of *K*-nearest neighbors or ε-radius, they are regarded as neighbors. A weighted graph *G* over all the data points is defined by this means.

(2) Calculate shortest paths. The shortest path distances $d_G(i, j)$ in the graph *G* defined above are calculated and used to approximate the true geodesic distances $d_M(i, j)$ between all pairs. This can be done by various graph analysis algorithms. Floyd's algorithm, for example, iteratively improves the estimate on the shortest paths by comparing all the possible paths between all point pairs through the graph, until the optimal value is obtained. Graph *G* is firstly initialized by

$$d_G(i, j) = \begin{cases} d_X(i, j) & i, j \text{ are linked by an edge} \\ \infty & \text{otherwise} \end{cases} \tag{17}$$

Then, for each $k = 1,2,\ldots,N$ in turn, $d_G(i, j)$ is replaced by the minimal value in $\{d_G(i, j),\ d_G(i,k)+d_G(k, j)\}$. Finally, the collection $D_G=\{d_G(i, j)\}$ represents the shortest paths between all point pairs in graph *G*.

(3) Construct *d*-dimensional embedding. The classical MDS is applied to construct an embedding of geodesic distance data

$D_G = \{d_G(i, j)\}$ into a d-dimensional Euclidean space Y. The vectors y_i in Y are those that can minimize the cost function

$$E = \|\tau(D_G) - \tau(D_Y)\|_{L^2} \tag{18}$$

in which $\| \ \|_{L^2}$ is the L^2 matrix norm and τ is an operator that converts distances to inner products. The number of dimensionality d is described by the top d eigenvectors in the matrix $\tau(D_G)$. Freely available MATLAB code which implements Isomap can be downloaded from http://isomap.stanford.edu.

The first principal dimensionality that accounts for the most original variance is taken as the comprehensive profile. Each core promoter is then described by this profile, which is a 251×1 matrix, or a 251-dimensional vector. We use a sliding window of varied size (5 bp, 10 bp and 20 bp) to smooth the comprehensive profile at the speed of 1 bp shift each time, and the average value within this window is taken as a physicochemical feature at the fist position of the window. So finally, each core promoter is described by three vectors of 246-, 241- and 231-dimensionality, respectively. Here, we use LibSVM – a library for SVM developed by Chang and Lin [70] to train on the collection of promoters and non-promoters. The RBF kernel is used. The 10-fold cross validation shows that the sensitivity and specificity are 0.54 and 0.58 for the profile averaged by a 5bp window, 0.56 and 0.61 for the profile averaged by a 10bp window, and 0.54 and 0.59 for the profile averaged by a 20 bp window. So the profiles smoothed by windows of different sizes achieve a similar performance. We also try the 5-fold and 7-fold cross validation and obtain a similar discriminating performance.

Our model is compared with two *Drosophila* promoter predictors McPromoter [58] and NNPP [20], which use the occurrence frequency of motifs as input. The average sensitivity and specificity for McPromoter are 0.43 and 0.47, while the average sensitivity and specificity for NNPP are 0.37 and 0.10. Chan and Kibler used 6-mer distribution for identifying *cis*-regulatory motifs in *Drosophila* and achieved a sensitivity and specificity of 0.39 and 0.94 [71]. Hence, our SVM-approach based on the comprehensive DNA physicochemical profile obtains an equivalent or better discriminating performance. On the other hand, compared with the good promoter prediction performances reported by Anwar *et al.* (sensitivity: 0.96 and specificity: 0.92) [72] and Down *et al.* (sensitivity: 0.90 and specificity: 0.97) [73], our method has its drawbacks, implying that promoter recognition based purely on the DNA physicochemical and structural properties has its limits. One major reason is the high noise in the physicochemical profile that may obscure the local important features and thus impair its discriminating ability. Anyhow, the structural features still serve as a good complement to the motif features in describing the intrinsic differences between transcriptional regulation elements, protein-coding regions and non-functional regions.

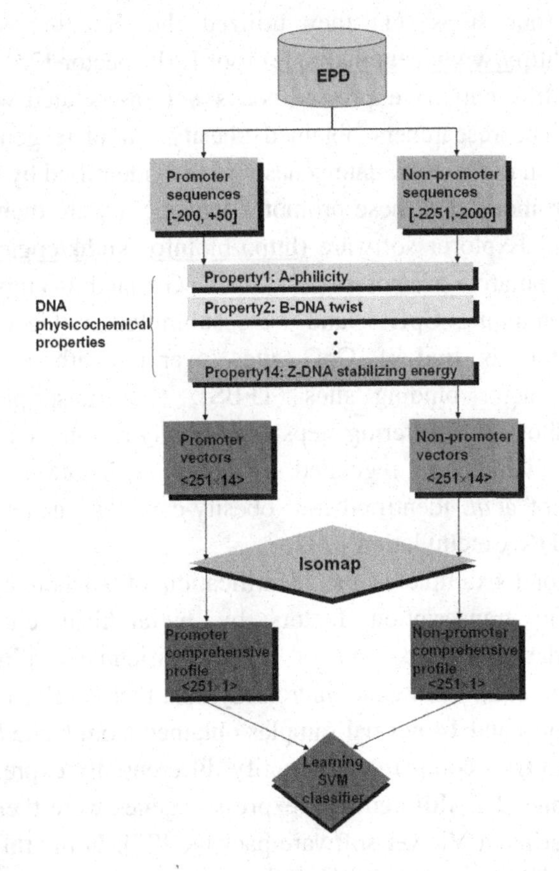

Fig. 5. The architecture of promoter prediction system based on Isomap and SVM.

3.5. *Computational identification of disease-related genes and regulatory regions*

The rapid development of bioinformatics methods and tools have greatly facilitated the identification of disease-related genes and regulatory regions. Here we provide some examples. The first example is the identification of epigenetically regulated genes associated with obesity by comprehensively using several online bioinformatics tools [74]. Turcot *et al.* firstly obtained gene expression microarrays from the peripheral blood RNA from children aged 11-13 years who were grouped according to their obesity degree. Genes showing no less than 2.0 fold differential expression were then selected for computational analysis that consists of four steps: (1) they utilized the literature search tool Genomatix (http://www.genomatix.de) tool LitInspector [75] to identify whether the differentially expressed genes were associated with obesity phenotypes. The researchers obtained obesity-candidate genes; (2) the promoter region of the candidate genes was then identified by Genomatix tool Gene2Promoter; (3) these promoter sequences were then submitted to CpG island Explorer software (http://bioinfo.hku.hk/cpgieintro.html) to find which putative promoters contain a CpG island; (4) those putative promoters containing CpG island were submitted to Genomatix tool ModelInspector to find if CpG sites overlap with at least one transcription factor binding sites (TFBS). The transcripts obtained through the above four filtering steps were finally recognized as obesity-related genes which are regulated by DNA methylation. Using this method, Turcot *et al.* identified four obesity-candidate genes putatively regulated by DNA methylation.

The second example is the identification of Alzheimer's disease (AD) specific transcription factors by hierarchical clustering of microarray data and analysis of TFBS enrichment [76]. Firstly, Krishnamurthy *et al.* used gene microarray data that consisted of 14 AD affected samples and 14 normal samples obtained from Gene Expression Omnibus (GEO) to compare and identify differentially expressed genes in the AD stage. The differentially expressed genes were then clustered by MultiExperiment Viewer software package [77]. In the third step, by using oPOSSUM program [78], each gene cluster was analyzed for enrichment of TFBS. For each transcript, the top 10% of conserved regions in the 2000 bp upstream and downstream range was scanned for TFBS with a matrix match threshold of 80% based on a position weight matrices algorithm. The TFBS profiles come from JASPAR database [79]. Krishnamurthy *et al.* finally determined five transcription factor binding sites as the most important regulator of Alzheimer's disease during the apoptosis pathway.

The third example is a computational study on prediction of both genetic and serum markers for seven cancer types conducted by Xu *et al.* [80]. Firstly, microarray gene expression data were downloaded for seven cancer types, namely, breast, colon, kidney, lung, pancreatic, prostate and stomach cancer from the GEO database of NCBI [81]. The dataset consisted of cancer and control samples from the same patients. Being similar as the first and second example, differentially expressed genes were then identified according to a series of statistical standards specified by the authors. Next, Xu *et al.* used a SVM-based classifier to predict if the proteins, as the product of differentially expressed genes, are blood-secretory. The input sequence-based features include signal peptides, transmembrane domains, glycosylation sites and polarity measures. They also applied the SVM-based classifier to identify marker genes. For each cancer type, all markers were ranked according to the 5-fold cross-validation performance on the training dataset and the top ones were recognized as discriminators for each cancer type. They also calculated the discerning power of the *k*-gene groups (combinations of *k* genes for *k*=1, 2, 3 and 4) across multiple cancer types. Xu *et al.* identified 19 genes that are differentially expressed in more than four cancer types and 11 *k*-gene discriminators with their proteins being blood secretory and highest discerning power. The knowledge found is helpful in elucidating the genetic alterations in various cancers.

Computational analysis is indispensable in almost all disease-related genomic studies. Application of in silico analysis on identification of disease-related genes and regulatory regions provides opportunities for new approaches in diagnostics and therapeutics. As a further refinement of various disease-related gene databases, the prediction accuracy of diverse computational approaches will also be improved.

4. Summary

The application of signal processing and machine learning techniques, on the whole, has achieved good results on the prediction of genomic functional regions. However, most computational methods are confronted with the weakness of high false positive rates, poor compromise between prediction sensitivity and specificity and conflicts between prediction precision and computational cost. The bottleneck of computational prediction of genes and promoters mainly lies in two aspects. One is the lack of experimentally determined functional regions, leading to a very limited number of training and testing samples. The second problem is the intrinsic weakness of various machine learning algorithms. Efforts to develop better methods for protein-coding and transcription regulatory region detection can be made in two directions: one is to find more accurate, informative and effective feature sets for exons and transcription binding sites. The other is to take advantage of new machine learning methods or give partial modifications on the frameworks established in previous studies to make the models more approximate to the true exon and transcription regulatory regions in the genome of a specific species.

Additionally, for a detailed list of popular gene and promoter prediction software, readers can refer to references [82-84].

Acknowledgement

This work is supported by the Hong Kong Research Grants Council (Project CityU 123809) and City University of Hong Kong (Project 7002843).

References

1. Wong, G.K.S., *et al.*, *Is "Junk" DNA Mostly Intron DNA?* Genome Research, 2000. 10(11): p. 1672-1678.
2. Rogozin, I.B., *et al.*, *Analysis of evolution of exon-intron structure of eukaryotic genes.* Briefings in bioinformatics, 2005. 6(2): p. 118-134.
3. Glanz, S. and U. Kück, T*rans-splicing of organelle introns–a detour to continuous RNAs.* Bioessays, 2009. 31(9): p. 921-934.

4. Down, T.A., *et al.*, *Large-scale discovery of promoter motifs in Drosophila melanogaster.* PLoS Computational Biology, 2007. 3(1): p. e7.
5. Zhang, Z.D., *et al.*, *Statistical analysis of the genomic distribution and correlation of regulatory elements in the ENCODE regions.* Genome Research, 2007. 17(6): p. 787-797.
6. Costa, F.F., L.S. Foly, and M.P. Coutinho, *DataGenno: building a new tool to bridge molecular and clinical genetics.* The Application of Clinical Genetics, 2011. 4: p. 45-54.
7. Beckmann, J.S., X. Estivill, and S.E. Antonarakis, *Copy number variants and genetic traits: closer to the resolution of phenotypic to genotypic variability.* Nature Reviews Genetics, 2007. 8(8): p. 639-646.
8. Safran, M., *et al.*, *GeneCards Version 3: the human gene integrator.* Database: the journal of biological databases and curation, 2010. baq020.
9. Theuns, J., *et al.*, *Promoter mutations that increase amyloid precursor-protein expression are associated with Alzheimer disease.* The American Journal of Human Genetics, 2006. 78(6): p. 936-946.
10. Bekri, S., *et al.*, *A promoter mutation in the erythroid-specific 5-aminolevulinate synthase (ALAS2) gene causes X-linked sideroblastic anemia.* Blood, 2003. 102(2): p. 698-704.
11. Haussler, D.K.D. and M.G.R.F.H. Eeckman, *A generalized hidden Markov model for the recognition of human genes in DNA.* Proceedings International Conference on Intelligent Systems for Molecular Biology, 1996. 4: p. 134-142.
12. Krogh, A., I.S. Mian, and D. Haussler, *A hidden Markov model that finds genes in E. coli DNA.* Nucleic Acids Research, 1994. 22(22): p. 4768-4778.
13. Audic, S. and J.M. Claverie, *Detection of eukaryotic promoters using Markov transition matrices.* Computers & chemistry, 1997. 21(4): p. 223-227.
14. Ohler, U., *et al.*, *Interpolated markov chains for eukaryotic promoter recognition.* Bioinformatics, 1999. 15(5): p. 362-369.
15. Won, K.J., *et al.*, *Modeling promoter grammars with evolving hidden Markov models.* Bioinformatics, 2008. 24(15): p. 1669-1675.
16. Xu, Y., R.J. Mural, and E. C. Uberbacher, *Constructing gene models from accurately predicted exons: an application of dynamic programming.* Computer Applications in the Biosciences, 1994. 10(6): p.613-623.
17. Grzegorczyk, M. and D. Husmeier, *Improvements in the reconstruction of time-varying gene regulatory networks: dynamic programming and regularization by information sharing among genes.* Bioinformatics, 2010. 27(5): p. 693-699.
18. Knudsen, S., *Promoter2.0: for the recognition of PolII promoter sequences.* Bioinformatics, 1999. 15(5): p. 356-361.
19. Arniker, S.B., *et al. Promoter prediction using DNA numerical representation and neural network: Case study with three organisms.* India Conference (INDICON), 2011 Annual IEEE, 2011. p.1-4.

20. Reese, M.G., *Application of a time-delay neural network to promoter annotation in the Drosophila melanogaster genome.* Computers & chemistry, 2001. 26(1): p. 51-56.
21. Davuluri, R.V., I. Grosse, and M.Q. Zhang, *Computational identification of promoters and first exons in the human genome.* Nature genetics, 2001. 29(4): p. 412-417.
22. Tiwari, S., *et al.*, *Prediction of probable genes by Fourier analysis of genomic sequences.* Bioinformatics, 1997. 13(3): p. 263-270.
23. Ouyang, Z., *et al.*, *Multivariate entropy distance method for prokaryotic gene identification.* Journal of Bioinformatics and Computational Biology, 2004. 2(2): p. 353-373.
24. Zhu, H., *et al.*, *MED: a new non-supervised gene prediction algorithm for bacterial and archaeal genomes.* BMC Bioinformatics, 2007. 8: p. 97.
25. Pedersen, A.G., *et al.*, *DNA structure in human RNA polymerase II promoters.* Journal of molecular biology, 1998. 281(4): p. 663-673.
26. Zeng, J., *et al.*, *Finding human promoter groups based on DNA physical properties.* Physical Review E, 2009. 80(4): p. 041917.
27. Cao, X.Q., J. Zeng, and H. Yan, *Structural properties of replication origins in yeast DNA sequences.* Physical biology, 2008. 5(3): p. 036012.
28. Yandell, M. and D. Ence, *A beginner's guide to eukaryotic genome annotation.* Nature Reviews Genetics, 2012. 13(5): p. 329-342.
29. Zhang, M.Q., *Computational prediction of eukaryotic protein-coding genes.* Nature Reviews Genetics, 2002. 3(9): p. 698-709.
30. Burge, C. and S. Karlin, *Prediction of complete gene structures in human genomic DNA.* Journal of molecular biology, 1997. 268(1): p. 78-94.
31. Korf, I., *et al.*, *Integrating genomic homology into gene structure prediction.* Bioinformatics, 2001. 17(suppl 1): p. S140-S148.
32. Fickett, J.W., *Recognition of protein coding regions in DNA sequences.* Nucleic Acids Research, 1982. 10(17): p. 5303-5318.
33. Tiwari, S., *et al.*, *Prediction of probable genes by Fourier analysis of genomic sequences.* Bioinformatics, 1997. 13(3): p. 263-270.
34. Akhtar, M., E. Ambikairajah, and J. Epps. *Optimizing period-3 methods for eukaryotic gene prediction.* IEEE International Conference on Acoustics, Speech and Signal Processing, 2008 (ICASSP 2008). p. 621-624.
35. Vaidyanathan, P. and B.J. Yoon. *Gene and exon prediction using allpass-based filters.* Workshop on Genomic Signal Processing and Statistics (GENSIPS), Raleigh NC, 2002.
36. Román-Roldán, R., P. Bernaola-Galván, and J.L. Oliver, *Sequence compositional complexity of DNA through an entropic segmentation method.* Physical Review Letters, 1998. 80(6): p. 1344-1347.
37. Yan, H. and T.D. Pham, *Spectral estimation techniques for DNA sequence and microarray data analysis.* Current Bioinformatics, 2007. 2(2): p. 145-156.

38. Choong, M.K. and H. Yan, *Multi-scale parametric spectral analysis for exon detection in DNA sequences based on forward-backward linear prediction and singular value decomposition of the double-base curves.* Bioinformation, 2008. 2(7): p. 273.
39. Jiang, R. and H. Yan, *Studies of spectral properties of short genes using the wavelet subspace Hilbert–Huang transform (WSHHT).* Physica A: Statistical Mechanics and its Applications, 2008. 387(16): p. 4223-4247.
40. George, T. and T. Thomas, *Discrete wavelet transform de-noising in eukaryotic gene splicing.* BMC bioinformatics, 2010. 11(Suppl 1): p. S50.
41. Nancy Yu, S. and Y. Hong, *Short exon detection in DNA sequences based on multifeature spectral analysis.* EURASIP Journal on Advances in Signal Processing, 2010. 2011.
42. Zhang, W.F. and H. Yan, *Exon prediction using empirical mode decomposition and Fourier transform of structural profiles of DNA sequences.* Pattern Recognition, 2012. 45(3): p. 947-955.
43. Wu, Y., *et al.*, *DB-Curve: a novel 2D method of DNA sequence visualization and representation.* Chemical Physics Letters, 2003. 367(1): p. 170-176.
44. Florquin, K., *et al.*, *Large-scale structural analysis of the core promoter in mammalian and plant genomes.* Nucleic Acids Research, 2005. 33(13): p. 4255-4264.
45. Ohler, U., *Computational promoter recognition in eukaryotic genomic DNA.*2002: Logos-Verlag.
46. Hawkin, J.D., *A survey on intron and exon lengths.* Nucleic Acids Research, 1988. **16**(21): p. 9893-9908.
47. Lang, S. and J. McClellan, *Frequency estimation with maximum entropy spectral estimators.* Acoustics, Speech and Signal Processing, IEEE Transactions on, 1980. 28(6): p. 716-724.
48. Cortes, C. and V. Vapnik, *Support-vector networks.* Machine learning, 1995. 20(3): p. 273-297.
49. Burges, C.J.C., *A tutorial on support vector machines for pattern recognition.* Data mining and knowledge discovery, 1998. 2(2): p. 121-167.
50. Gangal, R. and P. Sharma, *Human pol II promoter prediction: time series descriptors and machine learning.* Nucleic Acids Research, 2005. 33(4): p. 1332-1336.
51. Srivastava, S., M.R. Gupta, and B.A. Frigyik, *Bayesian quadratic discriminant analysis.* Journal of Machine Learning Research, 2007. 8(6): p. 1277-1305.
52. Scherf, M., A. Klingenhoff, and T. Werner, *Highly specific localization of promoter regions in large genomic sequences by PromoterInspector: a novel context analysis approach.* Journal of molecular biology, 2000. 297(3): p. 599-606.
53. Bucher, P., *Weight matrix descriptions of four eukaryotic RNA polymerase II promoter elements derived from 502 unrelated promoter sequences.* Journal of molecular biology, 1990. 212(4): p. 563-578.

54. Hannenhalli, S. and L.S. Wang, *Enhanced position weight matrices using mixture models.* Bioinformatics, 2005. 21(suppl 1): p. i204-i212.
55. Gershenzon, N.I., G.D. Stormo, and I.P. Ioshikhes, *Computational technique for improvement of the position-weight matrices for the DNA/protein binding sites.* Nucleic Acids Research, 2005. 33(7): p. 2290-2301.
56. Ben-Gal, I., *et al.*, *Identification of transcription factor binding sites with variable-order Bayesian networks.* Bioinformatics, 2005. 21(11): p. 2657-2666.
57. Prestridge, D.S., *Computer Software of Eukaryotic Promoter Analysis.* Methods in Molecular Biology, 2000. 130: p. 265-295.
58. Ohler, U., *et al.*, *Joint modeling of DNA sequence and physical properties to improve eukaryotic promoter recognition.* Bioinformatics, 2001. 17(suppl 1): p. S199-S206.
59. Xie, X., *et al.*, *PromoterExplorer: an effective promoter identification method based on the AdaBoost algorithm.* Bioinformatics, 2006. 22(22): p. 2722-2728.
60. Bajic, V.B., *et al.*, *Dragon Promoter Finder: recognition of vertebrate RNA polymerase II promoters.* Bioinformatics, 2002. 18(1): p. 198-199.
61. Zeng, J., *et al.*, *SCS: Signal, context, and structure features for genome-wide human promoter recognition.* Computational Biology and Bioinformatics, IEEE/ACM Transactions on, 2010. 7(3): p. 550-562.
62. Saxonov, S., *et al.*, *EID: the Exon–Intron Database—an exhaustive database of protein-coding intron-containing genes.* Nucleic Acids Research, 2000. **28**(1): p. 185-190.
63. Pesole, G., *et al.*, *UTRdb and UTRsite: specialized databases of sequences and functional elements of 5′ and 3′ untranslated regions of eukaryotic mRNAs. Update 2002.* Nucleic Acids Research, 2002. **30**(1): p. 335-340.
64. Wu, S., *et al.*, *Eukaryotic promoter prediction based on relative entropy and positional information.* Physical Review E, 2007. 75(4): p. 041908.
65. Down, T.A. and T.J.P. Hubbard, *Computational detection and location of transcription start sites in mammalian genomic DNA.* Genome Research, 2002. 12(3): p. 458-461.
66. Goñi, J.R., *et al.*, *Determining promoter location based on DNA structure first-principles calculations.* Genome Biology, 2007. 8(12): p. R263.
67. Larkin, M., *et al.*, *Clustal W and Clustal X version 2.0.* Bioinformatics, 2007. 23(21): p. 2947-2948.
68. Cayton, L., *Algorithms for manifold learning.* University of California, San Diego, Tech. Rep. CS2008-0923, 2005.
69. Tenenbaum, J.B., V. De Silva, and J.C. Langford, *A global geometric framework for nonlinear dimensionality reduction.* Science, 2000. 290(5500): p. 2319-2323.
70. Chang, C.C. and C.J. Lin, *LIBSVM: a library for support vector machines.* ACM Transactions on Intelligent Systems and Technology (TIST), 2011. 2(3): p. 27.
71. Chan, B.Y. and D. Kibler, *Using hexamers to predict cis-regulatory motifs in Drosophila.* BMC bioinformatics, 2005. 6: p. 262.

72. Anwar, F., *et al.*, *Pol II promoter prediction using characteristic 4-mer motifs: a machine learning approach.* BMC bioinformatics, 2008. 9(1): p. 414.
73. Down, T.A., *et al.*, *Large-scale discovery of promoter motifs in Drosophila melanogaster.* PLoS Computational Biology, 2007. 3(1): p. e7.
74. Turcot, V., *et al.*, *Bioinformatic selection of putative epigenetically regulated loci associated with obesity using gene expression data.* Gene, 2012. 499(1):p. 99-107.
75. Frisch, M., *et al.*, *LitInspector: literature and signal transduction pathway mining in PubMed abstracts.* Nucleic Acids Research, 2009. 37: p. W135-W140.
76. Krishnamurthy, V., N.S. Sweety, J. Natarajan, *Computational Identification of Alzheimer's Disease Specific Transcription Factors using Microarray Gene Expression Data.* Journal of Proteomics and Bioinformatics, 2009. 2: p. 505-508.
77. Saeed, A.I., *et al.*, *TM4: a free, open source system for microarray data management and analysis.* Biotechniques, 2003. 34(2): p. 374-378.
78. Ho Sui, S.J., *et al.*, *oPOSSUM: integrated tools for analysis of regulatory motif over-representation.* Nucleic Acids Research, 2007. 35(Web Server issue): p. W245-252.
79. Sandelin, A., *et al.*, *JASPAR: an open-access database for eukaryotic transription factor binding profiles.* Nucleic Acids Research, 2004. 32(Database issue): p. D91-94.
80. Xu, K., *et al.*, *A comparative analysis of gene-expression data of multiple cancer types.* PLoS One, 2010. 5(10): p. e13696.
81. Edgar, R., M. Domrachev, A.E. Lash, *Gene Expression Omnibus: NCBI gene expression and hybridization array data repository.* Nucleic Acids Research, 2002. 30(1): p. 207-210.
82. Wang, Z., Y. Chen, Y. Li, *A brief review of computational gene prediction methods.* Genomics, Proteomics and Bioinformatics, 2004. 2(4): p. 216-221.
83. Mathé, C., *et al.*, *Current methods of gene prediciton, their strengths and weaknesses.* Nucleic Acids Research, 2002. 30(19): p. 4103-4117.
84. Abeel, T., Y. Van de Peer, Y. Saeys, *Toward a gold standard for promoter prediction evaluation.* Bioinformatics, 2009. 25: i313-i320.

Chapter 3

Mining Genomic Sequence Data for Related Sequences Using Pairwise Statistical Significance

Yuhong Zhang[1] and Yunbo Rao[2]

[1] *College of Information Science and Engineering, Henan University of Technology, Zhengzhou, P.R. China, 450001, zhangyuhong001@gmail.com*

[2] *School of Information and Software Engineering, University of Electronic Science and Technology of China, Chengdu, 610054, P.R. China, cloudrao@gmail.com*

Next generation sequencing has led to a deluge of biological sequence data in the public domain. Almost everything in bioinformatics depends on the inter-relationship between sequence, structure and function (all encapsulated in the term *relatedness*), which is far from being well understood. In particular, pairwise statistical significance has been recognized to be able to accurately identify related sequences (homologs), which is a very important cornerstone procedure in numerous bioinformatics applications. However, it is both computationally and data intensive. To prevent it from becoming a performance bottleneck, we resort to high performance computation techniques for accelerating the computation. In this chapter, we first present the algorithm of pairwise statistical significance, then describe several high performance algorithms, which enable significant acceleration of pairwise statistical significance estimation.

1. Introduction

The past decades have witnessed dramatically increasing trends in the quantity and variety of publicly available genomic and proteomic sequence data. As of August 15, 2013, for example, there are approximately 154,192,921011 bases from 167,295,840 reported sequences in GenBank, which is maintained by National Center for Biotechnology Information (NCBI). In addition, GenBank still continues to grow at an exponential rate, doubling in size every 18 months (Moore's Law at work again). Thus, how to deal with these data, make sense of them, and enable them accessible to biologists working on a wide diversity of problems is a big challenge in bioinformatics.[1]

The overall aim of the data mining is to identify patterns and establish relationships from a raw data set and transform it into a comprehensible form for further use. In particular, the "raw data" in bioinformatics mainly refers to genomic sequence. There exist many complex data mining tasks, which usually cannot be handled directly by standard data mining algorithms. Genomic data mining and knowledge extraction for biological problems has become one of top 10 challenging problems in data mining research.[2]

Pairwise sequence alignment (PSA)[3] is one of the most widely used technique that tries to mine such *relatedness* from proteomic and genomic data. Many bioinformatics applications have been developed on the basis of pairwise sequence alignment, such as BLAST,[4] PSI-BLAST,[5] and FASTA.[6]

PSA generates a score for an alignment as a measure of the similarity between two sequences. Generally, the higher the score, the more related the sequences. However, the alignment score depends on various factors such as alignment methods, scoring schemes, sequence lengths, and sequence compositions.[7] As a result, judging the relationship between two sequences solely based on the scores may often lead to a wrong conclusion. For instance, it is possible that two related sequences of length 100 have an alignment score of 50, but two other unrelated sequences of length 500 have an alignment score of 200, suggesting that alignment score by itself is not sufficient to comment on homology. Therefore, it is more appropriate to measure the quality of PSA using the statistical significance of the score rather than the score itself.[8] Statistical significance of sequence alignment scores is very important to know whether an observed sequence similarity could imply a functional or evolutionary link, or is a chance event.[7] Homology (i.e., common evolutionary ancestry) can be reliably deduced for proteins that share statistically significant sequence similarity.[9] Accurate estimation of statistical significance of gapped sequence alignment has attracted a lot of investigation in recent years.[10–25]

1.1. *Biological sequence*

There exist millions of different species inhabiting the Earth. At the molecular level, however, all these living entities basically consist of three major macromolecules, i.e., deoxyribonucleic acid (DNA), ribonucleic acid (RNA), and proteins, which are essential for all known forms of life, serving as the building blocks of life.

DNA is a nucleic acid which carries the genetic instructions used in the development and functioning of all known organisms (with the exception of some viruses). DNA is capable of self-replication and also responsible for the transmission of hereditary characteristics from one generation to its offspring.

In essence, the DNA is a double chain of molecules called nucleotides, twisted together into a helical structure known as the double helix. The two chains (called strands) are complementary. In such a way, it is possible to infer one strand from the other. The nucleotides are differentiated by a nitrogen *base* that is made up of four types: adenosine, cytosine, guanine, and thymine. These bases tie the two strands together. Adenosine always bonds to thymine, whereas cytosine always bonds to guanine, resulting in forming base pairs (bp) (Fig. 1), which are the most common units for measuring the length of a DNA. In fact, a DNA can be determined uniquely by listing its sequence of nucleotides, or base pairs. Consequently, for practical purposes, the DNA is abstracted as a responding text over a four-letter alphabet, each representing a different nucleotide. Adenosine, cytosine, guanine and thymine are abbreviated as A, C, G and T, respectively.

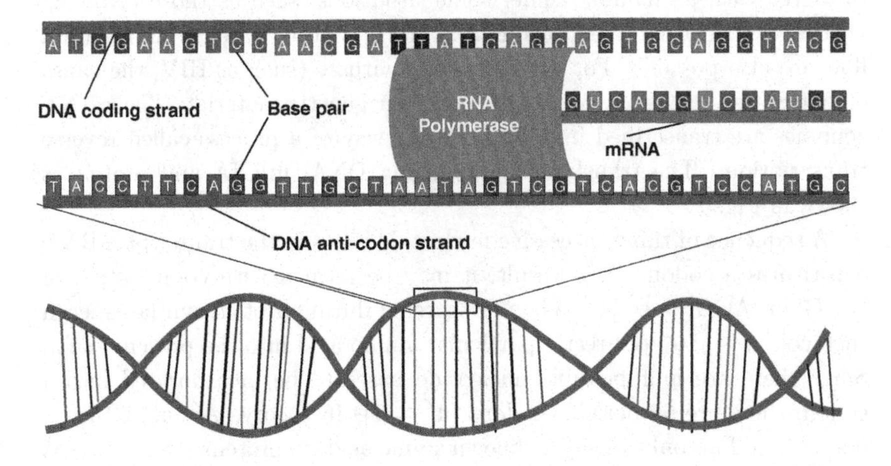

Fig. 1. The structures of DNA and RNA

Although both RNA and DNA are nucleic acids, unlike double-stranded DNA, RNA is a single-stranded molecule in many of its biological roles. It has ribose instead of deoxyribose and has a much shorter chain of nucleotides compared to DNA. In addition, the complementary base in RNA sequence to adenine is not thymine, but rather uracil (abbreviated as U).

Generally speaking, proteins are accountable for what a living being is and does in a physical sense.[26] They accomplish most of the functions of a living cell and determine its shape and structure. A protein is a linear sequence of molecules named *amino acids*.

The DNA segment that carries genetic information is called a *gene*. In the simplest sense, expressing a gene means that manufacturing its corresponding protein is a must. To to this, two major steps should be done.[27] In the first step, the genetic information in DNA is transferred to a messenger RNA (mRNA) molecule. This process is called *transcription*, in which RNA polymerase is used to construct RNA chains using DNA genes as templates. The resulting mRNA is a single-stranded copy of the gene, which next is translated into a protein molecule. In the second step, the mRNA is "read" according to genetic code. On the basis of the instructions from mRNA, the DNA sequence can be translated into the amino acid sequence in proteins. This process is named *translation*.

According to "central dogma" first formulated by Francis Crick,[28] there exists one flow in one direction: from DNA to mRNA and from mRNA to proteins in biological systems. However, this dogma does not always hold (disagree with its name). Under some situations, such as those involving viruses or special interventions in a laboratory, other types of information flow are also possible. For example, some viruses (such as HIV, the cause of AIDS) use RNA instead of DNA as their genetic material. Their RNA genomes are transcribed into DNA by the way of a process called reverse transcription. The translation flow among DNA, mRNA and protein is shown in Fig. 2.

A sequence of three successive nucleotide bases in the transcript mRNA constitutes a codon. As a result, it may be formed with combination of 'AUG' or 'ACG' or so on. When interpreted during protein synthesis, each such codon is able to direct a particular amino acid into the protein chain. Since there are four possible nucleotide bases to be incorporated into a codon, there are 64 possible codons ($4^3 = 64$) in theory. Actually, the 61 out of 64 codons only signify 20 known amino acids in proteins, which means that more than one codon specify the same amino acid. For instance, in addition to 'CGU', five additional codons specify the amino acid arginine. The 3 out of 64 condos (i.e., 'UAA', 'UAG' and 'UGA') are the stops which signal the end of a protein sequence.

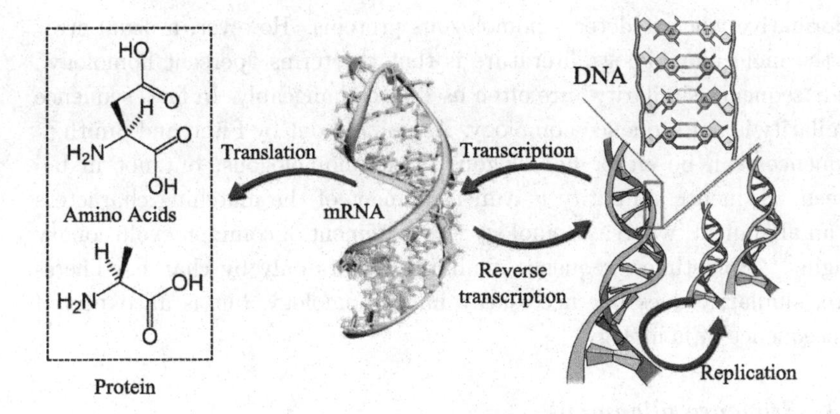

Fig. 2. Translation flow among DNA, mRNA and protein

For simplification, the 20 different amino acids have their individual three-letter code as well as single-letter code (SLC). For example, alanine, one the most frequently appearing amino acids, is represented by three-letter code 'Ala' or the SLC 'A'. Similar to DNA, proteins also can be readily represented as a string of letters describing its sequence of amino acids using their SLCs. Note that the four base letters (i.e., 'A', 'C', 'G' and 'T') that constitute of DNA sequences also appear in the set of 20 residue characters that form protein sequence, but have respectively different meanings in the two encodings.

1.2. *Homology and similarity*

Given a newly sequenced gene and its protein product, how can we predict its potential functions? Sequence comparison is often used to answer this question. It aims at finding important sequence similarities that would allow one to infer homology.

Homology between DNA (RNA) or protein sequences is defined in terms of shared ancestry. In general, homologs display conserved functions, which means that homologous sequences usually have the same or similar functions. Thus, the functions of a new sequence can be deduced relatively reliably from its homologous sequences if the homologous sequences with known functions can be identified.

Homology is usually inferred based on the similarity between sequences. Intuitively, the similarity of sequences refers to the degree of match between corresponding positions in sequence. Sequence comparison is most informative when it detects homologous proteins. However, a usual error in the molecular biology literature is that the terms "percent homology" and "sequence similarity" are often used interchangeably. In fact, sequence similarity is not sequence homology. As pointed out by Fitch and Smith,[29] sequences can be either homologous or non-homologous, but not in between. Sequence similarity is a measurement of the matching characters in an alignment, whereas homology is a statement of common evolutionary origin.[30] Sometimes, sequence similarity occurs only by chance. Therefore, similarity does not necessarily imply homology, but is an expected consequence of homology.

1.3. *Sequence alignment*

Protein sequence comparison is the most powerful tool for characterizing protein sequences since the enormous amount of information is preserved throughout the evolutionary process. For many protein sequences, an evolutionary history can be traced back 1-2 billion years.[31] By contrast, DNA sequences tend to be less informative than protein sequences.

With the development of genome analysis and large-scale sequence comparisons, sequence alignment becomes essential to recognize sequence similarity, which may be a meaningful indicator of homology.

Definition 1. Sequence alignment *is the comparison of two or more sequences by searching for a series of individual characters or character patterns that are in the same order in the sequences.*[30]

In pairwise alignment, two sequences are aligned through writing them in two rows. Identical or similar nucleotide or amino acid residues are placed in the same column. Otherwise, non-identical characters are placed either in the same column as a mismatch or opposite a gap in the other sequence so that identical characters are aligned in successive columns. The two sequences may be "broken" into smaller pieces by inserting necessary spaces in themselves, which guarantees that identical subsequences are finally aligned in a one-to-one correspondence. Clearly, there is no use for inserting spaces in both sequences at the same position. At last, the two sequences end up with equal size.

In the simplest sense, the following example illustrates an alignment between the sequences $Seq1$ ="CTGTCGCTGC" and $Seq2$ ="TGCCGTG", as shown in Fig. 3. In this example, four matches are highlighted with vertical bars. By contrast, non-identical characters are placed either in the same column as a mismatch or opposite a gap in the other sequence so that identical characters are aligned in successive columns. Two mismatches can be identified in the example: a 'T' of *Seq*1 aligned with a 'C' of *Seq*2, and a 'G' of *Seq*1 aligned with a 'C' of *Seq*2 . The insertion of spaces generated gaps in the two sequences. They are important to allow a good alignment between the sub-sequence "TG" of both sequences.

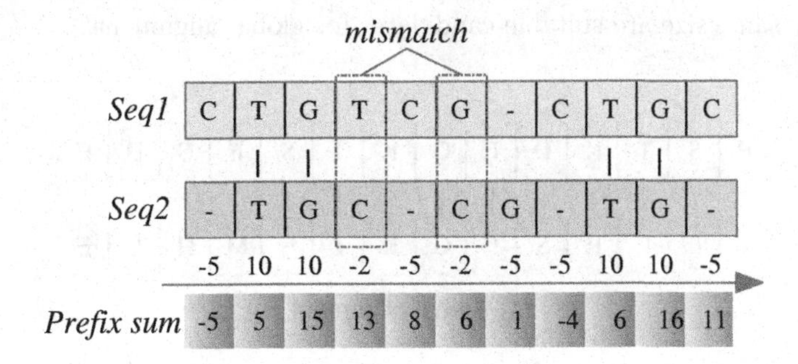

Fig. 3. Scoring pairwise sequence alignment

Once the alignment is done, we need to differentiate good alignments from poor ones. As a result, alignment score is needed to evaluate the quality of an alignment.

Definition 2. ***Sequence alignment score*** *is the sum of the individual log odds scores which are assigned to each pair of aligned characters on the basis of a given scoring scheme in the alignment*

In sequence alignment, matches are usually rewarded, whereas mismatches and gaps are penalized. The overall score of the alignment can then be obtained by adding up the score of each pair of characters. For instance, using a scoring scheme that gives a '+10' value to matches and '-2' to mismatches and -5 to gaps, the alignment of Fig. 3 scores $-5+10+10-2-5-2-5-5+10+10-5 = 11$. In general, the higher the score, the better the alignment.

There are two kinds of pairwise sequence alignment(i.e., global and local). In global alignment, an attempt is made to align every residue in every sequence up to both ends of each sequence. This is to say, global alignment tries to find best match of both sequences in their entirety. For the two assumed protein sequence fragments in Fig. 4-(a), the global alignment is extended over the whole sequence length to include as many matching residues as possible. In this tiny example, $Seq1$ = "PSTKDFGKISESRE" (whose length is 14) and $Seq2$ = "LERSFGKINMRLE" (whose length is 13), and the vertical bars between the sequences indicate that the presence of same residue along the whole entire sequence. Global alignment tries to match them to each other from end to end, even if parts of the alignment are not very convincing. This alignment can be made possible by including gaps within one or both sequences. The sequences that are similar and of roughly same size are suitable candidates for global alignment.

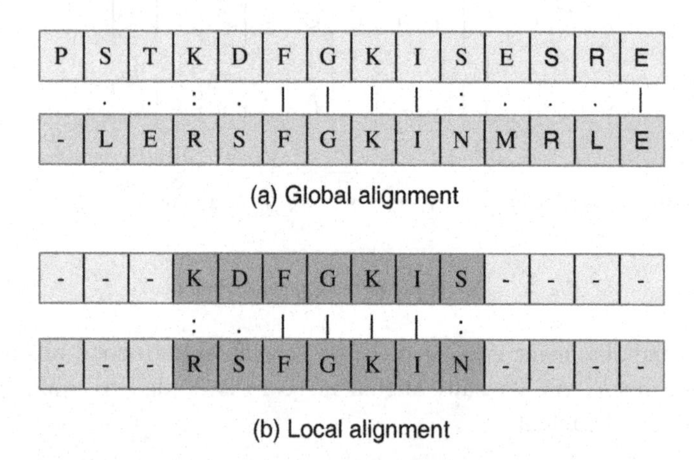

Fig. 4. Distinction between global and local alignment (*Symbols: '|' denotes identical character, ':' denotes conserved substitutions, '.' denotes semi-conserved substitution, and '-' denotes a gap in the sequence.*)

In local alignment, alignment will stop at the borderlines of regions of strong similarity, and finding these local regions has a much higher priority than extending the alignment to include more neighboring residue pairs. One or more of the most similar region(s) within the sequences are aligned, thus generating one or more islands of matches or sub-alignments in the aligned sequences, as shown in Fig. 4-(b). Dashes in the figure signify that the sequence is not included in the alignment. Local alignment is more suitable for aligning sequences that differ in length, or share a conserved region of domain. In the cases of the analysis of protein sequence, local alignment is expected as a more useful tool for dissimilar sequences, which may include possible similar sequence motifs or regions of similarity within their larger sequence context.

Obviously, alignment scores are not precise enough if only "*match*" or "*mismatch*" are used to estimate the similarity of sequences. Thus, we need a more accurate scoring scheme (also called scoring system) to find the optimal alignment among sequences.

Definition 3. ***Scoring scheme*** *is the system that defines the score for aligning two characters of a given alphabet, and the score for gaps (or gap extensions) within alignments.*

Given a scoring scheme and an alignment between two sequences, a more precise alignment score can be calculated as the sum of the scores for aligned character pairs plus the sum of the scores for all gaps. Levenshtein distance is a vivid example for a scoring scheme. During the computation of Levenshtein distance, each *mismatch* between two aligned characters costs 1 and each character that is aligned with a gap costs 1. Converted into scores instead of costs, *mismatches* obtain a score of -1 and gaps obtains a score of -1 per character, whereas *matches* costs nothing. This scoring scheme makes a distinction between between "*match*", "*mismatch*", and gaps. The score for a gap of length k is gap open + $(k-1)\times$ gap extend. If gap open is equal gap extend, the score scheme uses linear gap costs, otherwise it uses affine gap costs.

For the convenience of calculation, scoring schemes can store a score value for each pair of characters in a substitution matrix. Substitution matrix is a key factor in evaluating the quality of a pairwise sequence alignment, which assigns a score for aligning any possible pair of (protein or DNA) residues. The theory of amino acid substitution matrices is described in 32, and also applied to DNA sequence comparison in 33. Generally, different substitution matrices are *custom-tailored* to detecting similarities among sequences, which are diverged by different degrees.[32–34] Examples for this kind of scoring scheme are PAM250 and BLOSUM62.

The similarly score that an alignment program outputs depend strongly on the scoring scheme it uses. No single scoring scheme is the best for all purposes. Also, even if the alignment score is optimal for a given scoring scheme, it does not mean the aligned sequences are biologically related.

To determine whether the similarly score between two sequences implies an evolutionary link or not, one has to know how probable it is that we could have obtained the same score by aligning completely unrelated sequences.[35] Therefore, statistical approaches are often used to estimate the significance of the alignment score.

2. Statistical significance

In this section, we discuss how to verify whether or not one given alignment between two sequences is significant in terms of potential homology. This task can boil down to asking an equivalent question, i.e., when the same scoring system is used, how often two unrelated sequences or two different random sequences of about the same length as the test sequences can be aligned to obtain as high an alignment score as the two test sequences.[30] The task related to estimating the significance of an alignment score, in essence, is estimating data reliability and how well the data support a specific hypothesis or prediction.

2.1. *Why statistical significance?*

Biologists usually use pairwise alignment programs to identify similar, or more specifically, related sequences or homologs (having common ancestor). In general, more related sequences will have higher similarity scores. However, there are many factors, such as the compositions of sequences, the lengths of sequences and the user-defined scoring system, that can affect an alignment score to different extents.[7] Thus, inferring the relationship between two sequences solely based on the scores may lead to a wrong conclusion.

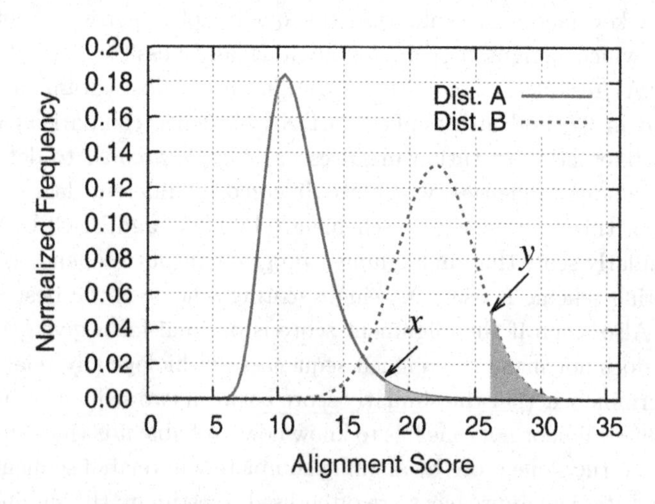

Fig. 5. Two different distributions of alignment score

Fig. 5 shows that two alignment score distributions (probability density functions) "Dist. A" and "Dist. B". Assuming that scores x and y are in the score distributions A and B, respectively. Obviously, $x < y$, but x is more statistically significant than y, since x lies more in the right tail of its distribution, meaning that it is less probable to have occurred by chance. The shaded region in this figure presents the probability which a score equal or higher could have been obtained by chance. Thus, the biological significance of a pairwise sequence alignment or the potential relatedness of the two sequences being aligned is better gauged by the statistical significance of the alignment score rather than by the alignment score alone.[8]

Although statistically significant patterns have attracted much investigation, it is necessary to note that statistical significance is not equivalent to biological significance.[36,37] The main aim of the statistical significance estimation is to accentuate the targets for experiments with potentially meaningful results. Whether such highlighted sequences are indeed worth further research or not, relies upon the quality of the *P-value* estimation method.[38]

2.2. *P-value in statistical significance*

In statistical hypothesis testing, the *P-value* can be defined by:

Definition 4. ***P-value*** *is the probability of obtaining an observed result as extreme as, or more extreme than the one that was actually observed, assuming that the null hypothesis is true.*[39]

One often "rejects the null hypothesis" when the P-value is less than the predetermined significance level α, which is often 0.05 or 0.01, indicating that the observed result would be highly unlikely under the null hypothesis. Herein the null hypothesis is a hypothesis about a population parameter, which proposes that no statistical significance exists in a set of given observations. Depending on these observations, the null hypothesis either will or will not be rejected as a viable possibility. Generally, the smaller the P-value, the more strongly the test rejects the null hypothesis, that is, the hypothesis being tested. The vertical coordinate in Fig. 6 is the probability density of each outcome, computed under the null hypothesis. The *P-value* is the area under the curve past the observed data point, which is shown as a shaded area in Fig. 6.

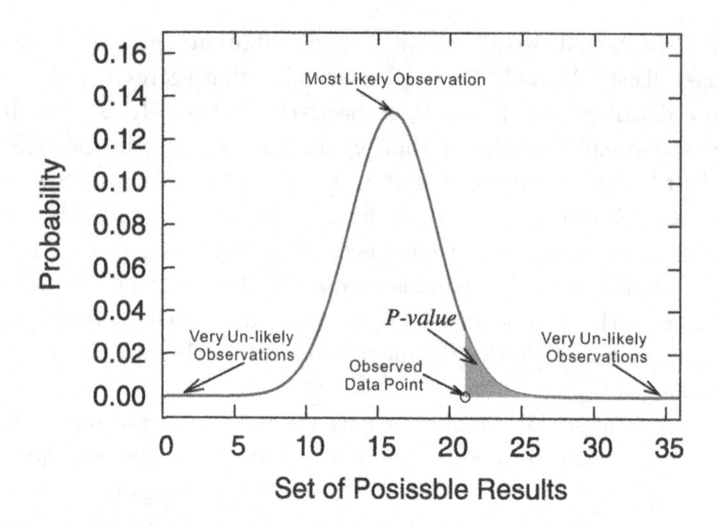

Fig. 6. Example of a P-value computation

In particular, the statistical significance of the resulting sequence alignment score between the sequences can be presented by the probability (i.e., *P-value*) that random or unrelated sequences could be aligned to generate the same score. If an alignment score has a low probability of occurring by chance, the alignment is considered statistically significant, and hence potentially biologically significant.

Statistical approaches can be based upon theoretical models or upon permutation reconstructions of the observed sequences.[40] However, it is only possible when the distribution (or statistical model) of the alignment scores is known precisely, which can be obtained by aligning many pairs of random sequences.

2.3. *Modeling statistical for local sequence alignment*

2.3.1. *Coin-Toss model*

Originally, the significance of sequence alignment scores was estimated under the assumption that these scores follow a normal statistical distribution. If the sequences to be aligned are randomly produced by Monte Carlo strategy, as in producing a sequence by picking balls representing 20 amino acids or 4 bases (residues) out of box, and the number of each type is proportional to the frequency that is found in sequences, the distribution of alignment scores seems to be normal at first glance. However, this statistical model might not be appropriate since it is not suitable for predicting the number of sequentially matched positions to expect between random sequences of a given length. To estimate this number, coin-tossing experiments are used to model a DNA or protein sequence.[41]

Random alignments normally comprise mixture of mismatches and matches, just as a series of coin tosses generate a mixture of tails and heads. The chance of generating a series of matches in a sequence alignment with no mismatches is similar to the chance of tossing a coin and coming up with a series of heads.[30] The numbers of interest are the highest possible score that can be obtained in a certain number of trials.

To utilize the coin model, an alignment of two random sequences is converted into a series of heads and tails. Here the two sequences are noted as $\mathbf{a} = a_1a_2 \ldots a_n$, $\mathbf{b} = b_1b_2 \ldots b_n$, respectively. Each of the two sequences $\mathbf{a}$ and $\mathbf{b}$ is the same length n. if $a_i = b_i$, the equivalent toss result is an 'H', otherwise the result is a 'T'. The conversion of an alignment to a series of 'H' and 'T' tosses is illustrated in the following example:

$$\left.\begin{matrix} a_1a_2a_3a_4 \cdots a_n \\ b_1b_2b_3b_4 \cdots b_n \end{matrix}\right] \Rightarrow HTTH \cdots$$

where $a_1 = b_1$ and $a_4 = b_4$.

In such model, coins are usually considered that the probability of a head is equal to that of a tail. In this example, the coin has a certain probability p of scoring a head (H). The expected number of runs of heads of length l in n coin tosses is $E(l) \cong np^l$. This relationship results from the logic that the expectation (number of heads) is the product of the number of tosses n, times the probability of p^l heads at each toss. If the longest run R_l is expected once, $1 = np^{R_l}$ and thus $R_l = \log_{1/p}(n)$.

In some sense, finding the highest scoring segment in a single protein or DNA sequence using a scoring matrix can be equivalent to finding the longest run of heads in the coin-tossing sequence. In the scoring matrix, a positive value is allocated to some of the residues, whereas $-\infty$ is allocated to all of the others. By the coin-tossing model, the average longest run of matches in the alignment should be possible to use the Erdös-Rényi law to predict the longest run of matches. If two random sequences of length m and n (where $m \neq n$) are aligned in the same manner, the same law still applies, but the length of the predicted matches is $\log_{1/p} mn$.

A more precise formula for the expectation value or mean of the longest match, $E(M)$ has been derived by Arratia and Waterman:[42,43]

$$E(M) \cong \log_{1/p}(mn) + \log_{1/p}(q) + \gamma \log(e) - 1/2\ , \qquad (1)$$

where $q = 1 - p$ and $\gamma = 0.577$ is the Euler's number. Equation 1 can be further simplified to

$$E(M) \simeq \log_{1/p}(Kmn) , \tag{2}$$

where K is a constant which relies on the base composition in sequence. Equation 2 is very important for computing the statistical significance of local alignment scores. Basically, it describes that as the lengths of unrelated or random sequences increase , the mean of the highest possible local alignment score will be proportional to the logarithm of the product of the sequence lengths.

2.3.2. *Assessing the statistical significance using alignment scores*

In sequence analysis, it is more useful to use alignment scores instead of the match lengths for comparing alignments as discussed previously. The mean alignment length two unrelated (or random) sequences given by Equations 2 can be easily extended to a mean alignment score just by using scoring matrix values. Not limited to predict match lengths of sequences $E(M)$, Equations 2 can also be used to predict the mean alignment score $E(S)$ between random sequences of lengths m and n. In another word, assessing statistical significance then is equivalent to computing the probability which $E(S)$ between two random (unrelated) sequences will be greater than an observed alignment score between two test sequences. Based on the above changes, the expected (or mean extreme) score between random sequences can be given by :

$$E(S) = [\log_e (Kmn)]/\lambda . \tag{3}$$

Karlin and Altschul[40] offered a natural extension of the problem of head-runs, or match-runs to the more general case of local similarity scores for non-intersecting alignments without gaps. To ensure the scores are local, the requirement that $E(s_{i,j}) = \sum_{i,j} p_i p_j s_{i,j} < 0$ should first be met. The remarkable Karlin-Altschul model[40] interprets the number of highest scoring matching regions above a threshold by a Poisson distribution. Under this assumption, the expected number of distinct local alignments with score values of at least x is approximately Poisson distributed with mean

$$E(x) \simeq Kmne^{-\lambda x} , \tag{4}$$

where λ and K are easily calculated parameters, and are dependent upon the scoring scheme (substitution matrix and gap costs) employed.[40,44] Eq. 4 is also known as "*E-value*"(Expectation value). In Eq. 4, the coefficient

mn in can be regarded as a "*space factor*" for the fact that there are not really mn independent places that could have generated score $S \geq x$. Thus, $K < 1$ is a proportionality constant that corrects the mn. Compared to λ, K has a small effect during estimation of the statistical significance of a similarity score.

For sequence analysis, the parameters λ and μ rely upon the composition and length of the sequences being aligned. For those alignments that do not include any gaps, parameter λ and μ can be computed from their scoring matrix.[30] The parameter λ provides the scale factor by which a similarity score is multiplied to determine its probability. For alignment without gaps, λ is the unique positive root to the following equation:

$$\sum p_i p_j e^{s_{i,j}\lambda} = 1 , \tag{5}$$

where p_i and p_j are the respective fractional representations of residues i and j in the sequences, and $s_{i,j}$ is the score for a match between i and j, taken from a log odds scoring matrix, which is defined by[32,40]

$$s_{i,j} = ln \frac{p_{i,j}}{p_i p_j} , \tag{6}$$

where $p_{i,j}$ is the probability that the residues i and j form a pair in the alignment, thus, $p_{i,j} = p_{j,i}$.

In some sense, λ can be regarded as a factor that converts pairwise match scores to probabilities, so that $e^{-\lambda x}$ is similar to p^l in the coin tossing case. Therefore, just as in the coin-tossing example discussed above, the expectation of a run of heads (or an alignment run that generates a score S) is the product of a "*space-factor*" term, Kmn, and a probability term $e^{-\lambda S}$.[45]

The following step is to determine the probability that a given alignment score S reaches a greater score x. As in the single sequence case, we can convert the problem from the probability of the longest match run to the probability of score $S_l \geq x$ by considering the probability $P(S \geq x)$ when a pair of residues a_i, b_j is matched with positive score $s_{i,j}$ and all negative scores are $-\infty$. For local pairwise alignment scores with no gaps and a mismatch score of $-\infty$, the expected number of runs of score $S \geq x$ can have a general form: $E(S \geq x) \propto mnp^x$, or equivalently $E(\geq x) \propto mne^{xlnp}$ or $mne^{-\lambda x}$ where $\lambda = -lnp$.[40,45]

The probability $P(S \geq x)$ can be predicted by the Poisson distribution where the mean of the Poisson distribution is x.[46] The Poisson distribution applies when the number of trials is large but the probability of *success* in a single trial is small, as in comparing many random (unrelated) sequences and finding a rare positive alignment which has a high score. When the average number of times is μ, the Poisson distribution provides the probability P_n of the number *successes* (i.e., $0, 1, 2, 3, \ldots, n$), $P_n = e^{-\mu}\mu^n/n!$. Generally, we are interested in the probability that an event occur $\geq n$ times. At the same time, we usually ask for the probability which a high score happens one or more times ($n \geq 1$) in the case of sequence comparisons. In this case, one can easily compute the probability that the event happened not zero times: $P(n \geq 1) = 1 - P(0)$, so $P(S \geq x) = 1 - P(n = 0) = 1 - e^{-\mu}\mu^0/0!$. Since $\mu = E(x) = Kmne^{\lambda x}$ (see Eq. 4), $0! = 1$, and $\mu^0 = 1$, the probability of seeing an optimal alignment score $S(\geq x)$ (i.e., *P-value*) can be given by:

$$P(S \geq x) \simeq 1 - exp(-\mu) = 1 - exp(-Kmne^{-\lambda x}) \; , \tag{7}$$

Equation 7 describes the probability of having a similarity score $S \geq x$ in a single pairwise comparison, in which a query sequence of length m aligns against a subject sequence of length n. In fact, Eq. 7 is also identical to an extreme-value distribution (EVD) or Gumbel distribution, which will be further discussed below.

2.4. *Gumbel extreme value distribution*

Karlin and Altschul[40] and Waterman and Arratia[42] proved that local similarity scores, at least for alignments without gaps, are expected to follow a Gumbel distribution (a type I extreme value distribution(EVD).[47]

The EVD, to some extent, is like a normal distribution but with a positively skewed tail in the higher score range. It is a limiting distribution for the maximum or the minimum of a large collection of random observations from the same arbitrary distribution. In particular, the maximum observations in this chapter refer to the optimal sequence alignment scores.

The probability density function (PDF) of EVD is given by

$$f(x) = \lambda exp\left[-\lambda(x - \mu) - e^{-\lambda(x-\mu)}\right] \; , \tag{8}$$

and its corresponding distribution function can be described as

$$P(S < x) = exp\left[-e^{-\lambda(x-\mu)}\right] \; , \tag{9}$$

where $-\infty < x < +\infty$, $\lambda > 0$ and $-\infty < \mu < +\infty$. Thus, the probability of a score $S \geq x$ (i.e., *P-value*) can be given by

$$P(S \geq x) = 1 - exp\left[-e^{-\lambda(x-\mu)}\right] \; , \tag{10}$$

where μ is the mode (highest point or characteristic value of the distribution), and λ is the decay of scale parameter. After strict theoretical analysis, Altschul and Gish[48] gave the representation of the characteristic parameter μ:

$$\mu = (lnKmn)/\lambda \, , \tag{11}$$

where m and n are the sequence lengths, and K is a either computed or estimated constant.

Combining Eq. 10 and Eq. 11 eliminates parameter μ and obtains the following relationship (i.e., P-value), which can be written as seen in Eq. 7:

$$P(S \geq x) \simeq 1 - e^{-Kmne^{-\lambda x}} = 1 - e^{-E(x)} \; .$$

The distribution of optimal alignment scores not only can be deduced from Karlin-Altschul statistical model,[40] but also can double confirmed by Bastien and Marechal using reliability theory (see Ref. 49 for more details).

For local alignment without gaps, rigorous statistical theory about the alignment score distribution has been developed, as discussed previously. There no perfect theory currently exists for gapped local alignment score distribution. However, many computation experiments or simulations[34,46,50,51] also strongly imply that the gapped local alignment scores, which are generated by the Smith-Waterman algorithm[52] and approximated by Gapped BLAST[34] or FASTA,[53] still follow an extreme value distribution.

3. Pairwise statistical significance

In general, there are two primary methods to estimate the statistical significance of local sequence alignment. One is called database statistical significance (DSS) reported by many popular database search programs, such as BLAST,[54] FASTA,[6] and SSEARCH (using full implementation of Smith-Waterman algorithm[52]). For a given sequence pair, pairwise local sequence alignment programs provides the optimal alignment. In the case of database searches, the second sequence is the database consisting of lots of sequences. Clearly, E-values and P-values in DSS rely on the average sequence features like length, amino acid composition of the database being searched.

To estimate statistical significance of a high similarity score obtained during a database search, DSS needs to calculate a probability (i.e., *P-value*) given by Eq. 7. To complete the calculation, it is fundamental to obtain the statistical parameters K and λ in Eq. 7 beforehand. The two parameters are dependent on the background amino acid frequencies of the sequences being compared and the scoring scheme. FASTA[6] estimates the two parameters from the scores generated by actual database searches (real unrelated sequences), whereas BLAST2[54] estimates them in advance for a given scoring scheme via comparing many artificial random sequences. To finish the assessment of statistical significance, this method needs to utilize sequence database. Thus, this method is named "database statistical significance".[8]

The other method is called the pairwise statistical significance (PSS)[8,13,22,55] , which is specific to the sequence-pair being aligned, and independent of any database. Below we will be introduced PSS in detail.

3.1. *The definition of pairwise statistical significance*

As Pearson pointed out,[31] the differences in the performance of sequence comparison algorithms are insignificant compared to the loss of information that occurs when one compares DNA sequences. If the biological sequence of interest encodes a protein, protein sequence comparison is always the method of choice. At the same time, the distribution of global similarity scores are rarely utilized to infer homology since they are not well understood, and thus it is difficult to assign a statistical significance to a global similarity score.[45]

Therefore, in this chapter, we use protein sequences as experimental data to estimate the performance of pairwise statistical significance of local sequence alignment. On the other hand, there is a common fact that statistical analysis approaches should be based on enough samples. Waterman and Vingron[46] showed that, for protein sequences, the estimation results using sequences selected from databases are in good agreement with those using random sequence models with independent residues. As a result, permuting the subject protein sequence is a convenient method producing enough sequence samples.

The significance of a specific pairwise similarity score in each of the two sequences, the query and library sequence, can also be estimated using a Monte-Carlo method. The two sequences are aligned, and then one or both of the sequences is shuffled thousands of times to produce a sample of random sequences with the same residue composition and length. Similarity scores are computed for these alignments between the query sequence and each of the shuffled sequences (also called subject sequences). The two parameters λ and K parameters, which are used to compute P-value in Eq. 7, can then be computed from this distribution of scores using maximum likelihood. This strategy initially was done by the *prss* program in the FASTA package.[56]

Recently, a novel strategy about pairwise statistical significance (PSS) is designed by Agrawal et al,[8,13,22,55] which is an attempt to make the statistical significance estimation procedure more specific to the sequence pair being compared. In addition to not needing a sequence database to estimate the statistical significance of an alignment, PSS is shown to be more accurate than database statistical significance reported by popular database search programs like BLAST, PSI-BLAST, and SSEARCH.[8] This brings us to the formal definition of pairwise statistical significance. Consider that the scoring scheme SC (substitution matrix, gap opening penalty, gap extension penalty), and the number of permutations N, the PSS[8] of two sequences is represented as:

$$PSS(Seq1, Seq2, m, n, SC, N) \; . \tag{12}$$

Through permuting $Seq2$ N times randomly, the function 12 generates N scores by aligning $Seq1$ against each of the N permuted sequences and then fits these scores to an EVD[6,48,57] using censored maximum likelihood.[58]

The scoring scheme SC is very important to obtain an optimal alignment score, which can be extended from sequence-pair-specific distanced substitution matrices[24] or multiple parameter sets.[14,23] In addition, a sequence-specific/position-specific scoring scheme SC_1 or SC_2 specific to one of the sequences (responding to $Seq1$ or $Seq2$) can be used to estimate its PSS using sequence-specific/position-specific substitution matrices.[13,17] Non-conservative pairwise statistical significance PSS_{nc} was defined in 19 when the above described function 12 is used two times with different ordering of sequence inputs. Let

$$Sig1 = PSS(Seq1, Seq2, m, n, SC_1, N) \; ,$$

$$Sig2 = PSS(Seq2, Seq1, n, m, SC_2, N) \; .$$

Correspondingly,

$$PSS_{nc} = min\{Sig1, Sig2\} \; . \tag{13}$$

Using the function 12 two times in this way ensures that both sequences are permuted to built two different alignment score distributions, which subsequently results in two different PSS estimates for the same sequence pair (*Seq*1 and *Seq*2), and the final reported PSS estimate is the minimum of the two individual estimates. A good review of the recent developments in sequence-specific sequence comparison using PSS can be found in Ref. 55.

PSS has been used to reorder the hits from a fast database search program like PSI-BLAST.[12] It has also been used for DNA alignment in some cases as well.[18,20] There also exist some other approaches to estimate PSS, including a clustering-classification approach,[21] which is a lookup-based approach and therefore not very accurate.

Note that the EVD distribution only applies to a gapless alignment. However, local sequence alignment methods are used to find the best-matching pairwise alignments with gap penalties. For the cases of gapped alignment, although no asymptotic score distribution has yet been established analytically, computational experiments strongly indicate these scores still roughly follow Gumbel law after pragmatic estimation of the λ and K parameters.[6,7,48,57] In the sense of gapless alignment, the parameters of the EVD can be obtained by theoretical computation. However, we cannot get these parameters through the same way. As a thumb of rule, parameter fitting is commonly used. A good fit plays a crucial role in performance improvement of pairwise statistical significance in terms of homology detection.

3.2. *Parameters fitting of pairwise statistical significance*

The accurate estimation of the statistical parameters is the key issue of using equation 4 and 7. The most direct approach to estimate parameters (that is, λ and K) is to generate large scale pairs of random sequences of equal length n, and find the optimal local alignment score for each pair. From these scores, these parameters can be estimated by simulation and empirical curve-fitting.[14,38]

There are various approaches to fitting the distribution parameters of pairwise statistical significance, such as ARIADNE,[7] PRSS,[59] censored-maximum likelihood fitting,[58] and linear regression fitting.[60] Agrawal et al.[8] found that the maximum likelihood fitting with censoring left of peak (described as type-I censoring ML) is the most accurate method for PSS.

Since the Type I censored maximum likelihood (ML) fitting contributes a lot to the performance of PSS in terms of homology detection, herein we give it a brief description. According to Eddy's technical report,[58] the Type

<table>
<tr><td>Query
sequence A</td><td>> 1repC1|123|2.600
SPRIVQSNDLTEAAYSLSRDQKRMLYLFVDQI
RKSHDGICEIHVAKYAEIFGLTSAEASKDIRQA
LKSFAGKEVVFYESFPWFIKPAHSPSRGLYSV
HINPYLIPFFIGLQNRFTQFRLS</td></tr>
<tr><td>Subject
sequence B</td><td>> 101m00|154|2.070
MVLSEGEWQLVLHVWAKVEADVAGHGQDILI
RLFKSHPETLEKFDRVKHLKTEAEMKASEDL
KKHGVTVLTALGAILKKKGHHEAELKPLAQS
HATKHKIPIKYLEFISEAIIHVLHSRHPGNFGAD
AQGAMNKALELFRKDIAAKYKELGYQG</td></tr>
</table>

I censored ML fit for 100 data points is better than the linear regression fit for 10,000 data points. That is to say, the censored-maximum likelihood fitting is significantly superior to linear regression for fitting EVD. In addition, because of the intrinsic nature of long right tail in EVD, one may wish to fit only the right tail of fitting histogram, rather than the whole one. The left (low scoring) tail may be *contaminated* with poor-scoring sequences that do not conform to the feature of EVD, so they are not included in the fit. If *a priori* value c can be given, any data $x_i < c$ will be cut off in the fit. After this purification, the data can achieve a much better fitting.

In the following we use a real pair of protein sequences to verify this fitting strategy. Assuming that the query sequence is A (named "1repC1") and the subject sequence is B (named "101m00"). To obtain enough samples of alignment score for fitting, subject sequence B is first shuffled N times (where $N = 1000$), then these shuffled copies are aligned against query sequence A using Smith-Waterman algorithm, finally 1000 optimal alignment scores are available for fitting.

Figure 7 shows the visualization diagram of the EVD fitting using those 1000 alignment scores. For simplicity, the diagram uses symbols '=' to present the histogram of those scores, whose length represents the actual number of observed sample scores, noted by "obs" in this figure. The curve, which is plotted by star symbol '*', is the one of extreme value distribution using the fitting parameters K and λ. Correspondingly, the expected number of sequences on the basis of the fitted EVD is noted by "exp" in this figure. To make this diagram a better visibility, one symbol '=' is determined to represent two entities dynamically. For example, in the bin whose the optimal alignment score is 31, the number of actual observed sequences "obs" is 93, whereas the expected one based on the EVD is 90.

```
score  obs   exp (one = represents 2 sequences)
-----  ---   ---
  23     4     2|*=
  24     5     7|==*
  25    13    18|======   *
  26    43    34|=================*====
  27    46    52|=======================   *
  28    76    69|==================================*===
  29    83    82|=========================================*=
  30    87    89|============================================*
  31    93    90|=============================================*==
  32    87    86|===========================================*=
  33    75    78|======================================*
  34    63    69|===============================  *
  35    65    59|=============================*===
  36    50    50|========================*
  37    47    41|====================*===
  38    27    33|=============   *
  39    23    27|===========  *
  40    24    21|==========*=
  41    18    17|========*
  42    10    13|===== *
  43    17    11|=====*===
  44     8     8|===*
  45     8     6|==*=
  46     8     5|==*=
  47     5     4|=*=
  48     2     3|=*
  49     2     2|*
  50     1     2|*
  51     3     1|*=
> 52     7     -|====

% Statistical details of theoretical EVD fit:
          μ = 31.1048
          λ =  0.2473
```

Fig. 7. Two different distributions of alignment score

It can be seen that the fitted EVD basically is able to cover the actual observations of the alignment score sample. Once μ is known via fitting, parameter K can be easily obtained according to Eq. 11.

3.3. *Evaluation of pairwise statistical significance*

Further studies on pairwise statistical significance using multiple parameter sets,[14] and sequence-specific substitution matrices (PSSS) and position-specific substitution matrices (PSSM)[13] have demonstrated it to be a promising method capable of producing much more biologically relevant estimates of statistical significance than database statistical significance (DSS).

Furthermore, in order to evaluate the accuracy of different approaches of statistical significance, our software library offers the popular evaluation measure, *Error per Query* (*EPQ*) versus Coverage, which can be plotted to visualize and compare the results. The EPQ is defined as

$$EPQ = F_{num}/Q_{num} \ , \tag{14}$$

where F_{num} is the total number of non-homologous sequences detected as homologs (i.e., false positives) and Q_{num} is the total number of queries. The *Coverage* can be given by

$$Coverage = H_d/H_t \ , \tag{15}$$

where H_d and H_t are the number of homologous pairs detected and the total number of homologous pairs presented in the sequence database, respectively. In order to plot the curves of *Errors per Query* versus *Coverage*, the list of pairwise comparisons is sorted based on increasing P-values (say, decreasing statistical significance). The higher position in the sorted list implies the higher probability that the compared pairwise sequences are homologs. Therefore, while traversing the sorted list from top to bottom, the coverage count is incremented if the two sequences of the pair are true homolog, else the error count is increased by one. The *Errors versus Coverage* curves illustrate how much coverage is obtained at a given error level. Obviously, at a same *EPQ* level, a curve shows a better performance if it is more to the right of x-axis (representing *Coverage*).

We next compare the retrieval accuracy of pairwise statistical significance against database statistical significance and the retrieval accuracy of pairwise statistical significance using different substitution matrices. In our experiments, the coverage of each of the 86 queries at the 1st, 3rd, 10-th, 30-th, and 100-th error is reported, and the median coverage at each error level is compared among different sequence-comparison schemes. As shown in Fig. 8-(a), pairwise statistical significance performs significantly better than database statistical significance (used by BLAST, PSI-BLAST, and SSEARCH) in term of retrieval accuracy (a percentage of 35-40% according

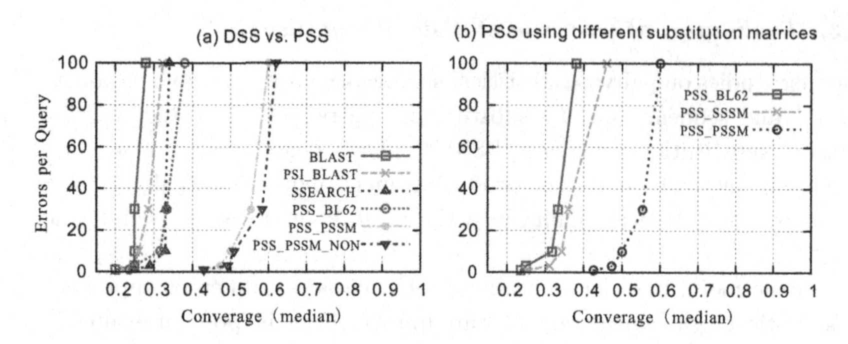

Fig. 8. *Errors per Query* vs.*Coverage* plots for different methods of statistical significance

to different schemes). The non-conservative pairwise statistical significance achieves the best performance. As for the PSS using different substitution matrices, the methods using PSSM win the best performance, followed by using SSSM. At the same level of *Error per Query*, the *Coverage* of the approach using position-specific scoring matrix is higher than the one using standard substitution matrix (such as BLOSUM62) about at a percentage of 25%, as shown in the Fig. 8-(b).

4. HPC solutions for accelerating pairwise statistical significance estimation

Although pairwise statistical significance estimation (PSSE) has been shown to be accurate, it involves thousands of such permutations and alignments, which are enormously time consuming and can be impractical for estimating pairwise statistical significance of a large number of sequence pairs. For instance, for our experiments with 86 query sequences and 2771 subject sequences, the sequential implementation takes more than 32 hours. Bigger data sets demand more computing power.

Careful analysis of the data pipelines of PSSE shows that the computation of PSSE can be decomposed into three computation kernels: (1) Permutation: It generates N random sequences by permuting $Seq2$, as shown in Fig. 9. In this kernel, in addition to produce enough random sequence for alignment, permutation also keeps the compositions and lengths of these copies unchanged. Permutation strategy depends on the assumption that the similarity scores of real unrelated protein sequences behave like the similarity scores of randomly generated sequences. (2) Alignment: the N

permuted sequences of $Seq2$ (also known as subject sequence) align with $Seq1$ (known as query sequence) using Smith-Waterman algorithm[52] and (3) Fitting: It obtains statistical constants K and λ by fitting the scores produced in (2) into an EVD, then returning the PSS between sequence pair $Seq1$ and $Seq2$ according to Eq. 7.

P	S	T	K	D	F	G	K	I	S	Seq2

(a) original (subject) sequence

S	P	T	D	K	G	F	I	K	S	Seq2*
K	S	T	P	D	F	K	G	S	I	Seq2*
...	...	...	...	...	...	...	...	...	...	...
T	S	P	K	F	D	G	S	I	K	Seq2*

(b) Permuted copies of original sequence (denoted by symbol '*')

Fig. 9. Permutation: produce N copies ($Seq2*$) by randomly shuffling the residues (characters) in the original sequence of $Seq2$

Reference 15 shows that permutation and alignment comprise the overwhelming majority (more than 99.8%) of the whole execution time. It hints us that efforts should be spent to optimize the two kernels to achieve high performance.

In addition, the kernel of permutation shows high degrees of data independency that are naturally suitable for single instruction, multiple threads/processes architectures, and therefore can be mapped very well to task parallelism models of multi-core CPU and many-core GPU. Applying high performance computing (HPC) techniques provides more opportunities of accelerating the estimation in reasonable time.

Over the past several years, high performance computation techniques have become extremely popular and have been applied in various domains like material science,[61] power systems,[62–66] linear solvers,[67–69] data mining,[70,71] multimedia security and compression[72–74] and bioinformatics.[75,76]

Multi-core computers and clusters have become increasingly ubiquitous and more powerful. Therefore, it is of interest to unlock the potential of multi-core desktop and clusters using high performance technologies. On the other hand, with general purpose graphics processing units (GPGPUs) becoming increasingly powerful, inexpensive, and relatively easy to program, it has become a very attractive hardware acceleration platform. Those strongly motivate the use of multi-core CPUs,[77,78] many-core GPUs[16,79] and FPGA[80,81] to accelerate the PSSE.

4.1. *Parallel paradigms of HPC techniques*

Message Passing Interface (MPI) is *de facto* standard for message-passing specification supporting parallel program running on computer clusters and supercomputers. It works on both shared and distributed memory machines, providing a convenient solution to parallel programming among different machines and hardware topologies. In MPI paradigm, every MPI task corresponds to one process and every process has its own private address space. Therefore, data transfers from one address space to another via explicit message passing, which means that extra data copies and/or duplication are usually required.

OpenMP is an application programming interface (API) that may be used to implement a multi-threaded, shared memory parallel algorithm. It supports multi-platform shared memory multiprocessing programming in C/C++ and Fortran. It consists of a set of compiler directives, environment variables, and library routines that determine run-time execution. OpenMP directives can be added to a sequential source code to instruct the compiler to execute the corresponding block of code in parallel. The main advantages of this technique are relative ease of use and portability between serial and multi-processor platforms, as for a serial compiler the directives are ignored as comments.

Note that the current OpenMP paradigm only applies one multi-core node which shares main memory among the multiple cores. Therefore, the number of cores can be used in parallel is limited. If one wants to further enhance computation performance, borrowing a hand from hybrid paradigm (using OpenMP and MPI together) usually is a necessity.

Although GPUs were originally designed as highly effective graphics accelerator, the modern programmable GPUs have evolved into highly parallel, many core processors with tremendous computational power and huge memory bandwidth. This makes them more effective compared to general-purpose CPUs for a wide range of applications. Using CUDA, the system consists of a host that is a traditional CPU and one or more compute devices (namely, GPUs) that can be seen as (a) coprocessor(s) of CPU, specialized for compute-intensive, highly parallel computation. Accordingly, a CUDA program is divided into two parts: a host program running on the host CPU and one or more parallel *kernels* running on the GPUs.

Even if the current generation of GPUs provides much higher computational parallelism than the CPUs, memory management is still more complex compared to CPU. To obtain high performance, a developer needs to have a good understanding of the hierarchy and features of the GPU memory. An insightful discussion on memory management and optimizations of GPUs could be found in Ref. 82.

4.2. *Implementations*

In the following, MPI, OpenMP and GPU implementations are presented briefly. Our MPI and OpenMP implementations about PSSE are carried out on Cray XE6 machines (provided by Hopper system at NERSC), each with 24 cores. Each node contains two twelve-core AMD® "Magny-Cours" @2.1 GHz processors. As for GPU implementations, they are carried out using Intel© Core™ i7 CPU 920 processors, and dual Tesla C2050 GPU (each with 448 CUDA cores). All the performance in terms of speedup is computed over the corresponding sequential implementation.

The sequences data used in this work comprise of a non-redundant subset of the CATH 2.3 database.[83] This dataset consists of 2771 domain sequences as our subject sequences library, which represents 1099 homologous superfamilies and 623 topologies and includes 86 CATH queries serving as our query set.

As discussed in the previous section, the kernels of permutation and alignment are independent of each other during the PSSE procedure, which therefore map very well to programming models capable of expressing MPI task parallelism. The PSSE of single-pair sequences processes only one pair of query and subject sequences. The basic idea of computing single-pair PSSE can be found in Refs. 84 and 15

The multi-pair PSSE processes multiple queries and subject sequences in parallel. In the rest of chapter, we mainly discuss multi-pair PSSE since the benefits of HPC are harvested usually when the bigger data are available. A coarse-grained parallelism, called tiling strategy, can be employed. Tiling strategy partitions the sequence database into different disjoint sets of subject sequences and assigns one set to each process. As result, each process has a subset of subject sequences and all the query sequences, thus estimating the PSS of each pair independent of other processes. This enhances data locality on single node and reduces the overhead of communication. See Ref. 78 for more details about the MPI implementation.

The basic idea of parallelizing the PSSE using OpenMP is similar to MPI implementation. In general, using more OpenMP threads can potentially save memory usage while it may also increase the synchronization cost (implicit or explicit) among these threads and the activation/deactivation overhead. It is hard to generalize a common rule for setting of the optimized number of OpenMP threads to deliver the best performance, because it also depends on specific computer architectures, platforms and problem size. The hybrid scheme (using OpenMP and MPI paradigms together) is also possible, Refs. 85 and 77 provide more details about this issue.

As for GPU implementation, to harvest a good performance from GPU, one is expected to have a good understanding of the hierarchy and features of GPU memory. It is especially important to optimize global memory access as its bandwidth is low and its latency is hundreds of clock cycles. Moreover, global memory coalescing is the most critical optimization for GPU programming.[82] Since the kernels of PSSE usually work over large numbers of sequences, which reside in the global memory, the performance is highly dependent upon hiding memory latency. At the same time, hiding global memory latency is also very important to achieve high performance on the GPU. This can be done by creating enough threads to keep the CUDA cores always occupied while many other threads are waiting for global memory accesses.[82] In another word, a higher occupancy of acceleration strategy using GPU will harvest more computing power of GPU.

Through analyzing experimental results of the single-pair PSSE, it is found that inter-task parallelism performs better than intra-task parallelism (results shown later). Thus, inter-task parallelism is considered to compute multi-pair PSSE.

In addition, the acceleration strategy will not work well when the size of subject sequence database becomes too big to be fitted into GPU global memory. This prohibits transfer of all the subject sequences to GPU at the same time. Therefore, an optimal number of subject sequences is needed to be shipped to GPU keeping in mind that the subject sequences and their permuted copies fit in global memory.

Moreover, the number of new generated sequences should be enough to keep all CUDA cores busy, i.e., keep a high occupancy of GPU, which is very important to harness the GPU power. Herein a memory tiling technique is developed, which is able to self-tuning based on the hardware configuration and can achieve a close-to-optimal performance. In out-of-core fashion, the data in the main memory is divided into smaller chunks called *tiles* and transferred to the GPU global memory. In this case, the *tile* refers to the number of subject sequences to be transferred to the GPU at a time.

Due to a high occupancy, the tile-based strategy using single (dual) GPU(s) achieves significant speedups of 240.31 (338.63)×, 243.51 (361.63)×, 240.71 (363.99)×, and 243.84 (369.95)× for 64, 128, 256, and 512 threads per block, respectively. More details about the GPU implementation can be found in Ref. 79.

In short, low occupancy interferes with the ability to hide latency on memory-bound kernels, causing performance degradation. However, increasing occupancy does not necessarily increase performance. In general, once a 50% occupancy is achieved, further optimization to gain additional occupancy has little effect on performance.

All the implementations of the proposed method and related programs in OpenMP, MPI and CUDA are available for free academic use at `http://cucis.ece.northwestern.edu/projects/PSSE/`.

4.3. *Summary*

In this chapter, we first discuss a recently developed method of mining genomic sequence data for related sequences using pairwise statistical significance. Then we give some high performance solutions to accelerate the estimation of the pairwise statistical significance of local sequence alignment, which supports standard substitution matrix like BLOSUM62 as well as position-specific substitution matrices. Our accelerator harvests computation power of many-core GPUs and multi-core CPU by using CUDA and OpenMP/MPI programming paradigms, respectively, which results in high end-to-end speedups for PSSE. As the size of biological sequence databases increase rapidly, even more powerful high performance computing accelerator platforms, comprising of heterogeneous components like multi-core CPU along with general-purpose GPGPUs (or GPU clusters) and possibly FPGAs, etc., are expected become more and more common and imperative for sequence mining.

Acknowledgement

This work is supported in part by Talent Introduction Fund of Henan University of Technology (No. 2013BS003), National Natural Science Foundation of China (Contract No. 61203265), Key Project of Henan Province (Contract No. 122102110106) and Research Fund of Henan University of Technology (No. 10XZR001). This research used resources of the National

Energy Research Scientific Computing Center, which is supported by the Office of Science of the U.S. Department of Energy under Contract No. DE-AC02-05CH11231. We also are thankful to Dr. Ankit Agrawal, Prof. Alok Choudhary, and Prof. Xiaoqiu Huang for offering their code, data, papers, and for helpful discussions.

References

1. D. S. Roos, COMPUTATIONAL BIOLOGY: Bioinformatics-trying to swim in a sea of data, *Science.* **291**, 1260–1261 (2001).
2. Q. Yang and X. Wu, 10 challenging problems in data mining research, *International Journal of Information Technology and Decision Making.* **5**(4), 597–604 (2006).
3. R. Giegerich and D. Wheeler, Pairwise sequence alignment, *BioComputing Hypertext Coursebook.* **2** (1996).
4. C. Camacho, G. Coulouris, V. Avagyan, N. Ma, J. Papadopoulos, K. Bealer, and T. Madden, BLAST+: architecture and applications, *BMC Bioinformatics.* **10**, 421 (2009).
5. A. Schäffer, L. Aravind, T. Madden, S. Shavirin, J. Spouge, Y. Wolf, E. Koonin, and S. Altschul, Improving the accuracy of PSI-BLAST protein database searches with composition-based statistics and other refinements, *Nucleic Acids Research.* **29**(14), 2994–3005 (2001).
6. W. R. Pearson, Empirical statistical estimates for sequence similarity searches, *Journal of Molecular Biology.* **276**, 71–84 (1998).
7. R. Mott, Accurate formula for P-values of gapped local sequence and profile alignments, *Journal of Molecular Biology.* **300**, 649–659 (2000).
8. A. Agrawal, V. Brendel, and X. Huang. Pairwise statistical significance versus database statistical significance for local alignment of protein sequences. In *Bioinformatics Research and Applications*, vol. 4983, *LNCS(LNBI)*, pp. 50–61, Springer Berlin/Heidelberg (2008).
9. M. L. Sierk, M. E. Smoot, E. J. Bass, and W. R. Pearson, Improving pairwise sequence alignment accuracy using near-optimal protein sequence alignments, *BMC Bioinformatics.* **11** (2010).
10. S. F. Altschul, R. Bundschuh, R. Olsen, and T. Hwa, The estimation of statistical parameters for local alignment score distributions, *Nucleic Acids Research.* **29**(2), 351–361 (2001).
11. S. Sheetlin, Y. Park, and J. L. Spouge, The gumbel pre-factor k for gapped local alignment can be estimated from simulations of global alignment, *Nucleic Acids Research.* **33**(15), 4987–4994 (2005).
12. A. Agrawal and X. Huang, PSIBLAST_PairwiseStatSig: reordering PSI-BLAST hits using pairwise statistical significance, *Bioinformatics.* **25**(8), 1082–1083 (2009).
13. A. Agrawal and X. Huang, Pairwise statistical significance of local sequence alignment using sequence-specific and position-specific substitution matrices,

IEEE Transactions on Computational Biology and Bioinformatics. **8**(1), 194–205 (2011).

14. A. Agrawal and X. Huang, Pairwise statistical significance of local sequence alignment using multiple parameter sets and empirical justification of parameter set change penalty, *BMC Bioinformatics.* **10**(Suppl 3), S1 (2009).
15. A. Agrawal, S. Misra, D. Honbo, and A. Choudhary. Mpipairwisestatsig: Parallel pairwise statistical significance estimation of local sequence alignment. In *ECMLS proceedings of HPDC 2010*, pp. 470–476 (2010).
16. Y. Zhang, S. Misra, D. Honbo, A. Agrawal, W. Liao, and A. Choudhary. Efficient pairwise statistical significance estimation for local sequence alignment using GPU. In *IEEE 1st International Conference on Computational Advances in Bio and Medical Sciences (ICCABS)*, pp. 226–231 (2011).
17. A. Agrawal, A. Choudhary, and X. Huang. Non-conservative pairwise statistical significance of local sequence alignment using position-specific substitution matrices. In *Proceedings of International Conference on Bioinformatics and Computational Biology*, pp. 262–268 (2010).
18. A. Agrawal and X. Huang. Pairwise DNA alignment with sequence specific transition-transversion ratio using multiple parameter sets. In *International Conference on Information Technology*, pp. 89–93 (dec., 2008).
19. A. Agrawal and X. Huang. Conservative, non-conservative and average pairwise statistical significance of local sequence alignment. In *Proc. of IEEE International Conference on Bioinformatics and Biomedicine, BIBM*, pp. 433–436 (2008).
20. A. Agrawal and X. Huang. DNAlignTT:pairwise DNA alignment with sequence specific transition-transversion ratio. In *Proceedings of IEEE International Conference on EIT* (2008).
21. A. Agrawal, A. Ghosh, and X. Huang. Estimating pairwise statistical significance of protein local alignments using a clustering-classification approach based on amino acid composition. In eds. I. Mandoiu, R. Sunderraman, and A. Zelikovsky, *Bioinformatics Research and Applications*, vol. 4983, *LNCS(LNBI)*, pp. 62–73, Springer Berlin/Heidelberg (2008).
22. A. Agrawal, V. P. Brendel, and X. Huang, Pairwise statistical significance and empirical determination of effective gap opening penalties for protein local sequence alignment, *International Journal of Computational Biology and Drug Design.* **1**(4), 347–367 (2008).
23. A. Agrawal and X. Huang. Pairwise statistical significance of local sequence alignment using multiple parameter sets. In *Proceedings of ACM 2nd International Workshop on Data and Text Mining in Bioinformatics, DTMBIO*, pp. 53–60 (2008).
24. A. Agrawal and X. Huang. Pairwise statistical significance of local sequence alignment using substitution matrices with sequence-pair-specific distance. In *Proceedings of International Conference on Information Technology*, pp. 94–99 (2008).
25. A. Agrawal, A. Choudhary, and X. Huang. Derived distribution points heuristic for fast pairwise statistical significance estimation. In *ACM International Conference On Bioinformatics and Computational Biology*, pp. 312–321 (2010).

26. J. Setubal and J. Meidanis, *Introduction to computational molecular biology*. PWS Pub. (1997).
27. S. Clancy and W. Brown, Translation: DNA to mRNA to protein, *Nature Education*. **1**(1) (2008).
28. F. Crick, On protein synthesis, *Symposia of the Society for Experimental Biology*. **12**, 138–163 .
29. W. Fitch and T. Smith, Optimal sequence alignments, *Proceedings of the National Academy of Sciences*. **80**(5), 1382 (1983).
30. D. W. Mount, *Bioinformatics: sequence and genome analysis (2nd ed.)*. Cold Spring Harbor Laboratory Press (2004).
31. W. R. Pearson. Protein sequence comparison and protein evolution. Tutorial - ISMB98 (1998).
32. S. F. Altschul, Amino acid substitution matrices from an information theoretic perspective, *Journal of Molecular Biology*. **219**, 555–565 (1991).
33. D. J. States, W. Gish, S. F. Altschul, Improved sensitivity of nucleic acid database searches using application-specific scoring matrices, *Methods*. **3**(1), 66–70 (1991).
34. S. F. Altschul, T. L. Madden, A. A. Schäffer, J. Zhang, Z. Zhang, W. Miller, and D. J. Lipman, Gapped BLAST and PSI-BLAST: a new generation of protein database search programs, *Nucleic acids research*. **25**(17), 3389–3402 (1997).
35. R. Bundschuh, Rapid significance estimation in local sequence alignment with gaps, *Journal of Computational Biology*. **9**(2), 243–260 (2002).
36. S. Altschul, M. Boguski, W. Gish, J. Wootton, Issues in searching molecular sequence databases, *Nature genetics*. **6**(2), 119–129 (1994).
37. M. Waterman, Consensus patterns in sequences, *Mathematical Methods for DNA sequences*. pp. 93–116 (1989).
38. A. Y. Mitrophanov and M. Borodovsky, Statistical significance in biological sequence analysis, *Briefings in Bioinformatics*. **7**(1), 2–24 (2006).
39. G. N. Goodman, Toward evidence-based medical statistics. 1: The p value fallacy, *Annals of internal medicine*. **130**, 995–1004 (1999).
40. S. Karlin and S. F. Altschul, Methods for assessing the statistical significance of molecular sequence features by using general scoring schemes, *Proceedings of the National Academy of Sciences*. **87**(6), 2264–2268 (1990).
41. R. Arratia and M. Waterman, The erdos-rényi strong law for pattern matching with a given proportion of mismatches, *The Annals of Probability*. **17**(3), 1152–1169 (1989).
42. R. Arratia, L. Gordon, and M. Waterman, An extreme value theory for sequence matching, *The annals of statistics*. **14**(3), 971–993 (1986).
43. M. Waterman, Mathematical methods for DNA sequences CRC Press Inc., Florida (1989).
44. A. Dembo, S. Karlin, and O. Zeitouni, Limit distribution of maximal non-aligned two-sequence segmental score, *The Annals of Probability*. pp. 2022–2039 (1994).
45. W. Pearson and T. Wood, Statistical significance in biological sequence comparison, *Handbook of statistical genetics* (2004).

46. M. Waterman and M. Vingron, Sequence comparison significance and poisson approximation, *Statistical Science.* **9**(3), 367–381 (1994).
47. E. J. Gumbel, *Statistics of Extremes.* Columbia University Press, New York (1958).
48. S. F. Altschul and W. Gish, Local alignment statistics, *Methods in Enzymology.* **266**, 460–480 (1996).
49. O. Bastien and E. Marechal, Evolution of biological sequences implies an extreme value distribution of type I for both global and local pairwise alignment scores, *BMC Bioinformatics.* **9**(1), 332 (2008).
50. R. F. Mott, Maximum-likelihood estimation of the statistical distribution of smith-waterman local sequence similarity scores, *Bulletin of Mathematical Biology.* **54**, 59–75 (1992).
51. W. R. Pearson, Empirical statistical estimates for sequence similarity searches, *Journal of molecular biology.* **276**(1), 71–84 (1998).
52. T. F. Smith and M. S. Waterman, Identification of common molecular subsequences, *Journal of Molecular Biology.* **147**(1), 195–197 (1981).
53. W. R. Pearson and D. J. Lipman, Improved tools for biological sequence comparison, *Proceedings of the National Academy of Sciences.* **85**(8), 2444–2448 (1988).
54. S. Altschul, T. Madden, A. Schäffer, J. Zhang, Z. Zhang, W. Miller, and D. Lipman, Gapped BLAST and PSI-BLAST: A new generation of protein database search programs, *Nucleic Acids Research.* **25**, 3389–3402 (1997).
55. A. Agrawal, A. Choudhary, and X. Huang. Sequence-specific sequence comparison using pairwise statistical significance. In *Software Tools and Algorithms for Biological Systems*, vol. 696, *Advances in Experimental Medicine and Biology*, pp. 297–306. Springer New York (2011).
56. W. R. Pearson, Effective protein sequence comparison, *Methods in enzymology.* **266**, 227 (1996).
57. M. S. Waterman and M. Vingron, Rapid and accurate estimates of statistical significance for sequence database searches, *Proceedings of the National Academy of Sciences.* **91**(11), 4625–4628 (1994).
58. S. R. Eddy, Maximum likelihood fitting of extreme value distributions (1997). URL `citeseer.ist.psu.edu/370503.html`. unpublished work.
59. W. R. Pearson, Flexible sequence similarity searching with the fasta3 program package, *Methods in Molecular Biology.* **132**, 185–219 (2000).
60. X. Huang and D. L. Brutlag, Dynamic use of multiple parameter sets in sequence alignment, *Nucleic Acids Research.* **35**(2), 678–686 (2007).
61. M. T. Lusk and A. E. Mattsson, High-performance computing for materials design to advance energy science, *MRS Bulletin.* **36**, 170–174 (2011).
62. S. K. Khaitan and J. D. McCalley. TDPSS: A scalable time domain power system simulator for dynamic security assessment. In *Networking and Analytics for the Power Grid*, pp. 323-332 (2012).
63. S. Khaitan and J. D. McCalley. EmPower: An efficient load balancing approach for massive dynamic contingency analysis in power systems. In *Networking and Analytics for the Power Grid*, pp. 289-298 (2012)

64. S. K. Khaitan and J. D. McCalley. High performance computing for power system dynamic simulation. In *High Performance Computing in Power and Energy Systems*, pp. 43–69. Springer-Verlag Inc. (2012).
65. S. K. Khaitan and J. D. McCalley. Dynamic load balancing and scheduling for parallel power system dynamic contingency analysis. In *High Performance Computing in Power and Energy Systems*, pp. 189–209. Springer-Verlag Inc. (2012).
66. S. K. Khaitan, C. Fu, and J. D. McCalley. Fast parallelized algorithms for online extended-term dynamic cascading analysis. In *IEEE PES Power Systems Conference and Exposition*, pp. 15–18 (2009).
67. M. Raju and S. K. Khaitan, Implementation of shared memory sparse direct solvers for three dimensional finite element codes, *Journal of Computing.* **1** (8), 699–706 (2010).
68. M. Raju and S. K. Khaitan, Domain decomposition based high performance parallel computing, *International Journal Computer Science Issues.* **5**, 691–704 (2009).
69. M. Raju and S. K. Khaitan, High performance computing using out-of-core sparse direct solvers, *International Journal of Mathematical, Physical and Engineering Sciences.* **3**(2) .
70. W. Hendrix, M. M. A. Patwary, A. Agrawal, W. keng Liao, and A. Choudhary. Parallel hierarchical clustering on shared memory platforms. In *19th Annual International Conference on High Performance Computing (HiPC'12)*, pp. 34–42 (2012).
71. M. M. A. Patwary, D. Palsetia, A. Agrawal, W. keng Liao, F. Manne, and A. Choudhary. A new scalable parallel dbscan algorithm using the disjoint set data structure. In *Proceedings of the International Conference on High Performance Computing, Networking, Storage and Analysisthe (Supercomputing, SC'12)*, pp. 62:1–62:11 (2012).
72. A. Pande and J. Zambreno, The secure wavelet transform, *Journal of Real-Time Image Process.* **7**, 131–142 (2012).
73. A. Pande and J. Zambreno, Poly-dwt: Polymorphic wavelet hardware support for dynamic image compression, *ACM TECS.* **1**, 1–26 (2012).
74. A. Pande and J. Zambreno. Polymorphic wavelet architecture over reconfigurable hardware. In *Proceedings of Field Programming Logic and Applicaions*, pp. 471–474 (2008).
75. Y. Liu, D. L. Maskell, and B. Schmidt, CUDASW++: optimizing Smith-Waterman sequence database searches for CUDA-enabled graphics processing units, *BMC Research Notes.* **2**(1), 73 (2009). ISSN 1756-0500.
76. D. Hains, Z. Cashero, M. Ottenberg, W. Bohm, and S. Rajopadhye. Improving CUDASW++, a Parallelization of Smith-Waterman for CUDA Enabled Devices. In *IPDPS*, pp. 485–496 (2011).
77. Y. Zhang, F. Zhou, J. Gou, H. Xiao, Z. Qin, and A. Agrawal, Accelerating pairwise statistical significance estimation using numa machine, *Journal of Computational Information Systems.* **8**(9), 3887–3894 (2012).
78. Y. Zhang, M. M. A. Patwary, S. Misra, A. Agrawal, W. keng Liao, Z. Qin, and A. Choudhary, Par-PSSE: Software for pairwise statistical significance estimation in parallel for local sequence alignment, *International Journal of Digital Content Technology and its Applications.* **6**(5), 200–208 (2012).

79. Y. Zhang, S. Misra, A. Agrawal, M. Patwary, W. Liao, Z. Qin, and A. Choudhary, Accelerating pairwise statistical significance estimation for local alignment by harvesting GPU's power, *BMC Bioinformatics.* **13**(Suppl 3), S1 (2012).
80. D. Honbo, A. Pande, and A. Choudhary. Hardware architecture for pairwise statistical significance estimation in bioinformatics problems. International Journal of High Performance Systems Architecture (2013). To appear.
81. D. Honbo, A. Agrawal, and A. N. Choudhary. Efficient pairwise statistical significance estimation using FPGAs. In *Proceedings of International Conference on Bioinformatics and Computational Biology*, pp. 571–577 (2010).
82. S. Ryoo, C. Rodrigues, S. Baghsorkhi, S. Stone, D. Kirk, and W. Hwu. Optimization principles and application performance evaluation of a multithreaded GPU using CUDA. In *ACM SIGPLAN Symposium on Principles and Practice of Parallel Programming*, pp. 73–82 (2008).
83. C. Orengo, A. Michie, S. Jones, D. Jones, M. Swindells, and J. Thornton, CATH - a hierarchic classification of protein domain structures, *Structure.* **5** (8), 1093 – 1109 (1997).
84. A. Agrawal, S. Misra, D. Honbo, and A. N. Choudhary, Parallel pairwise statistical significance estimation of local sequence alignment using message passing interface library, *Concurrency and Computation: Practice and Experience.* **23**(17), 2269–2279 (2011).
85. Y. Zhang, M. M. A. Patwary, S. Misra, A. Agrawal, W. keng Liao, and A. Choudhary. Enhancing parallelism of pairwise statistical significance estimation for local sequence alignment. In *The Workshop on Hybrid Mutl-core Computing*, pp. 1–8 (2011).

2. Biological Network Mining

Chapter 4

Indexing for Similarity Queries on Biological Networks

Günhan Gülsoy[1], Md Mahmudul Hasan[1], Yusuf Kavurucu[1,2] and Tamer Kahveci[1]

[1] *Computer and Information Science and Engineering, University of Florida, Gainesville, FL 32611, USA*
{ggulsoy, mmhasan, tamer}@cise.ufl.edu
[2] *Turkish Naval Academy, Istanbul, Turkey*
ykavurucu@dho.edu.tr

Biological networks demonstrate how different biochemical entities interact with one another to perform vital functions in an organism. Due to recent advances in high throughput computing and experimental techniques, several network databases, each containing large number of biological networks, have emerged. We need efficient methods for accessing and querying these databases. As these networks are generally represented as graphs in theory, several graph indexing methods are developed for answering queries on them. We group these methods in three categories, namely feature, tree and reference based indexing. In this chapter, we describe these network indexing categories on large biological network databases using example algorithms. Moreover, we provide a review of the literature and research directions on this topic.

1. Introduction

Recent advances in high throughput computing and experimental techniques in biological sciences led to rapidly growing amount of interaction data among molecules. *Biological networks* constitute a key model to represent the information governed by these interactions. Depending on the types of interactions, different types of biological networks exist, such as metabolic,[10] gene regulatory[17] and protein-protein interaction[11] networks.

Biological networks hold the information on how molecules work collectively to perform key functions. Because of this, extracting knowledge from biological networks has been an important goal in computational biology.

One way to do this is through comparative analysis that aligns networks to identify the similarities between them. Comparative analysis of biological networks has been successfully used for finding functional annotations of molecules,[6] identifying drug targets,[24] reconstructing metabolic networks from newly sequenced genome[10] and building phylogenetic trees.[5]

Alignment of biological networks is a computationally challenging problem. Existing methods often map the global and local network alignment problems to graph and subgraph isomorphism problems respectively. These two problems however have no known polynomial time solution.[7,21] In the literature, two approaches exist to solve the network alignment problem. First one either ensures optimality or at least a user supplied confidence in the optimality.[8,22,23] The second one encompasses the heuristic approaches which do not provide any optimality guarantees.[1,2,18] Both of these approaches are computationally intensive, and thus require a significant amount of running time and memory space.

Given a database that contains a large set of biological networks, a similarity query returns all database networks that have a higher alignment score with the query network compared to a user-specified similarity threshold. Rapid growth in the size of biological network databases coupled with the costly network alignment necessitates efficient methods for accessing and querying these databases. Exhaustively aligning each query network with all the networks in a large database one by one is neither practical nor feasible. Therefore, it is imperative to find alternative techniques that reduce the number of network alignment. Database indexing has been successfully used for similarity queries on traditional databases, such as relational, multi-dimensional or time series databases. Biological network databases have inherent properties that distinguish them from traditional databases. *In this chapter, we discuss network indexing strategies on biological network databases.*

There are a number of methods in the literature for network database indexing. A number of these solutions have been adapted to biological network databases. We group these methods mainly in three categories, namely feature, tree and reference based indexing. Next, we briefly summarize these categories.

The first category, *feature based indexing*, extracts sets of predefined features from all database networks. Thus, the methods in this category summarize the database networks with these features. Given a query network, they extract same or similar features for that query network. They compare those features of the query with the feature set of the entire database.

Using this comparison, they filter some of the networks in the database quickly.

The second category, *tree based indexing*, arranges the database networks hierarchically at different nodes of a tree. Thus, each node of a tree is a summary of a subset of the database networks. For a given query network, the methods in this category start aligning the query network to each node starting from the root node. Then, they progressively move down through the tree and filter out branches (i.e., subsets of networks) in the process.

The final category, *reference based indexing*, summarizes the database networks using a small set of networks called reference networks. The methods in this category align all database networks with all of the references as a preprocessing step. Given a query network, instead of aligning it with the database networks, they align it with the references. Using these and precomputed alignments they filter a substantial subset of the database.

2. Preliminaries

In this section, we start by presenting some of the common terminology used in this chapter. We then formally define the problem considered in this paper.

2.1. *Definitions*

Throughout this chapter, we utilize a number of abbreviations and symbols to simplify the explanation of methods. Table 1 lists the most frequently used symbols.

Table 1. Commonly used symbols in this chapter.

Symbol Name	Description
$\mathcal{D}$	A biological network database
d_i	i^{th} database network
Q	Query network
$\mathcal{R}$	Reference network set
r_j	j^{th} reference network
λ	A mapping between two networks
$sim(Q, d_i)$	Alignment (similarity) score of networks Q and d_i.
ϵ	Similarity cutoff

We list the terminology needed to understand the rest of this chapter as follows.

Definition 1 (Biological Network).
A biological network $N = (V_N, E_N)$ *consists of a set of nodes* V_N *and a set of interactions* $E_N \subseteq V_N \times V_N$.

Depending on the type of network (e.g., protein-protein interaction network, metabolic network, etc.), nodes can be genes, proteins, enzymes, etc. Also depending on the type of the network, an interaction can indicate various relationships among molecules. For example, gene regulatory networks are usually represented by directed graphs, whereas protein-protein interaction networks are modeled using undirected graphs.

Definition 2 (Subnetwork). *A subnetwork* N' *of a network* N *is defined as:*

$$N' = (V', E') \text{ where}$$
$$V' \subseteq V_N \text{ and}$$
$$E' \subseteq (V' \times V') \cap E_N$$

In the rest of this chapter, we will use the terms graph and subgraph interchangeably with network and subnetwork respectively.

Definition 3 (Network Alignment). *Let* $N_1 = (V_1, E_1)$ *and* $N_2 = (V_2, E_2)$ *be two biological networks. An alignment* λ *between* N_1 *and* N_2 *is mapping between* v_1 *and* v_2*, where* $v_1 \subseteq V_1$ *and* $v_2 \subseteq V_2$.

As the above definition implies, not all the nodes in the networks need to be aligned to a node in the other network. The unaligned nodes are referred to as *insertions* and *deletions.* Insertion and deletions are referred to as *indels* in the literature. We will use the term indel throughout this chapter to refer to any node insertions or deletions.

Definition 4 (Alignment Score). *Given two networks* (N_1, N_2) *and an alignment* λ *between the two, alignment score between* N_1 *and* N_2 $(sim(N_1, N_2))$ *is the sum of the following:*

(1) Similarity scores of the aligned nodes. Depending on the alignment method used, the scores may be discrete (1 or 0, match or mismatch) or more continuous such as protein sequence alignment scores.
(2) Indel penalties for such nodes.
(3) Based on λ*, edge alignment scores. A common definition of aligned edges is if two ends of the edges are mapped to each other.*

2.2. *Problem Formulation*

To extract knowledge from huge amount of data stored in biological network databases, we need efficient methods to perform queries on them. There are a number of different types of queries that can be performed on biological network databases. In this section we will focus on similarity queries. Next, we define the problem we consider in this chapter:

Problem definition: *Assume that we have a database of* n *biological networks denoted by* $\mathcal{D} = \{d_1, d_2, ..., d_n\}$, *where* d_i *is the* i^{th} *network in our database. Also, we are given a query network denoted by* Q. *Let us denote the alignment score between* Q *and* d_i $(d_i \in \mathcal{D})$ *with* $sim(Q, d_i)$. *Given a similarity cutoff* ϵ, similarity search *returns all the database networks* d_i *that satisfy* $sim(Q, d_i) \geq \epsilon$. *The aim is to reduce the processing time of similarity searches in biological network databases to a practical level.*

3. Feature Based Indexing

Several indexing methods exist for similarity queries in network databases. Majority of these methods can be classified as *feature based indexing* methods.[12,19,25,27,28] These methods use specific features (i.e., nodes, common subnetworks, etc.) for filtering purposes. They start by extracting these features from all the database networks. When the database receives a query, they process the query network using the same procedure as the database network to extract the same set of features. In order to filter database networks, they compare the set of features extracted from the query network with those of the database networks. Finally, they report the remaining networks as the results of the query. For a number of applications, verification of the results using a costly alignment method is necessary.

A number of different methods exist for feature based indexing. GraphGrep chooses paths as index feature.[12] gIndex uses frequent subnetworks for the same purpose.[27] Both methods apply exact subnetwork matching which have limited usage on biological networks. SAGA is a recent study that is developed for biological networks.[25] It uses fragments of networks as its features. Grafil[28] extends gIndex to support approximate matching for modeling similarity on biological networks. A recent feature based indexing method is SIGMA which concentrates on the problem of inexact matching in network databases.[19]

We will use SAGA as an example to study feature based indexing methods.

SAGA: A Case Study. *Substructure Index-based Approximate Graph Alignment (SAGA)*[25] is a recent study that applies approximate matching on biological networks. SAGA uses fragments of database networks as features and tries to combine them together to find larger matches. In this section, we describe SAGA in detail as an example of feature based indexing method.

Each node in a network has both a label and a unique id in SAGA.[25] This unique id is used to establish a total order among the nodes. In its applications SAGA uses network alignment where each node is either a match or a mismatch.

Definition 5 (Subgraph Distance under λ). *Let $G_1 = (V_1, E_1)$, $G_2 = (V_2, E_2)$ be two graphs. The subgraph distance (SGD) between G_1 and G_2 under a mapping $\lambda : V_1^* \leftrightarrow V_2^*$ is:*

$$\begin{aligned} SGD_\lambda(G_1, G_2) = {} & \omega_e \times StructDist_\lambda \\ & + \omega_n \times VertexMismatches_\lambda \\ & + \omega_g \times VertexGaps_\lambda \end{aligned}$$

where

$d_G(u, v)$ = *Length of shortest path between nodes u and v in G.*

$StructDist_\lambda = \sum_{u,v \in V_1^*, u<v} \mid d_{G_1}(u, v) - d_{G_2}(\lambda u, \lambda v) \mid$

$Mismatch(u, v) = 0$ *if u and v have same label (or group),* 1 *otherwise.*

$VertexMismatches_\lambda = \sum_{u \in V_1^*} Mismatch(u, \lambda u)$

$Gap_\lambda(u) = 0$ *if u has mapping under λ,* 1 *otherwise.*

$VertexGaps_\lambda = \sum_{u \in V_1 - V_1^*} Gap_\lambda(u)$

ω_e, ω_n and ω_g are the weights for each component in the matching under λ.

Definition 6 (Subgraph Distance). *Let $G_1 = (V_1, E_1)$, $G_2 = (V_2, E_2)$ be two graphs. The subgraph distance (SGD) between G_1 and G_2 is the minimum distance among all matchings:*

$SGD(G_1, G_2) = \min_\lambda \{SGD_\lambda(G_1, G_2)\}$

If the query network is subnetwork-isomorphic to the target network, then the subgraph distance between them is 0, and vice versa.

The main idea behind SAGA is to generate an index, namely *FragmentIndex* on small structures of the database networks and then use it for matching fragments of the query with the fragments in the database networks. After that, those matching fragments are used for larger alignments.

A fragment is a set of k (user specified parameter) nodes in the database networks. The selected nodes do not need to be connected in the source network. However, enumerating all possible size-k fragments is computationally expensive. Moreover, it is not meaningful to select two nodes which have large distance between them. Because of this, a parameter $d_{\max}$ is used to specify the maximum possible distance between two nodes in the source network. After selecting k nodes, two nodes having distance less than $d_{\max}$ are connected with a pseudo edge. If the generated fragment is a connected network, then it is indexed. However, this fragment does not need to be a connected subnetwork in the source database network which allows node insertion or deletions in the following matching steps.

An entry in the *FragmentIndex* has five components. The first one is a sequence of node IDs for the nodes in it (*vertexSeq*). The next one is the sequence of group labels associated with the nodes (*groupSeq*). The third component is the sequence of pairwise distances between the nodes (*distSeq*). The next one is the sum of these pairwise distances (*sumDist*). The final one is a unique network ID (*gid*).

SAGA also maintains a second index, namely *DistanceIndex* to evaluate the subnetwork distance between the query network and target database network efficiently.

Given a query network, SAGA enumerates the fragments in the query in a similar way to database networks. It probes *FragmentIndex* for each query fragment. Then, it first uses *groupSeq* and *sumDist* values in each *FragmentIndex* to filter out unmatched index entries. After that, it uses *distSeq* values to filter additional false positives. At the end of filtering step, SAGA produces a set of small fragment hits. In the following step, SAGA assembles those smaller hits into bigger matches. Finally, it examines each candidate match and produces a set of real matches.

The parameter, k, affects the performance of SAGA. A larger k value results in a larger *FragmentIndex* which increases the index probe cost. However, this may result in fewer false positives in the matching algorithm and reduce the cost of remaining step. Due to this fact, it is better to define a small fragment size if queries are expected to have many matches in the database networks so that it may not process many false positives. On the other hand, it is better to define a large fragment size if queries tend to have very few matches.

4. Tree Based Indexing

The second indexing strategy that we will describe is *tree based indexing.* Tree based indexing method builds a tree structure and associates the database networks to different parts of this tree. When processing a query, it traverses the tree starting from the root node. As moving down towards the leaf nodes of the tree, it tries to filter out subtrees; resulting in exclusion of large portions of the database networks. It reports the remaining nodes as the results as it reaches the leaf nodes in the tree.

There exist a few algorithms that fall under this indexing category. Berretti *et. al.* applied metric trees to attributed relational graph (ARG) databases for content-based image retrieval.[3] ARGs are clustered hierarchically according to their mutual distances and indexed by M-trees.[4] This technique is also used to model graphical representations of foreground and background scenes in videos.[16] We continue our discussion on this indexing strategy by explaining a representative algorithm, namely *Closure-Tree (CTree).*[14]

CTree: A Case Study. One of the popular methods for tree based indexing is Closure-Tree(CTree). CTree supports both subgraph and similarity queries. It organizes the networks in the database using a binary tree (named as *C-tree* in the corresponding literature) structure. Each leaf node represents a database network. Each internal node (named as *graph closure*) has structural information about its descendants in order to facilitate effective pruning. The internal nodes are actually hypothetical networks that are obtained by aligning the networks corresponding to their children nodes. An interesting property of the C-tree is the following: The score of the alignment of any query network with an internal node is at least as much as that with a leaf node rooted at that internal node. Following from this property, given a query network, CTree algorithm starts aligning query to the root node. It then proceeds to the children nodes. It prunes an entire subtree rooted at an internal node, if the alignment to that internal node has a score less than the given cutoff.

CTree defines a correspondence between two networks of unequal size by extending each network with dummy nodes and edges such that every node and edge has a corresponding element in the other graph. The extended network is represented as $G^* = (V^*, E^*)$ and a dummy node or edge has a special label ε.

Definition 7 (Graph Mapping). *Let $G_1 = (V_1, E_1)$, $G_2 = (V_2, E_2)$ be two graphs and $G_1^* = (V_1^*, E_1^*)$, $G_2^* = (V_2^*, E_2^*)$ be their extended graph versions, respectively. A graph mapping is a bijection $\phi : G_1^* \to G_2^*$, where (i)$\forall v \in V_1^*, \phi(v) \in V_2^*$ and at least one of v and $\phi(v)$ is not a dummy node, (ii)$\forall e = (u,v) \in E_1^*, \phi(e) = (\phi(u), \phi(v)) \in E_2^*$ and at least one of e and $\phi(e)$ is not a dummy edge.*

Edit distance between two graphs is the cost of transforming one graph into the other in terms of insertion and removal of vertices and edges, and the changing of attributes on them. CTree calculates the edit distance between two graphs under a mapping ϕ as follows:

Definition 8 (Edit Distance under ϕ). *Let G_1 and G_2 be two graphs. The edit distance between G_1 and G_2 under a mapping ϕ is the cost of transforming G_1 into G_2:*

$$d_\phi(G_1, G_2) = \sum_{v \in V_1^*} d(v, \phi(v)) + \sum_{e \in E_1^*} d(e, \phi(e))$$

where $d(v, \phi(v))$ and $d(e, \phi(e))$ are the node distance and the edge distance measures respectively.

The vertex and edge distance measures are application specific. But, in general, CTree defines the distance between two vertices or edges as 1 if they have different labels or 0 otherwise.

Definition 9 (Graph Distance). *Let G_1 and G_2 be two graphs. The distance between G_1 and G_2 is the minimum edit distance between them under all possible mappings.*

CTree defines similarity between two graphs ($sim(G_1, G_2)$) as similar to the definition of distance between them. But this time, it uses the maximum similarity under all possible mappings between them.

Recall that, each internal node (graph closure) in the C-tree structure has structural information about its descendants. Actually, CTree called it as the closure of its descendants.

Definition 10 (Vertex Closure and Edge Closure). *The closure of a set of vertices is a generalized vertex whose attribute is the union of attribute values of the vertices. Similarly, the closure of a set of edges is a generalized edge whose attribute is the union of attribute values of the edges.*

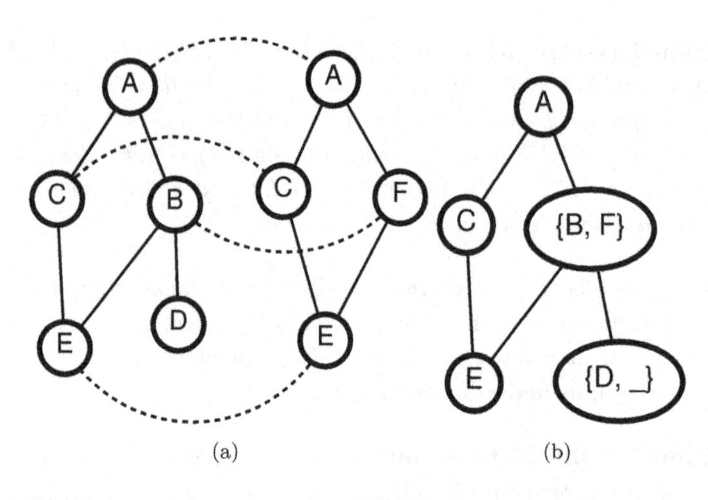

Fig. 1. (a) Alignment of two networks. Solid lines represent edges in the network. Dashed lines represent the mapping between the nodes of the two networks. (b) The closure. Note that, if there is a mismatch in the alignment, closure contains both of the nodes. If there is an indel in the alignment, the closure of the node contains a dummy node denoted by "_".

Definition 11 (Graph Closure under ϕ). *Let G_1 and G_2 be two graphs. The Closure(G_1, G_2) under a mapping ϕ is a generalized graph $G = (V, E)$ where V is the set of vertex closures of the corresponding vertices and E is the set of edge closures of the corresponding edges.*

Figure 1 presents a sample alignment between two networks, and a closure constructed using the given alignment.

CTree defines the distance and similarity between two closures similar to corresponding definitions between two graphs. In the case of multiple graphs, CTree calculates the total closure incrementally, i.e., $C_1 = Closure(G_1, G_2)$, $C_2 = Closure(C_1, G_3)$, and so on.

CTree uses a C-tree data structure for indexing in which each leaf node corresponds to a graph in the database and every internal node is the closure of its children nodes. Each node (except root) has at least $m(m \geq 2)$ and at most $M(\frac{M+1}{2} \geq m)$ children in the C-tree structure.

An insertion of a graph begins at the root node and iteratively chooses a child node until it reaches a leaf node. CTree chooses the child node that results in the minimum increase in the volume of the tree. If a node has more than M child nodes because of insertion, CTree applies linear partitioning on that node to split it into two nodes. For the deletion of a graph, CTree finds the leaf node where the graph is stored and delete that graph. If a node has less than m entries after deletion, then that node is deleted and its entries are reinserted. For the construction of tree, CTree uses hierarchical clustering instead of inserting each graph one by one stored in the database.

Due to NP-hard complexity of subgraph isomorphism, CTree uses an approximate technique called pseudo subgraph isomorphism. This is indigenous to the CTree method and for clarity and brevity, we will not discuss about it here. Interested readers are referred to the original paper.

CTree supports both subgraph and similarity queries. CTree processes a subgraph query in two steps. In the first step, it traverses the *C-tree* and prunes nodes according to pseudo subgraph isomorphism and returns a candidate answer set. In the second step, it applies exact subgraph isomorphism on each candidate answer and return the final answer set.

As exact similarity computing is expensive, CTree computes approximate graph similarity (or distance) using a heuristic graph mapping method. For this purpose, CTree contains a heuristic method, namely Neighbor Biased Mapping (NBM), in which the neighbors of a mapped vertex pair have higher chances to be mapped in the remaining iterations of the mapping process. CTree supports *K-NN* (K Nearest Neighbor) and range queries by using a priority queue that stores the nodes of *C-tree*.

5. Reference Based Indexing

The final indexing strategy we will describe is *reference based indexing*. Reference based indexing tries to summarize the database using a small set of references. In any application, references may be a small subset of the database itself or elements of the same type as the database elements. Using the set of references, any query received by the database is first run on the references. Using the result of the query on the reference set, reference based indexing estimates the results of the query on the whole database.

Reference based indexing is a relatively new approach in indexing biological network databases.[13] However, reference based indexing has been successfully used in databases of sequences[26] and elements in metric spaces.[9,15,29] The main difference between these methods is the task of selecting references. In this chapter, we will explore the details of reference based indexing using RINQ.

RINQ: A Case Study. *Reference-based Indexing for Biological Network Queries (RINQ)*[13] is a recent indexing method for similarity queries on biological networks. It uses a small set of fragments of database networks as reference networks. During the database setup, it aligns the reference networks with the database networks and stores all the alignment mappings and scores. During query processing, RINQ aligns the given query network only with the reference networks. Finally, according to these alignments, it computes a lower and an upper bound for the similarity value between query network and each database network. By using these lower and upper bound values, RINQ;

(1) Prunes some of the database networks directly,
(2) Selects some of the database networks as a part of result set without extra computation,
(3) Runs the costly alignment algorithm for the rest of the database networks.

In this section, we describe RINQ in detail as an example of reference based indexing in biological network databases.

RINQ has two major steps, namely index creation (which is performed only once before the database goes online) and query processing. The index creation step itself also has two phases. In its first phase, a large number of candidate reference networks are created from the database networks. In the second phase of index creation, a small subset of the candidate reference networks are selected to use as the actual reference set.

The success of RINQ depends on the quality of the reference set. The authors show that following properties of reference networks help RINQ perform better.

(1) Each reference network has a small number of nodes so that the query network aligns with the reference network quickly. For this purpose, RINQ sets the size of each reference network as the size of largest query allowed.
(2) Reference networks are non-redundant. No two reference networks have significant similarity.
(3) Reference networks comprehensively represent all the database networks.

The index creation stage of RINQ is designed to create a reference step that satisfies the above properties. In candidate reference creation, RINQ selects a random database network, and extracts a fixed size (Property 1) subnetwork using random walk. This subnetwork is a possible candidate reference. In order to avoid redundancy (Property 2), this possible candidate reference is aligned to all previous candidate references. If similarity above a given threshold is detected, this possible candidate reference is discarded. If no similarity is found, the subnetwork is added to the set of candidate references. Candidate reference creation is repeated until no more subnetworks can be added to the final set (Property 3).

In the second step of index creation, RINQ uses a set of training queries. In this step, RINQ starts with a random subset of the candidate reference set. Then it uses expectation maximization to improve the subset selection. Authors suggest the number of reference networks to be approximately 10 % of the number of database networks.

As described in Section 2.1, an alignment is a one to one mapping between the nodes of two networks. Let us use λ to represent the alignment between different networks as a function from the nodes of one network to the other. In order to calculate lower and upper bounds (LB and UB respectively) for the alignment score between a query network Q and ith database network d_i ($sim(Q, d_i)$), RINQ uses alignments of the references with both the given query and the database networks. Next, we describe how RINQ calculates these bounds.

In order to calculate $LB(Q, d_i)$, RINQ uses each reference network separately. Using each reference network r_j, and its alignments with Q and d_i ($\lambda_1 : Q \to r_j$, $\lambda_2 : r_j \to d_i$), RINQ calculates an exact lower bound ($LB_j(Q, d_i)$). Finally it picks the largest lower bound, as the best (tightest) lower bound, to use in the rest of the computations. In order to calculate $LB_j(Q, d_i)$, RINQ calculates an *indirect alignment* between Q and d_i using λ_1 and λ_2. Let us denote the indirect alignment function with λ'. RINQ calculates λ' as the composition of the functions λ_1 and λ_2:

$$\lambda' = \lambda_2(\lambda_1)$$

Finally, RINQ calculates the alignment score for the alignment λ' as $LB(Q, d_i)$. Note that RINQ works independently from the underlying alignment method used for defining similarities in the database. Therefore, calculation of the alignment score may change depending on the underlying alignment algorithm used.

RINQ calculates an approximate upper bound ($UB(Q, d_i)$) for the alignment score between Q and d_i using all the reference networks at once. In order to do this, RINQ calculates all the indirectly mapped nodes to each query node using all r_j. If it finds a node in d_i which no reference network nodes map to, it marks such nodes as mappable to all the nodes in the query. Finally, RINQ picks a node mapping between nodes of Q and d_i from the list of mappabilities just extracted, using maximum weight bipartite matching, disregarding all topological constraints. Finally, RINQ calculates the alignment score of this mapping as $UB(Q, d_i)$.

In order to speed up the similarity queries RINQ exploits the fact that aligning the query with all the references and calculating upper and lower bounds is faster than aligning the query with all the database networks by orders of magnitude. For each database network d_i, RINQ uses the $LB(Q, d_i)$ and $UB(Q, d_i)$ and the given similarity cutoff ϵ to make a decision to calculate the actual $sim(Q, d_i)$. Depending on the values of $LB(Q, d_i)$, $UB(Q, d_i)$ and ϵ, there are three possible cases, which RINQ treats as follows:

- Case 1: $UB(Q, d_i) < \epsilon$ implies $sim(Q, d_i) < \epsilon$, thus alignment not performed, d_i is filtered out.
- Case 2: $\epsilon \leq LB(Q, d_m)$ implies $\epsilon \leq sim(Q, d_i)$, thus alignment not performed, d_i is added to the result.
- Case 3: $LB(Q, d_m) < \epsilon \leq UB(Q, d_m)$ has no clear implications on result, therefore $sim(Q, d_i)$ has to be calculated.

For the first two cases above, RINQ decides not to run the network alignment algorithm for Q and d_i. Only for the networks satisfying the third case calculating the alignment between Q and d_i is necessary. Therefore, the number of alignments needed is highly reduced in this step.

6. Comparison between Indexing Strategies

In this section, we provide comparison of the three approaches we described in this chapter. In this comparison, we use the three model methods we used to describe all the approaches. For our comparison, SAGA[25] web server was used for the results concerning this method. For the tests involving CTree[14] and RINQ,[13] we used our own implementations in C++. As the underlying alignment strategy, we used the QNet algorithm.[8]

In order to test these methods, we used a database of gene regulatory networks we downloaded from KEGG PATHWAY database.[20] We used 297 gene regulatory networks, their sizes ranging from 15 nodes to 120 nodes. In this database, we have 21 different types of networks (e.g. MAPK signaling pathway) and 46 different organisms.

In order to test the database, we used 7 node queries. We limited the size of queries to 7 nodes because of the network size limitations in QNet.

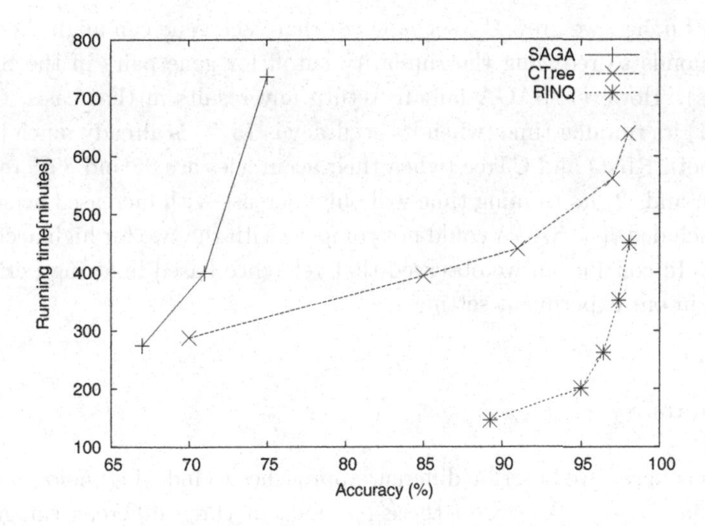

Fig. 2. Running time versus accuracy of SAGA, CTree and RINQ. Experiments are repeated using the same set of query networks with varying selectivities. Running time represents average query processing time. Accuracy is calculated over all test queries. Better performance is indicated by lower running time and higher accuracy.

We extracted the query networks using random walk on random database networks. We used 200 query networks for our experiments. We used the same query networks with similarity cutoff values that provide 2, 4, 8 and 16 % query selectivity values.

In order to compare the three algorithms, we used the average running time and average accuracy as the performance measures. In order to calculate accuracy, we compare the results returned by each method with the exhaustive comparison results with the query and the database. We report the number of true results returned by each method divided by the number of total true results as the accuracy value. For running time, we report the total filtering time plus the time it takes to verify the results as the total running time. However, it is worth to note that in all three methods, the time for filtering takes less than 1% of the total running time.

Figure 2 plots the comparison results. In our experiments RINQ performed much better in terms of both accuracy and speed than both CTree and SAGA. RINQ is up to 3 times faster than CTree for the same accuracy values. We could not achieve comparable accuracy with SAGA. We believe that this is probably because SAGA is well suited for exact matches and its accuracy drops quickly as the difference between the genes in the query and the database network grows. In order to improve the accuracy of

SAGA further, we grew the set of genes that each gene can align (i.e., this corresponds to reducing the similarity cutoff for gene pairs in the SAGA queries). However, SAGA fails to return any results in that case. Given that (1) its running time (when its accuracy is 75 %) is already much larger than both RINQ and CTree (when their accuracies are 85 and 89% respectively), and (2) its running time will only increase with increased accuracy, we concluded that SAGA could not compete with the two for high accuracy values. In conclusion, we observed that reference based indexing performs better in our experiment setting.

7. Summary

In this chapter, we describe different approaches to indexing biological network databases. We group these methods in three different categories. Then, we describe each of these groups in detail using a model algorithm in each group.

The First category, *feature based indexing* extracts sets of predefined features from all database networks. Thus, the methods in this category summarize the database networks with these features. Given a query network, they extract same or similar features for that query network. They compare those features of the query with the feature set of the entire database. Using this comparison, they filter some of the networks in the database quickly.

The second category, *tree based indexing*, arranges the database networks hierarchically at different nodes in a tree. Thus, each node of a tree is a summary of a subset of the database networks. For a given query network, the methods in this category start aligning the query network to each node starting from the root node. Then, they progressively move down through the tree and filter out branches (i.e., subsets of networks) in the process.

The third category, *reference based indexing*, summarizes the database networks using a small set of networks called references. The methods in this category align all database networks with all the references as a preprocessing step. Given a query network, instead of aligning it with the database networks, they align it with the references. Using these and precomputed alignments they filter a substantial portion of the database. Finally, we conclude our analysis by showing a comparison between the three approaches.

References

1. Ferhat Ay and Tamer Kahveci. SubMAP: Aligning metabolic pathways with subnetwork mappings. In *RECOMB*, pages 15–30, 2010.
2. Ferhat Ay, Tamer Kahveci, and Valérie de Crécy-Lagard. A fast and accurate algorithm for comparative analysis of metabolic pathways. *J. Bioinformatics and Computational Biology*, 7(3):389–428, 2009.
3. Stefano Berretti, Alberto Del Bimbo, and Enrico Vicario. Efficient matching and indexing of graph models in content-based retrieval. *IEEE Trans. Pattern Anal. Mach. Intell.*, 23(10):1089–1105, 2001.
4. Paolo Ciaccia, Marco Patella, and Pavel Zezula. M-tree: An efficient access method for similarity search in metric spaces. In *VLDB'97, Proceedings of 23rd International Conference on Very Large Data Bases*, pages 426–435. Morgan Kaufmann, 1997.
5. Jose C Clemente, Kenji Satou, and Gabriel Valiente. Reconstruction of phylogenetic relationships from metabolic pathways based on the enzyme hierarchy and the gene ontology. *Genome Inform*, 16(2):45–55, 2005.
6. Jose C Clemente, Kenji Satou, and Gabriel Valiente. Finding conserved and non-conserved reactions using a metabolic pathway alignment algorithm. *Genome Inform*, 17(2):46–56, 2006.
7. Stephen A. Cook. The complexity of theorem-proving procedures. In *STOC*, pages 151–158, 1971.
8. Banu Dost, Tomer Shlomi, Nitin Gupta, Eytan Ruppin, Vineet Bafna, and Roded Sharan. QNet: a tool for querying protein interaction networks. *J Comput Biol*, 15(7):913–925, Sep 2008.
9. R.F.S. Filho, A. Traina, Jr. Traina, C., and C. Faloutsos. Similarity search without tears: the OMNI-family of all-purpose access methods. In *Data Engineering, 2001. Proceedings. 17th International Conference on*, pages 623 –630, 2001.
10. Christof Francke, Roland J Siezen, and Bas Teusink. Reconstructing the metabolic network of a bacterium from its genome. *Trends Microbiol*, 13(11):550–558, Nov 2005.
11. L. Giot and J. S. Bader et al. A protein interaction map of Drosophila melanogaster. *Science*, 302(5651):1727–1736, 2003.
12. R. Giugno and D. Shasha. GraphGrep: A fast and universal method for querying graphs. In *Proc. 16th Int Pattern Recognition Conf*, volume 2, pages 112–115, 2002.
13. Günhan Gülsoy and Tamer Kahveci. RINQ: Reference-based indexing for network queries. *Bioinformatics [ISMB/ECCB]*, 27(13):149–158, 2011.
14. Huahai He and Ambuj K. Singh. Closure-Tree: An index structure for graph queries. *International Conference on Data Engineering*, 0:38, 2006.
15. H. V. Jagadish, Beng Chin Ooi, Kian-Lee Tan, Cui Yu, and Rui Zhang. iDistance: An adaptive B+-tree based indexing method for nearest neighbor search. *ACM Trans. Database Syst.*, 30(2):364–397, June 2005.
16. JeongKyu Lee, Jung-Hwan Oh, and Sae Hwang. STRG-Index: Spatio-temporal region graph indexing for large video databases. In *SIGMOD Conference*, pages 718–729, 2005.

17. Michael Levine and Eric H. Davidson. Gene regulatory networks for development. *Proceedings of the National Academy of Sciences of the United States of America*, 102(14):4936–4942, 2005.
18. C.S. Liao, K. Lu, M. Baym, R. Singh, and B. Berger. IsoRankN: spectral methods for global alignment of multiple protein networks. *Bioinformatics*, 25(12):253–238, 2009.
19. Misael Mongiovi, Raffaele Di Natale, Rosalba Giugno, Alfredo Pulvirenti, Alfredo Ferro, and Roded Sharan. SIGMA: a set-cover-based inexact graph matching algorithm. *J. Bioinformatics and Computational Biology*, 8(2):199–218, Apr 2010.
20. H. Ogata, S. Goto, K. Sato, W. Fujibuchi, H. Bono, and M. Kanehisa. KEGG: kyoto encyclopedia of genes and genomes. *Nucleic Acids Res*, 27(1):29–34, Jan 1999.
21. Sriram Pemmaraju and Steven Skiena. *Computational discrete mathematics: combinatorics and graph theory with mathematica.* Cambridge University Press, 2003.
22. Ron Y Pinter, Oleg Rokhlenko, Esti Yeger-Lotem, and Michal Ziv-Ukelson. Alignment of metabolic pathways. *Bioinformatics*, 21(16):3401–3408, Aug 2005.
23. Tomer Shlomi, Daniel Segal, Eytan Ruppin, and Roded Sharan. QPath: a method for querying pathways in a protein-protein interaction network. *BMC Bioinformatics*, 7:199, 2006.
24. Padmavati Sridhar, Tamer Kahveci, and Sanjay Ranka. An iterative algorithm for metabolic network-based drug target identification. *Pacific Symposium on Biocomputing*, pages 88–99, 2007.
25. Yuanyuan Tian, Richard C. McEachin, Carlos Santos, David J. States, and Jignesh M. Patel. SAGA: a subgraph matching tool for biological graphs. *Bioinformatics*, 23(2):232–239, 2007.
26. Jayendra Venkateswaran, Deepak Lachwani, Tamer Kahveci, and Christopher Jermaine. Reference-based indexing of sequence databases. In *Proceedings of the 32nd international conference on Very large data bases*, VLDB '06, pages 906–917. VLDB Endowment, 2006.
27. Xifeng Yan, Philip S. Yu, and Jiawei Han. Graph indexing: a frequent structure-based approach. In *SIGMOD*, pages 335–346, New York, NY, USA, 2004. ACM.
28. Xifeng Yan, Philip S. Yu, and Jiawei Han. Substructure similarity search in graph databases. In *SIGMOD Conference*, pages 766–777, 2005.
29. Peter N. Yianilos. Data structures and algorithms for nearest neighbor search in general metric spaces. In *Proceedings of the fourth annual ACM-SIAM Symposium on Discrete algorithms*, SODA '93, pages 311–321, Philadelphia, PA, USA, 1993. Society for Industrial and Applied Mathematics.

Chapter 5

Theory and Method of Completion for a Boolean Regulatory Network Using Observed Data

Takeyuki Tamura and Tatsuya Akutsu
Bioinformatics Center, Institute for Chemical Research, Kyoto University, Gokasho, Uji, Kyoto, Japan 6110011,
{tamura,takutsu}@kuicr.kyoto-u.ac.jp

In this chapter, we consider the knowledge completion problem, in which given existing knowledge is modified to be consistent with observed data. In particular, we focus on the knowledge completion problem on Boolean networks, and mathematically formalize the problem. Although basic versions of this problem are known to be NP-complete, they are solvable in polynomial time if the amount of observed data is not very large and the topology of given networks is a tree or a tree-like structure. We also consider the problem of detecting topological changes in signaling pathways after cell state alteration, and present an integer programming-based method. Our method is applied to a data set of gene expression profiles of colorectal cancer downloaded from the Gene Expression Omnibus and the signaling pathway data of colorectal cancer downloaded from the KEGG database.

1. Introduction

The basic idea of data mining is to extract important knowledge from a large amount of data. In some cases, almost no knowledge of the target topic exists, and the purpose of data mining is to obtain the basic knowledge as the first step. However, in many cases, some knowledge about the topic does exist, and the purpose of the data mining is to improve or enlarge this knowledge. Such a process is called *knowledge completion.* For example, when people eat bread, rice or pasta, glucose is assimilated by the small intestine and decomposed into pyruvate to provide us with necessary energy. This process is called the central metabolism and has been extensively studied in biochemical laboratories. Although much knowledge about the central metabolism is available, researchers have not yet discovered all the facts. Therefore, if results of a biological experiment are different from what was expected, the possibility of obtaining new knowledge via the knowledge completion approach exists.

In this chapter, we focus on the knowledge completion of biological networks using observed data. In particular, we assume a *Boolean network* model[1] as a model of biological networks, since it is a fundamental model and many theoretical and practical studies on the subject have been conducted. For example, a neurotransmitter signaling pathway was analyzed using Boolean networks in Gupta *et al.*[2]

In the Boolean network model of a signaling network, proteins (or protein complexes) are represented by nodes and the regulation among proteins is represented by edges. In Fig. 1(a), each of v_1, v_2, v_3 represents a protein, and the two edges indicate that both v_2 and v_3 regulate v_1. If there is an edge from v_a to v_b, v_a is called a *parent* of v_b. Therefore, v_2 and v_3 are parents of v_1. It should be note that the number of parents is not limited to two. States of nodes are represented by either 0 or 1. The assignment of 1 to a node indicates that the corresponding protein is activated. On the other hand, the assignment of 0 to a node indicates that the corresponding protein is suppressed. Readers may think that representing states of proteins only by 0 and 1 constitutes rough modeling. However, it is reasonable, since much of the existing knowledge on signaling pathways can be represented by Boolean functions.

In many cases, a relationship among a node and its parents can be represented by a combination of "AND," "OR," and "NOT." If v_1 is an "AND" node, v_1 is 1 only if values of all parents of v_1 are 1. For example, in Fig. 1(a), the relationship between v_1, v_2, v_3 is denoted by $v_1 = v_2 \wedge v_3$ if v_1 is an "AND" node. Therefore, $v_1 = 1$ holds only if $v_2 = v_3 = 1$ holds. On the other hand, if v_1 is an "OR" node, the relationship between v_1, v_2, v_3 is denoted by $v_1 = v_2 \vee v_3$. $v_1 = 1$ holds if one of v_2, v_3 is 1. Moreover, "NOT" is denoted by overlines. For example, $\overline{v_1}$ represents the negation of v_1, and $v_1 \neq \overline{v_1}$ always holds. Note that every variable takes only 0 or 1. Functions represented by a combination of "AND," "OR," and "NOT" are called *Boolean functions* (or AND/OR Boolean functions).

In signaling pathways, for example, the family of mitogen-activated protein kinase (MAPK) is known to activate the FBJ murine osteosarcoma virus oncogene (c-Fos).[3] MAPK1 and MAPK3 are subfamilies of MAPK,[4] both of which can activate cFOS.[3] Therefore, the condition of activation of c-FOS can be represented by cFOS= MAPK1 $\vee$ MAPK3, signifying that cFOS=1 if either MAPK1 or MAPK3 is active.

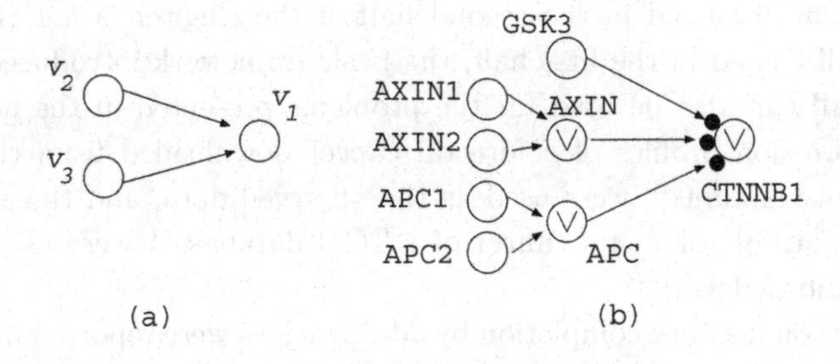

Fig. 1. (a) The relationship between v_1, v_2, v_3. v_1 is a child of v_2, v_3 and v_2, v_3 are parents of v_1. If v_1 is "AND," v_1 is 1 only if $v_2 = v_3 = 1$ holds. (b) The relationship between GSK3, AXIN1, AXIN2, APC1, APC2, and CTNNB1.

In the first half of this chapter, we discuss the theoretical aspects of the knowledge completion problem on Boolean networks in which we complete a given Boolean network so that the input/output behavior is consistent with given observed data,[5] where we consider only acyclic networks. In the theory of computational complexity, whether a problem is solvable in time proportional to a polynomial function of the problem size is very important. A detailed explanation is omitted here. However, in brief, that a problem is proved to belong to a class called NP-complete indicates it is almost impossible to solve the problem efficiently (in polynomial time). Thus, when a problem is considered, knowing whether it has a polynomial time algorithm is quite important. Readers interested in the details of computational complexity are advised to refer to papadimitrion,[6] and Arora and Barak.[7]

The knowledge completion problem is NP-complete in general and a basic version remains NP-complete even for tree-structured networks.[5] This implies that the knowledge completion problem with a large network and very large amount of data is almost not solvable, even when the given network has a simple structure such as trees. On the other hand, these problems can be solved in polynomial time for partial k-trees of bounded (constant) indegree if a logarithmic number of examples is given. This means that the problem can be efficiently solvable if the amount of data is not very large and the given network has a tree or tree-like structure.

One efficient method of handling the knowledge completion problem, which is NP-complete, is Integer Programming (IP). IP is known to be a useful method for solving NP-complete problems.[8] In the second half of the chapter, we consider the problem of detecting changes in signaling pathways after cell state alteration. To solve this problem, we developed an IP-based method and conducted a computer experiment.[9] Although the problem discussed in the second half of the chapter is not the same as those discussed in the first half, the basic framework introduced in the second half can also be used for the problems presented in the first half. Gene expression profiles of colorectal cancer downloaded from the Gene Expression Omnibus[10] were used as the observed data, and the signaling pathway data of colorectal cancer of KEGG database[11] were used as the existing knowledge.

Studies on network completion by adding edges were reported in clauset *et al.*[12] and Guimera and Sales-Pardo.[13] Kim and Leskovec developed the Metropolized Gibbs sampling approach for inferring missing nodes and edges simultaneously in networks.[14] Hanneke and Xing developed a method of deriving confidence bounds on the number of differences between the true and learned topologies.[15]

1.1. *Overview of Three Main Problems*

Although details are explained from Section 2, purposes of the three main problems discussed in this chapter are roughly summarized as follows: to obtain a consistent network with observed data,

- BNCMPL-1: assign Boolean functions to incomplete nodes,
- BNCMPL-2: change Boolean functions of minimum number of nodes,
- Problem of Section 5: detect deleted Boolean functions,

as shown in Fig. 2.

2. Network Completion Problem

Biological networks are often separately represented by each layer. In particular, genetic networks, protein-protein interaction networks, signaling pathways, and metabolic networks have attracted much research attention.[2,11,16,17] Inferring biological networks is one of the most important topics addressed by these studies. Extensive studies have also been conducted on inferring genetic networks.[18–20] The purpose of the Boolean version of this problem is, given a series of gene expression profiles, to infer a Boolean function, together with the input genes, that regulates each gene. It is often assumed in the inference of genetic networks that the states of all genes are observable in each environment and/or at each time step although there some noise exists. This assumption is reasonable because we

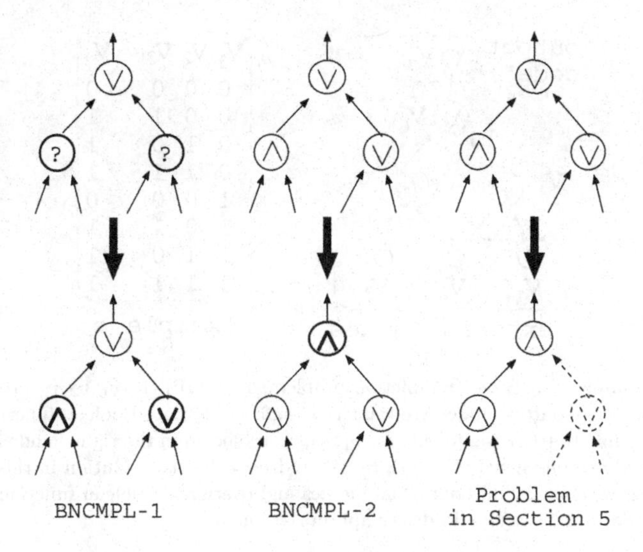

Fig. 2. Three main problems discussed in this chapter. Details are explained from Section 3.

can observe expression levels of all genes (or almost all genes) due to the technique of DNA chip and DNA microarray.

However, if some existing knowledge is available about the network that we want to reveal, it is more reasonable to take such existing knowledge into account in addition to the observed data. Furthermore, in the inference of *signaling networks* i.e., *signaling pathways*, in many cases, we need to measure the activity levels or quantities of proteins. Unfortunately, it is quite difficult to measure such data especially in living organisms. Reporter proteins (or reporter genes) are often used, each of which is associated with one or some kinds of protein.[21] However, both designing reporter proteins and introducing reporter proteins to cells are very difficult. In particular, introducing multiple types of reporter proteins is quite hard. Therefore, we can only assume in our analysis of signaling networks that the activity levels of one or a few kinds of proteins in various environments are observed. Thus, it is reasonable to assume that we have a preliminary network model of the target signaling pathway that includes unclear or invalid parts. Using observed data on the activity levels of a single or a few types of proteins in various environments, it may be possible to modify a preliminary network model so that it is consistent with the observed data. According to the well-known principle called Occam's razor in scientific discovery, it is reasonable to assume that the modification should be as small as possible. This leads to the study of *network completion* problems.

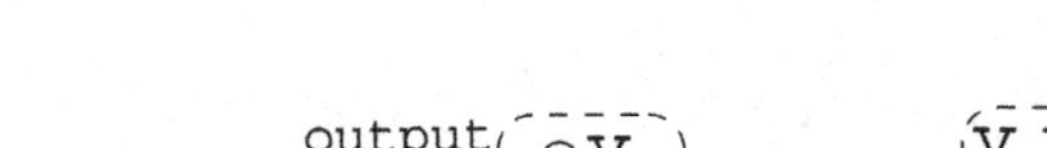
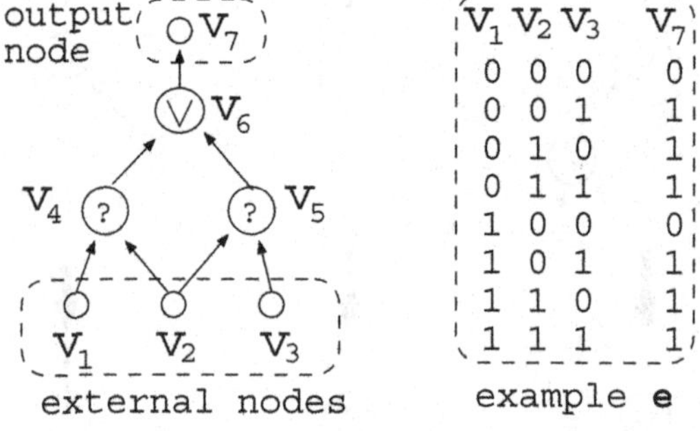

Fig. 3. Example of network completion problem, BNCMPL-1. v_1, v_2, v_3 are external nodes and v_7 is an output node. We want to assign appropriate Boolean functions to v_4 and v_5 according to 0/1 assignments for v_1, v_2, v_3, v_7 shown in the right-hand side. Note that "$\vee$" is already assigned to v_6. v_4 = "$\wedge$" and v_5 = "$\vee$" is a solution in this case. In BNCMPL-2, we should find appropriate nodes and overwrite Boolean functions. In the problem of Section 5, we should delete appropriate nodes.

To formalize the network completion problem, we employ the following mathematical model as shown in Fig. 3. We firstly assume that the network topology is given (i.e., a set of input nodes to each node is known) and Boolean functions are already assigned to a subset of nodes. We also assume that a set of nodes is divided into *external nodes*, *internal nodes* and *output nodes*, where only the activity levels of external and output nodes can be observed. Output nodes correspond to proteins whose activity levels are observed by reporter proteins, where we mainly consider the case that there exists only one output node because it is very difficult to introduce multiple reporter proteins. External nodes correspond to proteins whose activity levels are controlled by stimuli given from outside of the cell (e.g., environment), where these nodes can also be regarded as *input nodes* to the network. Furthermore, we assume that the network is acyclic because the state of the output node may not be determined uniquely if there exist cycles. Therefore, we can assume that the state of the output node is determined (through internal nodes) from the states of external nodes. Then, a basic version of the network completion problem is to determine Boolean functions for unassigned nodes so that the resulting network is consistent with the given set of examples (i.e., a series of external and output states). We also consider variants of the problem in which Boolean functions are assigned to all nodes but the minimum number of modifications (e.g., modification of Boolean functions, deletions of edges) are allowed. These problems are NP-complete and the basic version remains NP-complete even for tree structured networks. On the other hand, these problems can be solved in polynomial time for partial k-trees of bounded (constant) indegree if a logarithmic number of examples is given.[5]

3. Completing Boolean Functions

As the first step, we only consider acyclic Boolean networks as in Angluin *et al.*[22,23,24] Although the states of the nodes in a usual Boolean network are updated synchronously, we need not consider time steps because the states of all nodes in acyclic Boolean networks are determined uniquely from the states of external nodes and thus an acyclic Boolean network is equivalent to an acyclic Boolean circuit.

For modeling signaling networks, we define a Boolean network with *external*, *internal* and *output* nodes as follows.[5] A *Boolean network* $G(V, F)$ consists of a set $V = \{v_1, \ldots, v_n\}$ of nodes and a list $F = (f_1, \ldots, f_n)$ of *Boolean functions*, where each node takes a Boolean value (i.e., 0 or 1), and a Boolean function $f_i(v_{i_1}, \ldots, v_{i_l})$ with inputs from specified nodes $v_{i_1}, \ldots, v_{i_l}$ is assigned to each of internal and output nodes v_i. We use $\overline{x}$ to denote the negation of x, and use $\wedge$, $\vee$ and $\oplus$ to denote AND, OR and XOR, respectively. $IN(v_i)$ denotes the set of input nodes $v_{i_1}, \ldots, v_{i_l}$ to v_i. We allow that some v_{i_j} are not relevant (i.e., these v_{i_j} do not directly affect the state of v_i). For each $G(V, F)$, we associate a directed graph $G(V, E)$ defined by $E = \{(v_j, v_i) \mid v_j \in IN(v_i)\}$. We use $deg(v_i)$ to denote the indegree of v_i (i.e., $|IN(v_i)| = deg(v_i)$). In the analysis provided in this chapter, we assume that $G(V, E)$ is *acyclic*. We also assume that there exists only one output node, where some results can be extended for multiple output nodes. Since the network is acyclic, it is not necessary to consider state transition or/and time steps. We assume w.l.o.g. that $v_1, \ldots, v_h$ are *external nodes* (whose indegrees are 0) and v_n is the *output node*. Each node takes either 0 or 1 and the state of node v_i is denoted by $\hat{v}_i$. For an internal or output node v_i, $\hat{v}_i$ is determined by $\hat{v}_i = f_i(\hat{v}_{i_1}, \ldots, \hat{v}_{i_{l_i}})$.

In the basic setting, we assume that all f_i are known. However, f_i may not be known for some nodes v_i whereas $IN(v_i)$ are known. Such a node is called an *incomplete node*. A Boolean network is called *incomplete* if it contains an incomplete node, otherwise it is called complete. For example, the network of Fig. 3 is incomplete since the Boolean functions of v_4 and v_5 are not given.

Observed data are considered as a set of 0-1 assignments for external and output nodes. An $(h+1)$-dimensional 0-1 vector $\mathbf{e}$ is called an *example*, where the first h entries correspond to the external nodes and the last entry corresponds to the output node. In Fig. 3, h is 3 since v_1, v_2, v_3 are external nodes. An example $\mathbf{e}$ is called *positive* if $\mathbf{e}_{h+1} = 1$, otherwise it is called *negative.* $(0,0,0,0)$ and $(1,0,0,0)$ for (v_1, v_2, v_3, v_7) are negative examples and the others are positive examples in Fig. 3. A complete Boolean network $G(V, F)$ is *consistent* with $\mathbf{e}$ if $\hat{v}_n = \mathbf{e}_{h+1}$ holds under the condition that $\hat{v}_i = \mathbf{e}_i$ holds for $i = 1, \ldots, h$. If $v_4 =$ "$\vee$" and $v_5 = \wedge$ are given in advance in Fig. 3, it is consistent with the given examples. A basic version of the network completion problem is defined as follows.

Definition 1.[5] BNCMPL-1
Instance: An incomplete Boolean network $G(V, F)$, a set of examples $\{\mathbf{e}^1, \ldots, \mathbf{e}^m\}$,
Question: Is there an assignment of Boolean functions f_i to incomplete nodes so that the resulting network $G(V, F')$ is consistent with all examples?

An assignment that satisfies the above condition is called a *completion.* For example, suppose that a Boolean network as shown in Fig. 4 is given where v_5, v_6, v_7, v_8 are incomplete nodes, $v_1, \ldots, v_4$ are external nodes and v_8 is the output node. If $\mathbf{e}^1 = (v_1, v_2, v_3, v_4, v_8) = (0,0,0,0,1)$, $\mathbf{e}^2 = (0,0,1,1,1)$, $\mathbf{e}^3 = (1,1,0,0,1)$, $\mathbf{e}^4 = (1,0,0,1,0)$, $\mathbf{e}^5 = (0,1,0,0,0)$, $\mathbf{e}^6 = (1,1,0,1,0)$ are given, the answer for BNCMPL-1 is YES since $f_5 = v_1 \vee \overline{v_2} \vee v_3$, $f_6 = \overline{v_1} \vee v_3 \vee \overline{v_4}$, $f_7 = v_2 \vee v_3 \vee \overline{v_4}$, $f_8 = v_5 \wedge v_6 \wedge v_7$ satisfy all six examples.

In BNCMPL-1, a set of nodes to which Boolean functions are assigned is specified. However, there is a possibility that the existing knowledge about the target network contains mistakes. Therefore, it is useful to develop a method of modifying Boolean functions for the minimum number of nodes while retaining the network structure. To this end, we consider a variant of the network completion problem as follows.

Definition 2.[5] BNCMPL-2
Instance: A complete Boolean network $G(V, F)$, a set of examples $\{\mathbf{e}^1, \ldots, \mathbf{e}^m\}$, and a positive integer L,
Question: Is there an assignment of Boolean functions f_i to at most L nodes such that the resulting network $G(V, F')$ is consistent with all examples?

In BNCMPL-2, it is allowed to rewrite the complete nodes (i.e., other Boolean functions can be assigned to nodes for which Boolean functions are already assigned). For example, suppose that the previous six example

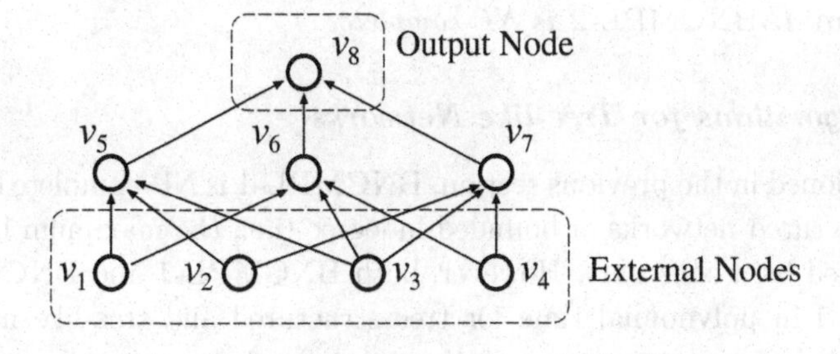

Fig. 4. Example for BNCMPL-1 and BNCMPL-2.

are given and $f_5 = v_1 \vee v_2 \vee v_3$, $f_6 = \overline{v_1} \vee v_3 \vee \overline{v_4}$, $f_7 = v_2 \vee v_3 \vee \overline{v_4}$, $f_8 = v_5 \wedge v_6 \wedge v_7$ are already assigned and $L = 1$. At this point, $G(V, F)$ is not consistent since $\mathbf{e}^1 = (v_1, v_2, v_3, v_4, v_8) = (0, 0, 0, 0, 1)$ is not satisfied. However, modifying f_5 from $v_1 \vee v_2 \vee v_3$ to $v_1 \vee \overline{v_2} \vee v_3$ results in all the examples being satisfied. Thus, in this case, the answer for BNCMPL-2 is also YES.

The problem of minimizing the number L of nodes for which other Boolean functions should be assigned can be regarded as a variant of BNCMPL-2. It is possible to solve this variant by considering BNCMPL-2 from $L = 0$ to n. It should be noted that *deletion of an edge* can be treated as a modification of the Boolean function and thus can be handled in BNCMPL-2 since it is allowed that some nodes in $IN(v_i)$ are not relevant[a].

Let D be a constant representing the maximum indegree of the given network. This assumption is reasonable because it is quite hard in general to learn Boolean functions with many inputs, and $O(2^n)$ bits are required to represent a Boolean function if an arbitrary Boolean function is allowed.

We can obtain the following hardness results. Since the details are described in Akutsu *et al.*,[5] the proofs are omitted here.

Theorem 1. BNCMPL-1 *is NP-complete even if one positive example is given and $D = 2$.*

Theorem 2. BNCMPL-1 *is NP-complete even if the network has a tree structure and $D = 2$.*

Theorem 3. BNCMPL-1 *is NP-complete even if only AND/OR nodes of $D = 2$ are allowed.*

[a] All the results in this section are valid even if all nodes in $IN(v_i)$ must be relevant.

Theorem 4. BNCMPL-2 *is NP-complete.*

3.1. *Algorithms for Tree-like Networks*

As mentioned in the previous section, BNCMPL-1 is NP-complete even for tree structured networks of bounded indegree (i.e., the maximum indegree is bounded by a constant). However, both BNCMPL-1 and BNCMPL-2 are solved in polynomial time for tree structured and tree-like networks of bounded constant indegree if the number of examples is small (i.e., $O(\log n)$). The algorithms are based on dynamic programming and are similar to that presented in Akutsu *et al.*[25] where additional ideas are introduced. Since the details are described in Akutsu *et al.*,[5] the proofs and algorithms are omitted in this chapter. Considering the case of a small number of examples is meaningful because a small number of data is usually available in the analysis of signaling networks.

Theorem 5. BNCMPL-1 *is solved in polynomial time if the network structure is a rooted tree of bounded indegree and the number of examples is* $O(\log n)$.

Theorem 6. BNCMPL-2 *is solved in polynomial time if the network structure is a rooted tree of bounded indegree and the number of examples is* $O(\log n)$.

The algorithms mentioned above can be extended for the case of *partial k-trees.* A partial k-tree is a graph with *treewidth* at most k, where the treewidth is defined via tree decomposition.[26] A *tree decomposition* of a graph $G(V, E)$ is a pair $\langle \mathcal{T}(\mathcal{V}_\mathcal{T}, \mathcal{E}_\mathcal{T}), (B_t)_{t\in\mathcal{V}_\mathcal{T}} \rangle$, where $\mathcal{T}(\mathcal{V}_\mathcal{T}, \mathcal{E}_\mathcal{T})$ is a rooted tree and $(B_t)_{t\in\mathcal{V}_\mathcal{T}}$ is a family of subsets of V such that (see also Fig. 5)

- For every $v \in V$, $B^{-1}(v) = \{t \in \mathcal{V}_\mathcal{T} | v \in B_t\}$ is nonempty and connected in $\mathcal{T}$,
- For every edge $\{u, v\} \in E$, there exists $t \in \mathcal{V}_\mathcal{T}$ such that $u, v \in B_t$.

$\max_{t\in\mathcal{V}_\mathcal{T}}(|B_t| - 1)$ is called the *width* of the decomposition and the *treewidth* is the minimum of the widths among all the tree decompositions of G.

Theorem 7. BNCMPL-1 *is solved in polynomial time if the network structure is a partial k-tree of bounded indegree and the number of examples is* $O(\log n)$.

This result can be extended for BNCMPL-2.

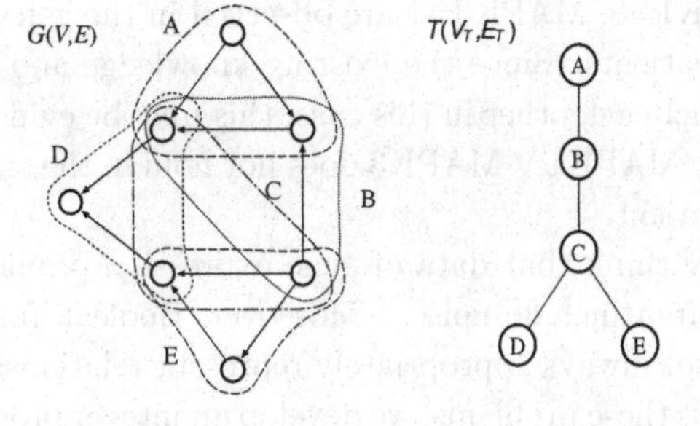

Fig. 5. Example of tree decomposition with treewidth 2.

Corollary 1. BNCMPL-2 *is solved in polynomial time if the network structure is a partial k-tree of bounded indegree and the number of examples is $O(\log n)$.*

4. Detecting Change of Signaling Pathway

In the previous section, we discussed the methods of completing Boolean functions when a complete (or incomplete) Boolean network and observed data are given. However, when genetic and epigenetic alteration occurs, the topology of signaling pathways in cells may also be affected. Of course, the degree of such change in pathways can be considered to be not very large since many functions of cells remain unchanged even when affected by cancers. Furthermore, in some cases, gene expression profiles are available for proteins related to signaling pathways. Therefore, in this section, we extend BNCMPL-2 and consider the problem of detecting a change in signaling pathways after cell state alteration by maximizing consistency. The content of Sections 4 to 6 is based on our previous paper.[9]

For this purpose, we apply gene expression profiles of cancer cells to signaling pathways representing the state before alteration. The values of gene expression profiles are encoded to 0 or 1 by using thresholds. Details are explained in Sec. 5.2. Since signaling pathways before alteration are represented by Boolean models and states of proteins after alteration are also represented by 0 or 1, assigned 0/1 values may sometimes contradict assigned Boolean functions. The part where such contradictions are detected can be considered the change by alteration.

For example, suppose that the relationship of cFOS= MAPK1 $\vee$ MAPK3 is given as the existing knowledge of the signaling pathway and cFOS=0, MAPK1=0, MAPK3=1 are observed in the gene expression profiles of cancer patients. Since the existing knowledge and gene expression profiles contradict each other in this case, this may be evidence that the relation of cFOS= MAPK1 $\vee$ MAPK3 does not hold in the signaling pathway of the cancer patient.

Readers may think that data of gene expression profiles and encoding by thresholds often include noises. Moreover, Boolean functions assigned to proteins do not always appropriately represent relationships among proteins. To address these problems, we develop an integer programming-based method where the noises incurred by the data of a small number of patients are automatically ignored.

Although polynomial time algorithms exist for BNCMPL-1 and BNCMPL-2 if the given network has a tree-like structure and the amount of observed data is not large as shown in the previous section, there was no algorithm for solving BNCMPL-1, BNCMPL-2 or their variants for the general case. Integer programming (IP) is a very common method in the field of operations research and often applied to complex problems[8] including NP-complete problems. To apply IP, problems must be formalized to maximize or minimize a given objective function which is a linear function of integer variables and constraints must also be given as linear equations or inequations of integer variables. Details of the proposed method are explained in Sec. 5.3.

We conducted computer experiments using the IP-based method mentioned above on the Boolean model of the signaling pathway of colorectal cancer in the KEGG database and gene expression profiles of colorectal cancer[27] from NCBI. These signaling pathway data include information about missing interactions among proteins. Missing interactions are relationships that are observed before, but not after, alteration. Therefore, if we apply data of gene expression profiles representing states after alteration, contradictions are expected to be detected on the part of missing interactions.

5. Integer Programming-based Method

5.1. *Data of Signaling Pathway*

The signaling pathway data of colorectal cancer was obtained from the Human Diseases section of the KEGG PATHWAY database.[11] The pathway can be regarded as a directed graph where each node corresponds to a protein and each edge corresponds to a relationship between proteins.

As described in Sec. 3, we represent the signaling pathway by a Boolean network. Protein-protein relationships can be classified into activation, inhibition, missing activation, and missing inhibition, where "missing" means that the protein-protein relationship is present in normal cells, but missing in cancer cells. Moreover, a relationship between a node and its parents can be represented by a combination of "AND", "OR" and "NOT". Thus the signaling pathway of colorectal cancer is represented by a network in which each node is assigned a Boolean function.

The assignment of a Boolean function to each node is conducted based on biological knowledge. For example, the Boolean functions shown in Fig. 1(b) can be obtained from the following biological knowledge. Note that "•" represents a negation in Fig. 1(b): CTNNB1 is degraded by a beta-catenin destruction complex, which includes Axin, adenomatosis polyposis coli (APC), and glycogen synthase kinase 3 (GSK3). From the view point of the Boolean network, this means that CTNNB1 is regulated by the existence of GSK3, AXIN and APC. GSK3 phosphorylates beta-catenin, resulting in beta-catenin recognition by an E3 ubiquitin ligase, and subsequent beta-catenin ubiquitination and proteasomal degradation.[28–30] Since AXIN has subfamilies AXIN1 and AXIN2, we can think that AXIN2 can be the alternative of AXIN1 and vice versa. Similarly, APC has subfamilies APC1 and APC2. Thus, using the Boolean network approach, the relationship between CTNNB1, GSK3, AXIN1, AXIN2, APC1 and APC2 can be represented as $\overline{CTNNB1} = GSK3 \wedge (AXIN1 \vee AXIN2) \wedge (APC1 \vee APC2)$. This implies that GSK3, AXIN and APC are all necessary to degrade CTNNB1. Since both AXIN1 and AXIN2 work as AXIN, AXIN is represented as $(AXIN1 \vee AXIN2)$. Similarly APC is represented as $(APC1 \vee APC2)$. By De Morgan's laws, $\overline{CTNNB1} = GSK3 \wedge (AXIN1 \vee AXIN2) \wedge (APC1 \vee APC2)$ is converted into $CTNNB1 = \overline{GSK3} \vee \overline{(AXIN1 \vee AXIN2)} \vee \overline{(APC1 \vee APC2)}$.

In this way, the Boolean model of the signaling pathway of colorectal cancer was constructed from the Human Diseases section of the KEGG PATHWAY database[11] of March 2010. The network, which contains 57 nodes and 154 edges, is available on "http://sunflower.kuicr.kyoto-u.ac.jp/~tamura/spcc.html".

5.2. *Gene Expression Data*

As for the observed data, gene expression profiles of colorectal cancer patients were obtained from the Gene Expression Omnibus (GEO).[31] This data set consists of 12 patients and 10 healthy controls, from which we use the data of only the 12 patients to draw a comparison with the signaling pathway of healthy people. In the original study, the authors performed an analysis of the normal-appearing colonic mucosa of early onset colorectal cancer patients.[27] Tumor specimens and adjacent grossly normal-appearing tissue were routinely collected and archived from patients undergoing colorectal resection. The healthy controls were those who underwent colonoscopic examination and were found to have no polyps and no known family history or previous colorectal cancer incidence.

Since these data consist of real values, some preprocessing is necessary to obtain 0/1 values. For this purpose, we first normalize the values as follows. For each gene, let $a_1, \ldots, a_{10}$ and $a_{11}, \ldots, a_{22}$ be the values of the gene expression profiles of healthy controls and colorectal cancer patients respectively. Let min and max be the minimum and maximum values in $a_1, \ldots, a_{22}$ respectively. Normalized values $a'_1, \ldots, a'_{22}$ are calculated by

$$a'_i = \frac{a_i - min}{max - min},$$

where each a'_i takes values in the range between 0 and 1. Then, a'_i is encoded to 0 or 1 by a threshold. In our experiment, 0.8 was used as the threshold. It should be noted that the data of gene expression profiles are not available for all the nodes in the signaling pathway.

5.3. *Representing by Linear Inequalities*

To represent the Boolean constraints of signaling pathways by IP, a method that we developed in Tamura *et al.*[32] may be applicable. However, since it is necessary to find a contradiction between the gene expression profiles and the signaling pathway in our problem, the extension is not very straightforward. Furthermore, it is preferable to develop a method that is tolerant for the noise included in gene expression profiles and differences between cancer types. To achieve this, in the first part of this subsection, we summarize how to formalize Boolean constraints as linear equations or inequations.

The Boolean relation of "AND" is converted into linear inequalities as follows. First,

$$x_1 = x_2 \wedge x_3 \wedge \cdots \wedge x_k$$

is converted into

$$(x_1 \vee \overline{x_2} \vee \overline{x_3} \vee \cdots \vee \overline{x_k}) \wedge (\overline{x_1} \vee x_2) \wedge (\overline{x_1} \vee x_3) \wedge \cdots \wedge (\overline{x_1} \vee x_k). \quad (1)$$

Then, (1) is converted into the following linear inequalities:

$$\begin{aligned} x_1 + \overline{x_2} + \overline{x_3} + \cdots + \overline{x_k} &\geq 1 \\ \overline{x_1} + x_2 &\geq 1 \\ \overline{x_1} + x_3 &\geq 1 \\ &\vdots \\ \overline{x_1} + x_k &\geq 1 \end{aligned}$$

where every variable takes only 0 or 1. On the other hand, for "OR" nodes,

$$x_1 = x_2 \vee x_3 \vee \cdots \vee x_k$$

is represented by

$$(\overline{x_1} \vee x_2 \vee x_3 \vee \cdots \vee x_k) \wedge (x_1 \vee \overline{x_2}) \wedge (x_1 \vee \overline{x_3}) \wedge \cdots \wedge (x_1 \vee \overline{x_k}). \quad (2)$$

Using a similar method for the case of "AND" nodes, (2) is converted into the following linear inequalities:

$$\begin{aligned} \overline{x_1} + x_2 + x_3 + \cdots + x_k &\geq 1 \\ x_1 + \overline{x_2} &\geq 1 \\ x_1 + \overline{x_3} &\geq 1 \\ &\vdots \\ x_1 + \overline{x_k} &\geq 1 \end{aligned}$$

where every variable takes only 0 or 1. Using the method explained above, the constraints of signaling pathways in the Boolean model can be represented by linear inequalities.

Note that the above constraints just represent the Boolean relations in the normal signaling pathway. In the signaling pathway of cancer patients, the function of some nodes for activations (or inhibitions) may be lost, which we call missing activations (or missing inhibitions). To detect such missing activations or/and missing inhibitions, we divide every node v of the original signaling pathway into v_1 and v_2 as shown in Fig. 6(a). v_1 and v_2 are connected by a directed edge e_1. The 0/1 value calculated by a Boolean function of v, which is determined by the values of the parents of v, is assigned to v_1. On the other hand, the 0/1 value encoded from gene expression profiles is assigned to v_2. Whichever "AND" or "OR" is assigned to v_2, $v_1 = v_2$ must be satisfied since v_2 has only one in-edge. If the value of v_1 is not the same as the value of v_2, we consider that the source of contradiction is a change in the signaling pathway. Although the source of this contradiction may be noise in the data, we assume that such noise can be ignored at this point.

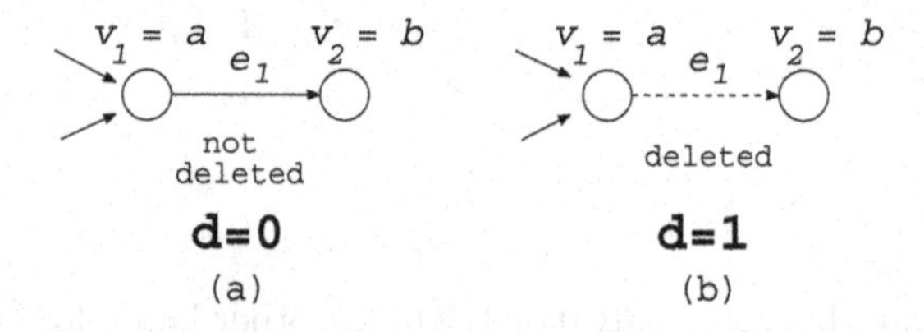

Fig. 6. To detect missing nodes, every node v is divided into two nodes v_1 and v_2. The 0/1 value calculated by the Boolean function corresponding to v is assigned to v_1. The 0/1 value encoded from gene expression profiles by a threshold is assigned to v_2. To handle a contradiction between v_1 and v_2, e_1 can be deleted if $d = 1$ holds.

If $v_1 \neq v_2$ holds, e_1 must be deleted to satisfy the linear constraints explained above. To represent whether e_1 is deleted or not, a parameter d is used in which $d = 0$ means that e_1 is not deleted and $d = 1$ means that e_1 is deleted. To this end, the relationship between v_1, v_2 and d is represented as

$$\overline{d}(v_1v_2 + \overline{v_1}\,\overline{v_2}) + d = 1 \tag{3}$$

where every variable and term take either 0 or 1. Not that $\overline{d}(v_1v_2 + \overline{v_1}\,\overline{v_2})$ means that $v_1 = v_2 = 1$ (represented by v_1v_2) or $v_1 = v_2 = 0$ (represented by $\overline{v_1}\,\overline{v_2}$) must be satisfied when $d = 0$ (represented by $\overline{d}$). On the other hand, d of (3) means that there is no constraint between v_1 and v_2 when $d = 1$.

Since (3) is not a linear inequality, it is converted into

$$\overline{d}ab + \overline{d}\overline{a}\overline{b} + d \geq 1$$

and further converted into

$$A + B + d \geq 1 \tag{4}$$

where

$$A = \overline{d} \wedge a \wedge b \tag{5}$$

$$B = \overline{d} \wedge \overline{a} \wedge \overline{b}. \tag{6}$$

Since (5) and (6) are not yet linear, (5) is converted into

$$\overline{A}+\overline{d}\geq 1 \tag{7}$$
$$\overline{A}+a\geq 1 \tag{8}$$
$$\overline{A}+b\geq 1 \tag{9}$$
$$A+d+\overline{a}+\overline{b}\geq 1 \tag{10}$$

and (6) is converted into

$$\overline{B}+\overline{d}\geq 1 \tag{11}$$
$$\overline{B}+\overline{a}\geq 1 \tag{12}$$
$$\overline{B}+\overline{b}\geq 1 \tag{13}$$
$$B+d+a+b\geq 1. \tag{14}$$

Thus, whether e_1 is deleted is represented by linear inequalities (4) and (7)-(14), where every variable takes either 0 or 1.

Now, we can check in which nodes the data of the gene expression profile do not satisfy the Boolean constraints for each patient. Since there are 12 patients of colorectal cancer in our dataset, the signaling pathway is duplicated into 12 copies $N_1,\ldots,N_{12}$ as shown in Fig. 7. It is to be noted that Fig. 7 describes only one node v_i although each N_j $(1\leq j\leq 12)$ originally contains 57 nodes and each node is duplicated into two nodes. Then for each N_j, 0/1 values of gene expression profiles are assigned. Since different parameters are necessary for each N_j, we use $d_{1\_i},\cdots,d_{12\_i}$ to represent whether e_{j_i} is deleted or not. If $v_{j\_i}\neq v'_{j\_i}$ holds, then $d_{j\_i}=1$ holds and e_{j_i} is deleted. Since the values of gene expression profiles differ among patients, $d_{j\_i}=1$ may hold for some j, but $d_{j\_i}=0$ holds for another j.

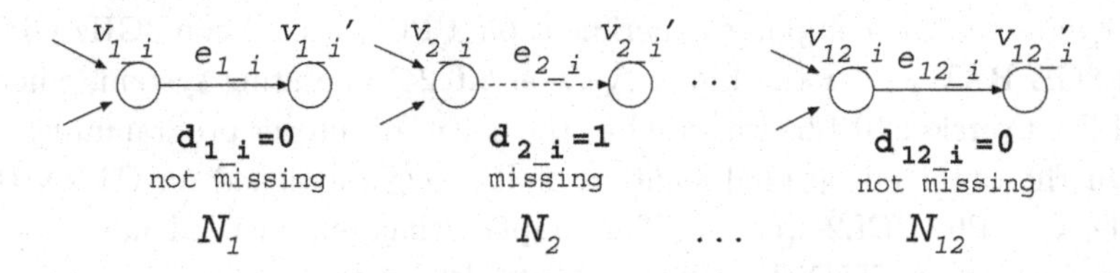

Fig. 7. There are 12 patients in the data of gene expression profiles and N_i corresponds to the network of the i-th patient of colorectal cancer. d_i represents whether e_{i_1} is deleted or not. It should be noted that each N_i contains $2n$ nodes although there are only two nodes for each N_i in this figure.

Furthermore, there are two major types of genomic instability for patients of colorectal cancer, chromosome instability (CIN) and microsatellite instability (MSI).[33] Therefore, some patients of colorectal cancer have relatively different signaling pathways when compared to those of other patients. According to the KEGG database, the signaling pathways of patients with MSI are partially different from those with CIN. However, since MSI occurs in less than 15%[33] of colorectal cancer, it is reasonable to assume that most data of gene expression profile is of CIN patients. In our method, data from MSI patients are treated as noise in addition to noises from gene expression profiles. To ignore noise appropriately, a parameter D_i determined by $d_{1\_i}, \ldots, d_{12\_i}$ is used to judge whether the function assigned to v_i is missing in the signaling pathway of CIN patients.

To take the effect of the type of colorectal cancer and the noise of gene expression profiles into account, we add

$$d_{1\_i} + d_{2\_i} + \cdots + d_{12\_i} - k \leq (12 - k)D_i$$

as a constraint of IP, where D_i takes either 0 or 1. Note that $D_i = 1$ means that the function of f_i is lost in more than k patients. It is also to be noted that D_i becomes 0 if at most k ds are 1 since the objective function of our method is

$$\min \sum_{i=1}^{n} D_i.$$

In our experiments, $k = 5$ was applied. This means that if contradictions are detected for at least six patients on v_i, our method considers that the the function of f_i is lost in the signaling pathway of patients of CIN type colorectal cancer.

6. Results of Computer Experiment

We conducted the computer experiment on a PC with a Xeon 3GHz CPU and 8GB RAM under the Linux (version 2.6.24) operating system, where CPLEX (Version 10.1.0) was used as the solver of integer programming.

In the obtained optimal solution of IP, "D_i"s assigned on CTNNB1, BAD, CASP9, BCL2 were 1. This implies that our method infers that the functions of CTNNB1, BAD, CASP9, BCL2 are lost in the signaling pathway of the CIN type colorectal cancer patients.

To evaluate the obtained results, we compared them with the KEGG database, in which there are missing activations or missing inhibitions on the functions of "CTNNB1", "CASP9" and "BCL2" in colorectal cancer. In fact, mutation of "CTNNB1" is often reported in human colorectal cancer.[34] "CASP9" is an essential downstream component for p53 to promote apoptosis as tumor suppressor function.[35] "BCL2" is known to suppress p53-dependent apoptosis in colorectal cancer cells.[36] Therefore, "CTNNB1", "CASP9" and "BCL2" can be considered as correctly predicted missing interactions. On the other hand, "BAD" can be considered as incorrectly predicted missing interactions because there is no missing interactions on "BAD" in the KEGG database of colorectal cancer. Although there is a possibility that incorrectly predicted missing interactions are newly found knowledge of missing interactions, we treat them as false prediction results in the estimation of the prediction accuracy.

To evaluate the prediction accuracy of our method, the following two measures[37] were used:

$$Sensitivity = \frac{TP}{TP + FN}, \quad Specificity = \frac{TP}{TP + FP}.$$

Here TP is the number of correctly predicted interactions (true positive). FP is the number of overpredicted interactions (false positive). FN is the number of underpredicted interactions (false negative).

In our experiment, since TP=3, FN=6 and FP=1 are obtained, sensitivity is 0.33 and specificity is 0.75. Although the level of specificity is relatively high, that of sensitivity is not satisfactory. However, for our purpose, specificity is much more important than sensitivity since in our method it may be possible to run iterations after adding newly found knowledge to the existing knowledge. It should be noted again that FP interactions may be newly found knowledge of missing interactions. The results of our computer experiment are summarized in the following table.

Gene	Normal cell	Predicted cancer cell	Cancer cell in database
CTNNB1	exist	deleted	deleted
BAD	exist	deleted	exist
CASP9	exist	deleted	deleted
BCL2	exist	deleted	deleted

7. Conclusion

In this chapter, the knowledge completion problem on Boolean network was discussed from the viewpoints of theory and application. Our theoretical contribution is the mathematically formalization of the knowledge completion problems BNCMPL-1 and BNCMPL-2, and analysis of their computational complexity. BNCMPL-1 is NP-complete in each of the following cases: (i) The number of examples is only one and $D = 2$; (ii) The given network has a tree structure and $D = 2$; (iii) Only AND/OR nodes of $D = 2$ are allowed. BNCMPL-2 is also NP-complete. However, both BNCMPL-1 and BNCMPL-2 can be solved in polynomial time if the network structure is a rooted tree or a partial k-tree of bounded indegree and the number of examples is $O(\log n)$. One of our ongoing studies addresses an extended version of BNCMPL-1, where we maximize the number of consistent nodes and a dynamic programming-based method with polynomial time running time may be developed. Another extension is that a Boolean function for each node is replaced with a quadratic function and real values are used for expression levels. In this version, least square fitting is used for finding the best completion on deleting or/and adding edges and we recently developed a dynamic programming-based method to achieve this.[38]

Our practical contribution is the development of an integer programming-based method for the knowledge completion problem on a Boolean network. We applied this method to detect changes in signaling pathways after cell state alteration. Gene expression profiles of colorectal cancer patients obtained from the Gene Expression Omnibus were used as the observed data whereas the topology and Boolean functions of the signaling pathway related to colorectal cancer downloaded from KEGG database were used as the existing knowledge. Applying our integer programming-based method to the above GEO and KEGG data, our method correctly detected the missing interactions of CTNNB1, CASP9 and BCL2 in signaling pathways after cell state alteration. In the above experiment, we only confirmed the accuracy of our method by comparison with existing knowledge. However, since our final objective is to find new knowledge about missing interactions after cell state alteration, this will be addressed in a future study.

References

1. S.A. Kauffman, *The Origins of Order: Self-organization and Selection in Evolution*, Oxford Univ. Press, NY (1993).
2. S. Gupta, S.S. Bisht, R. Kukreti, S. Jain, S.K. Brahmachari, Boolean network analysis of a neurotransmitter signaling pathway, *Journal of Theoretical Biology*, **244(3)**, 463–469 (2007).

3. J.Y. Fang, B.C. Richardson, The MAPK signalling pathways and colorectal cancer, *The Lancet Oncology*, **6(5)**, 322–327 (2005).
4. D. Kutz, M. Burg, Evolution of osmotic stress signaling via MAP kinase cascades, *The Journal of Experimental Biology*, **201**, 3015–3021 (1998).
5. T. Akutsu, T. Tamura, K. Horimoto, Completing networks using observed data, *Proc. 20th International Conference on Algorithmic Learning Theory*, 126–140 (2009).
6. C.H. Papadimitriou, *Computational Complexity*, Addison-Wesley, (1994).
7. S. Arora, B. Barak, *Computational Complexity: A Modern Approach*, Cambridge University Press, (2009).
8. J. Hromkovic, *Algorithmics For Hard Problems*, Springer-Verlag Berlin Heidelberg, (2001).
9. T. Tamura, Y. Yamanishi, M. Tanabe, S. Goto, M. Kanehisa, K. Horimoto, T. Akutsu, Integer Programming-based Method for Completing Signaling Pathways and its Application to Analysis of Colorectal Cancer, *Genome Informatics*, **24** (The 10th Int. Workshop on Bioinformatics and Systems Biology), pp. 193–203 (2010).
10. R. Edgar, M. Domrachev, A.E. Lash, Gene Expression Omnibus: NCBI gene expression and hybridization array data repository, *Nucleic Acids Research*, **30(1)**, 207–201 (2001).
11. M. Kanehisa, S. Goto, M. Furumichi, M. Tanabe, M. Hirakawa, KEGG for representation and analysis of molecular networks involving diseases and drugs, *Nucleic Acids Research*, **38**, D355–D360 (2010).
12. A. Clauset, C. Moore, M.E.J. Newman, Hierarchical structure and the prediction of missing links in networks, *Nature*, **453**, 98–101 (2008).
13. R. Guimera, M. Sales-Pardo, Missing and spurious interactions and the reconstruction of complex networks, *Proceedings of the National Academy of Sciences of the United States of America*, **106(52)**, 22073–22078 (2009).
14. M. Kim, J. Leskovec, The Network Completion Problem: Inferring Missing Nodes and Edges in Networks, *SIAM International Conference on Data Mining*, 47–58 (2011).
15. S. Hanneke, E.P. Xing, Network Completion and Survey Sampling, *Journal of Machine Learning Research - Proceedings Track* **5**, 209–215 (2009).
16. S. Kauffman, C. Peterson, B. Samuelsson, C. Troein, Genetic networks with canalyzing Boolean rules are always stable, *Proceedings of the National Academy of Sciences of the United States of America*, **101(49)**, 17102–17107 (2004).
17. J-D.J. Han, N. Bertin, T. Hao, D.S. Goldberg, G.F. Berriz, L.V. Zhang, D.D. Albertha, J.M. Walhout, M.E. Cusick, F.P. Roth, M. Vidal, Evidence for dynamically organized modularity in the yeast protein-protein interaction network, *Nature*, **430**, 88–93 (2004).
18. P. D'haeseleer, S. Liang, R. Somogyi, Genetic network inference: from co-expression clustering to reverse engineering, *Bioinformatics*, **16(8)**, 707–726 (2000).
19. T. Akutsu, S. Miyano, S. Kuhara, Identification of genetic networks from a small number of gene expression patterns under the Boolean network model,

Proceedings of Pacific Symposium on Bioinformatics, **(4)**, 17–28 (1999).
20. M. Hecker, S. Lambeck, S. Toepfer, E. van Someren, R. Guthke, Gene regulatory network inference: Data integration in dynamic models - A review, *Biosystems*, **96**, 86–103 (2009).
21. Y. Tokumoto, K. Horimoto, J. Miyake, TRAIL inhibited the cyclic AMP responsible element mediated gene expression, *Biochemical and Biophysical Research Communications*, **381**, 533–536 (2009).
22. D. Angluin, J. Aspnes, J. Chen, Y. Wu, Learning a circuit by injecting values, *Proc. 38th Annual ACM Symposium on Theory of Computing*, 584–593 (2006).
23. D. Angluin, J. Aspnes, J. Chen, L. Reyzin, Learning large-alphabet and analog circuits with value injection queries, *Machine Learning*, **72**, 113–138 (2008).
24. D. Angluin, J. Aspnes, J. Chen, D. Eisenstat, L. Reyzin, Learning acyclic probabilistic circuits using test paths, *Proc. 21st Annual Conference on Learning Theory*, 169–180 (2008).
25. T. Akutsu, M. Hayashida, W-K. Ching, M.K. Ng, Control of Boolean networks: Hardness results and algorithms for tree structured networks, *Journal of Theoretical Biology*, **244**, 670–679 (2007).
26. J. Flum, M. Grohe, *Parametrized Complexity Theory*, Springer, Berlin (2006).
27. Y. Hong, K.S. Ho, K.W. Eu, P.Y. Cheah, A susceptibility gene set for early onset colorectal cancer that integrates diverse signaling pathways: implication for tumorigenesis, *Clinical Cancer Research*, **13(4)**, 1107–1114 (2007).
28. R.Z. Karim, G.M.K. Tse, T.C. Putti, R.A. Scolyer, C.S. Lee, The significance of the Wnt pathway in the pathology of human cancers, *Pathology*, **36(2)**, 120–128 (2004).
29. Y. Komiya, R. Habas, Wnt signal transduction pathways, *Organogenesis*, **4(2)**, 68–75 (2008).
30. B.T. MacDonald, K. Tamai, X. He, Wnt/beta-catenin signaling: components, mechanisms, and diseases, *Developmental Cell*, **17(1)**, 9–26 (2009).
31. T. Barrett, T.O. Suzek, D.B. Troup, S.E. Wilhite, W.C. Ngau, P. Ledoux, D. Rudnev, A.E. Lash, W. Fujibuchi, R. Edgar, NCBI GEO: mining millions of expression profiles–database and tools, *Nucleic Acids Research*, **33**, D562–D566 (2005).
32. T. Tamura, K. Takemoto, T. Akutsu, Finding minimum reaction cuts of metabolic networks under a Boolean model using integer programming and feedback vertex sets, *International Journal of Knowledge Discovery in Bioinformatics*, **1**, 14–31, (2010).
33. W.M. Grady, Genomic instability and colon cancer, *Cancer and Metastasis Reviews*, **23**, 11–27 (2004).
34. S. Satoh, Y. Daigo, Y. Furukawa, T. Kato, N. Miwa, T. Nishiwaki, T. Kawasoe, H. Ishiguro, M. Fujita, T. Tokino, Y. Sasaki, S. Imaoka, M. Murata, T. Shimano, Y. Yamaoka, Y. Nakamura, AXIN1 mutations in hepatocellular carcinomas, and growth suppression in cancer cells by virus-mediated transfer of AXIN1, *Nature Genetics*, **24**, 245–250 (2000).

35. M.S. Soengas, R.M. Alarcon, H. Yoshida, A.J. Giaccia, R. Hakem, T.W. Mak, S.W. Lowe, Apaf-1 and Caspase-9 in p53-Dependent Apoptosis and Tumor Inhibition, *Science*, **284(5411)**, 156–159 (1999).
36. M. Jiang, J. Milner, Bcl-2 constitutively suppresses p53-dependent apoptosis in colorectal cancer cells, *Genes & Development*, **17**, 832–837 (2003).
37. D.W. Mount, *Bioinformatics, Sequence And Genome Analysis*, Cold Spring Harbor Laboratory Press, (2004).
38. N. Nakajima, T. Tamura, Y. Yamanishi, K. Horimoto, T. Akutsu, Network completion using dynamic programming and least-squares fitting, *The Scientific World Journal*, **2012(957620)**, 8 pages (2012).

Chapter 6

Mining Frequent Subgraph Patterns for Classifying Biological Data

Saeed Salem
Department of Computer Science,
North Dakota State University, Fargo, ND 58102, USA,
saeed.salem@ndsu.edu

Graphs have become widely adopted for representing biological data. Examples of biological data represented as graphs include chemical compounds, protein tertiary structure, protein-protein interaction networks, gene coexpression networks, etc. Graph mining techniques have proven to be powerful in discovering useful patterns in the data. The area of graph mining addresses the problem of discovering interesting subgraph patterns in a database of graphs or a single graph. The set of interesting subgraphs have immediate applications in graph clustering, graph classification, and graph indexing.

This chapter will focus on the problem of frequent subgraph pattern mining. Mining a summarized set of frequent patterns will also be discussed. We will present how frequent subgraphs have been successfully used in chemical compound classification and mining family-specific protein structural motifs.

1. Introduction

In today's world, graphs are commonly used to represent complex structures and their interactions. Examples include protein-protein interaction networks, metabolic networks, telecommunication networks, chemical networks, and social networks. The analysis of such complex patterns has gained a lot of interest in the research community in various fields. Among the different graph analysis techniques, mining frequent graphs, those occurring frequently in a graph database, has become one of the most widely used technique. Frequent subgraphs have a wide range of applications, e.g., discovering motifs in biological networks,[1,2] classifying chemical compounds,[3] discovery of functional modules from *gene coexpression* graphs,[4] and graph indexing using frequent pattern fragments.[5]

The problem of mining frequent graph patterns has received great attention from the data mining research community which resulted in several algorithms.[6–10] In this chapter, we present three different applications of graph mining techniques in biological data analysis.

Chemical compound classification is an important task in drug discovery that aims at identifying chemical compounds that exhibit the desired behavior.[11] One approach that has been successful in chemical compound classification is based on mapping each compound to a feature vector of the compound's physicochemical, geometric, and topological properties and then using traditional classification techniques.[12] Graph-based approaches for chemical compound classification are based on graph-representation of compounds.[3,13,14] Once chemical compounds are represented as graphs, discriminative subgraphs are extracted and each graph is then represented in the discriminative subgraphs space as a feature vector. Each vector represents a graph and each entry in the vector corresponds to the occurrence of the discriminative subgraph in the graph. Traditional classification techniques are then used to classify the graphs representing the chemical compounds.

Unknown protein functions can be inferred from other proteins that have high structural similarity.[15] Structural motifs are recurrent structural fragments that occur in a significant number of proteins in a given family. Family-specific motifs have direct applications in protein classification, function prediction, and folding. The tertiary structure of a protein is represented as a graph whose nodes represent the $C\alpha$-atom and edges represent the proximity between atoms based on various distance functions.[16]

In gene expression analysis, clustering genes that show high expression profile similarity has been proposed to predict the functions of unknown genes.[17] The effectiveness of the clustering approach is limited by the fact that some genes with similar expression profiles may not have the same function and the similarity in the profiles is attributed to the simultaneous perturbation of multiple biological pathways. Recent research have focused on integrating multiple gene expression datasets and discovering sets of genes that show similar expression profiles in a significant number of experiments.[18] Graph mining-based approaches have recently been employed to mine expression patterns from multiple cross-platform microarray data.[4]

In bioinformatics, interestingness measures that are based on the integration of frequency-based measures and density of the subgraph patterns have been proposed. One motivation to search for dense frequent subgraphs is that the number of frequent subgraphs can be very large, thus leading to an information overload problem. For example, in gene coexpression graphs, we can only search for sets of genes that are densely connected in at least *minsup* graphs. The CLAN[19] and Cocain [20] are two such algorithms which mine closed cliques and coherent closed quasi-cliques, respectively. In some applications, such as gene coexpression graphs, where nodes are uniquely identified, more efficient algorithms have been proposed to mine significant subgraphs. The Crochet and Crochet$^+$ algorithms[21,22] mine, respectively, cross-all-graph quasi-cliques and frequent cross-graph quasi-cliques from graphs with unique labels.

This chapter is organized as follows: In the next section, we introduce preliminary definitions and introduce the graph mining problem. Section 3 will give an overview of different frequent subgraph mining algorithms. Section 4 will present the problem of graph classification and discuss the pattern-based graph classification approach using chemical compound classification as a case study. Mining family-specific protein structural motifs will be discussed in Section 5. Finally, recent approaches for addressing the challenges and limitations of graph mining algorithms will be discussed in Section 6.

2. Frequent Graph Mining

First, we will introduce preliminary definitions that are used throughout the chapter. Then, we will introduce the problem of frequent subgraph mining and other related problems.

2.1. *A Primer on Graphs*

Graphs and Subgraphs: A graph $G = (V, E, \Sigma, \mathcal{L})$ consists of a set of vertices $V = \{v_1, v_2, ..., v_n\}$, and a set of edges $E \subseteq V \times V = \{(v_i, v_j) : v_i, v_j \in V\}$, Σ is the set of vertex and edge labels, and $\mathcal{L}$ is the labeling function that assigns labels to vertices and edges of the graph such that $\mathcal{L} : V \cup E \to \Sigma$. The vertex set and edge set of a graph G are denoted by $V(G)$, $E(G)$, respectively. The size of the graph G is the number of edges ($|E(G)|$) and is denoted as $|G|$.

Subgraph Isomorphism: A graph $G' = (V', E', \Sigma', \mathcal{L}')$ is a *subgraph* of another graph $G = (V, E, \Sigma, \mathcal{L})$, denoted $G' \subseteq G$, if there exists an injective mapping $\Phi : V' \rightarrow V$, such that for all $v \in V'$, $\Phi(v) \in V$ and $\mathcal{L}'(v) = \mathcal{L}(\Phi(v))$, and for all $(v_i, v_j) \in E'$, $(\Phi(v_i), \Phi(v_j)) \in E$ and $\mathcal{L}'((v_i, v_j)) = \mathcal{L}((\Phi(v_i), \Phi(v_j)))$. In other words, Φ is a vertex and edge label-preserving injective mapping function. For example, in Figure 1, g_{15} (in (c)) is a subgraph of G_1, G_2 , and G_3 (in (a)) but is not a subgraph of G_4. Throughout this chapter, we will use this definition for subgraph isomorphism.

Induced Subgraph Isomorphism: The induced subgraph isomorphism is more restrictive than the simple subgraph isomorphism defined above. A graph $G'(V', E', \Sigma', \mathcal{L}')$ is an induced subgraph of a graph $G(V, E, \Sigma, \mathcal{L})$, denoted $G' \subset G$, if there exists an injective mapping $\Phi : V' \rightarrow V$, such that for all $v \in V'$, $\Phi(v) \in V$ and $\mathcal{L}'(v) = \mathcal{L}(\Phi(v))$, and for all $v_i, v_j \in V'$, $(\Phi(v_i), \Phi(v_j)) \in E \implies (v_i, v_j) \in E'$ and $\mathcal{L}'((v_i, v_j)) = \mathcal{L}((\Phi(v_i), \Phi(v_j)))$. The induced subgraph definition requires all the edges present in G between the mirror vertices of $V(G')$ to appear in the subgraph G'. Note that the simple subgraph definition does not require this condition. In Figure 1, g_{15} (in (c)) is an induced subgraph of G_1, G_2 , and G_3, and while graph g_{14} is a simple subgraph of G_1, G_2 , and G_3, it is not an induced subgraph of any of the graphs in the database.

Graph Support: For a graph dataset, $\mathcal{G} = \{G_1, G_2, \cdots, G_n\}$, the support of a graph g is the number of graphs which has a subgraph isomorphic to g. Let $t(g, \mathcal{G}) = \{i : g \subseteq G_i \text{ and } G_i \in \mathcal{G}\}$ be the set of graph identifiers that contain a subgraph isomorphic to g. The support of g, is defined as the cardinality of the set of identifiers, i.e., $support(g, \mathcal{G}) = |t(g, \mathcal{G})|$. For the graph database shown in Figure 1(a), the support of g_{10} is 3 since g_{10} occurs in three graphs, G_1, G_2 and G_3, i.e., $t(g_{10}, \mathcal{G}) = \{1, 2, 3\}$.

2.2. *Mining Frequent Subgraphs*

A subgraph g is called frequent in a graph dataset $\mathcal{G}$ if it occurs in at least *minsup* graphs, where *minsup* is a user-specified minimum support threshold; i.e., g is frequent if $support(g, \mathcal{G}) \geq minsup$.

Given a graph dataset $\mathcal{G}$, and a minimum support threshold *minsup*, the problem of mining frequent subgraphs is to find the set $\mathcal{F} = \{g_1, g_2, \cdots, g_{|\mathcal{F}|}\}$ such that each $g_i \in \mathcal{F}$ is frequent, i.e., $support(g_i, \mathcal{G}) \geq minsup$.

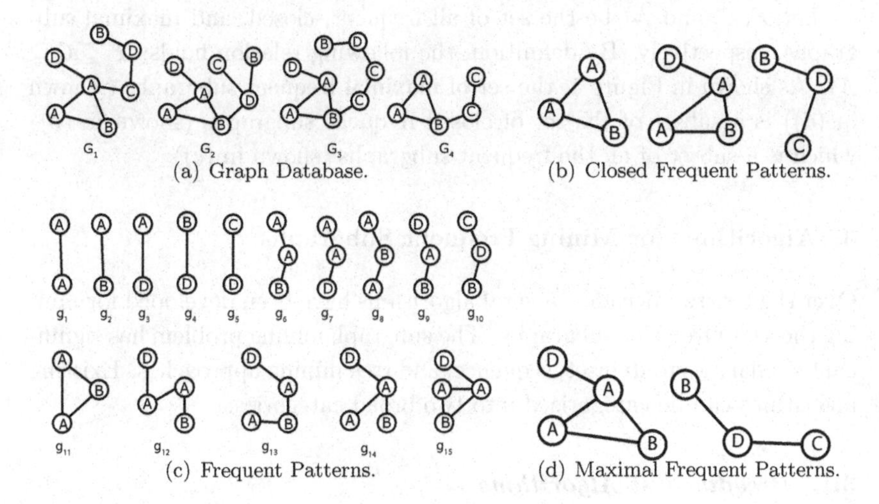

(a) Graph Database. (b) Closed Frequent Patterns.

(c) Frequent Patterns. (d) Maximal Frequent Patterns.

Fig. 1. A graph database and different types of frequent patterns. In this example, a subgraph is frequent if its support is at least 3.

Figure 1 is an illustrating example of finding frequent subgraphs from a graph dataset. Figure 1(a) is a graph dataset that has 4 graphs and the set of $\{A, B, C, D\}$ is the set of vertex labels and all edges are unlabeled. Figure 1(c) shows the set of frequent subgraph patterns (15 patterns) with the minimum support threshold set to 3.

2.3. *Mining Closed Frequent Subgraphs*

A frequent graph is called *closed* if it has no frequent super-graph with the same support, i.e., $\nexists g' : g \subseteq g'$ and $support(g', \mathcal{G}) = support(g, \mathcal{G})$. Given a graph dataset $\mathcal{G}$, and a minimum support threshold *minsup*, the problem of mining closed frequent subgraphs is to find the set $\mathcal{C} = \{g_1, g_2, \cdots, g_{|\mathcal{C}|}\}$ such that each $g_i \in \mathcal{C}$ is closed. Figure 1(b) shows the set of closed frequent subgraph patterns (3 patterns) with the minimum support set to 3.

2.4. *Mining Maximal Frequent Subgraphs*

A frequent graph is called *maximal* if it has no frequent super-graph, i.e., $\nexists g' : g \subseteq g'$ and g' is frequent. Given a graph dataset $\mathcal{G}$, and a minimum support threshold *minsup*, the problem of mining maximal frequent subgraphs is to find the set $\mathcal{M} = \{g_1, g_2, \cdots, g_{|\mathcal{M}|}\}$ such that each $g_i \in \mathcal{M}$ is maximal. Figure 1(b) shows the set of maximal frequent subgraph patterns (2 patterns) with the minimum support set to 3.

Let $\mathcal{F}$, $\mathcal{C}$, and $\mathcal{M}$ be the set of all frequent, closed, and maximal subgraphs, respectively. By definition, the following relation holds: $\mathcal{F} \supseteq \mathcal{C} \supseteq \mathcal{M}$. As shown in Figure 1, the set of maximal frequent subgraphs (shown in (d)) is a subset of the set of closed frequent subgraphs (shown in (b)) which is a subset of all the frequent subgraphs (shown in (c)).

3. Algorithms for Mining Frequent Subgraphs

Over the last two decades, several algorithms have been developed for mining the set of frequent subgraphs. The subgraph mining problem has significant similarities to itemset, sequence, and tree mining approaches. Existing algorithms can be categorized into two broad categories.

3.1. *Breadth-first Algorithms*

Methods in the first category (AGM[6,23] and FSG[7]) follow the level-wise approach that was first adopted in the Apriori algorithm for mining frequent itemsets.[24] The AGM algorithm mines all frequent subgraphs, and the FSG algorithm mines all frequent induced-subgraphs. The two main steps in these two algorithms are: 1) candidate generation and 2) support counting. The AGM algorithm employs a vertex-growth approach to generate a candidate (k+1)-subgraph (has $k+1$ vertices) in level $k+1$ by joining two frequent induced k-subgraphs from level k that has a common (k-1)-subgraph core. Once all candidate (k+1)-subgraphs are generated, the frequency of each candidate induced-subgraph is calculated and infrequent subgraphs are pruned. For each level, the database is scanned and the frequency of each candidate subgraph is computed. The algorithm stops when level $k+1$ is empty which indicates that we cannot generate any more patterns.

Candidate Generation in FSG: The FSG algorithm grows subgraphs by adding one edge at a time and level k has subgraphs with k edges. Candidate $(k+1)$-subgraphs in level $k+1$ are generated by joining two frequent k-subgraphs that share common cores. In subgraph mining and unlike candidate generating in itemsets, two frequent k-subgraphs can generate more than one candidate $k+1$ subgraph. The three major steps in candidate generation in the FSG algorithm are core identification, joining two subgraphs, and candidate elimination of $(k+1)$-subgraphs that has infrequent subgraphs. For each candidate subgraph in level $k+1$, the algorithm checks if the subgraph has been generated in the same level before.

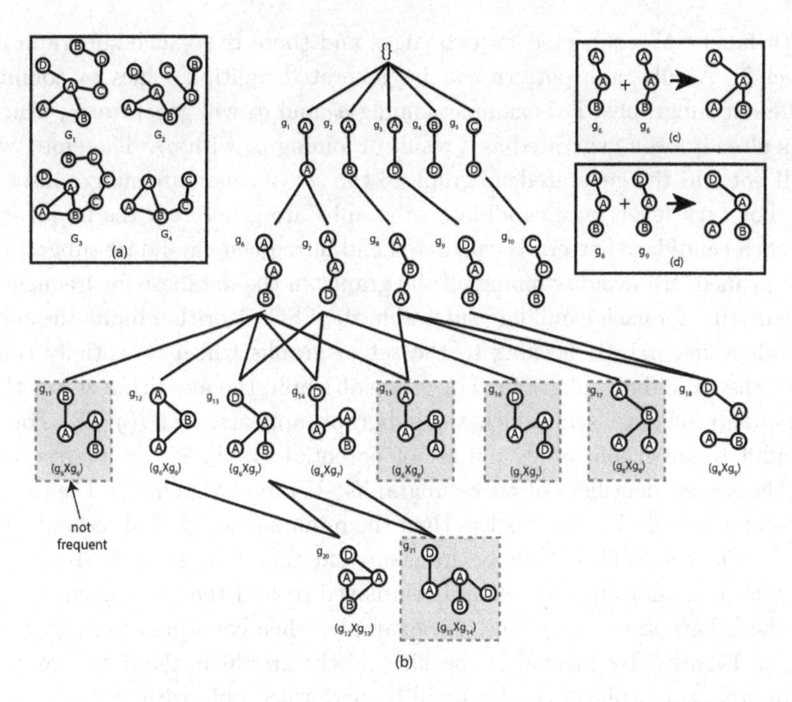

Fig. 2. Enumerating frequent patterns. (a) shows the graph database and (b) shows the enumeration of frequent subgraphs (15 frequent subgraphs). A subgraph is frequent if its support is at least 3. Shaded subgraphs are candidates that are not frequent and are pruned. (c) shows an example of joining two subgraphs that results in a non-candidate subgraph, and (d) shows an example of joining two subgraphs that results in a subgraph that has been generated before.

This is equivalent to graph isomorphism checking. The FSG algorithm employs an efficient canonical labeling that is based on flattened representation of the adjacency matrix to check if a particular subgraph is already in the candidate set.

Figure 2(b) shows an example of how the FSG algorithm enumerates frequent subgraphs from a database of 4 graphs (shown in (a)) with the *minsup* threshold set to 3. The FSG algorithm first enumerates all the frequent single and double edge graphs. The algorithm iteratively generate $(k+1)$-subgraphs by joining $k-$subgraphs. Candidate subgraphs shown in level 3 (3-subgraphs) are generated from joining subgraphs in level 2. For example, joining g_6 with itself (called self join) will generate three subgraphs. Two of these subgraphs are candidate subgraphs (g_{11} and g_{12}), and the third subgraph (shown in (c)) is not because it has a subgraph that is not frequent. The reason for this is that this subgraph has three nodes with label "A" connected by two edges and there is no such subgraph in level 2. A subgraph pattern can be generated multiple times by joining different subgraphs. For example, joining g_6 and g_9 will generate g_{13} which has already been generated as a result of joining g_6 with g_7. Therefore, we will not add the generated subgraph to the set of candidate subgraphs.

For each level, once candidate subgraphs are generated, the frequency of each candidate pattern is calculated and infrequent candidate subgraphs are pruned. To avoid scanning all the graphs in the database for frequency calculating for each candidate subgraph, the FSG algorithm limits the subgraph isomorphism checking to the set of graphs which potentially contain the candidate subgraph. For each subgraph, the algorithm stores the identifiers of graphs in which the subgraph appears, i.e., $t(g, \mathcal{G})$. For a candidate subgraph, g^{k+1}, the set of potential graphs is the intersection of the set of identifiers of all k-subgraphs. If the cardinality of the set of potential set of identifiers is less than the minimum support threshold, the candidate subgraph cannot be frequent and thus it is pruned. If not, a subgraph isomorphism checking is employed to find the exact occurrences of the subgraph in the graphs. For example, when counting the support of g_{13} in Figure 2(b), instead of checking all the graphs in the database, the subgraph isomorphism checking will be performed only with three graphs, G_1, G_2, and G_3. This is because the set $\{1, 2, 3\}$ is the intersection of the occurrence sets of all the subgraphs of g_{13}; we have $t(g_6, \mathcal{G}) = \{1, 2, 3, 4\}$, $t(g_7, \mathcal{G}) = \{1, 2, 3\}$, and $t(g_9, \mathcal{G}) = \{1, 2, 3\}$.

Experiments on a database of 340 graphs representing chemical compounds show that the FSG algorithm is efficient in mining the set of frequent subgraphs. The algorithm took few seconds when the support threshold was set to higher than 10% and the running time and the number of frequent subgraphs increase exponentially for lower support thresholds. It took 600 seconds when the support threshold was set to 7%. This is a significant improvement over the running time of the AGM algorithm which took eight days for 10% support and 400 seconds for 20%. The running time for level-wise mining algorithm depends largely on the size of frequent subgraphs (related to subgraph isomorphism checking) and the size of the largest subgraphs (related to database scans). Another factor that impacts the running time of the level-wise algorithm is the size of graphs in the database, the subgraph isomorphism takes longer for larger graphs. To summarize, the major problems in the level-wise frequent subgraph mining algorithms are: 1) costly candidate generation and false subgraph isomorphism testing, and 2) multiple scans of the database.

3.2. *Depth-first Algorithms*

The second category of frequent subgraph mining algorithms (e.g., gSpan,[8] FSSM[9]) adopt a depth-first search that results in improved running time performance and better memory utilization.

The major improvements of the gSpan algorithm include the design of a pattern growth approach that does not require any candidate generation step. A potentially frequent $(k+1)$-subgraph ($k+1$ edges) is grown by extending a k-subgraph with an edge. The gSpan algorithm developed a new canonical labeling approach, called DFS lexicographic order, to support the depth-first search. The search tree of the frequent subgraphs is built based on the DFS codes of the frequent subgraphs.[8] Experiments on synthetic data show that gSpan significantly outperforms the FSG algorithm by an order of magnitude.

The second algorithm that follows the depth-first search strategy is the FSSM algorithm.[9] The algorithm develops a canonical label that is based on the adjacency matrix in which each diagonal entry has label of the the corresponding vertex and each off-diagonal entry is filled with the label of the corresponding edge. The canonical code of the adjacency matrix is the sequence of the lower triangular entries of the matrix. The set of the canonical orders of all the adjacency matrices of a given graph is a total ordered set and the relation for the standard lexicographic order on sequences is used to define the relation between two canonical codes. The adjacency matrix with the maximum code is called the canonical adjacency matrix (CAM) of the graph and the canonical code of this matrix is called the canonical form of the graph.

The canonical codes of all the canonical adjacency matrices representing the set of all frequent subgraphs can be organized as a rooted tree with the empty matrix as a root. The FSSM avoids the computational overhead of the subgraph isomorphism checking by maintaining the set of all embeddings of each frequent subgraph. Experiment on several datasets with various support thresholds indicate that the running time of the FSSM algorithm is competitive to that of the gSpan algorithm.

4. Classifying Chemical Compounds

A major task in drug discovery is the identification of chemical compounds that exhibit the desired behavior. Recent advances in combinatorial chemistry and high-throughput screening have enabled scientists to synthesize and screen large number of chemical compounds.[11] The number of chemical

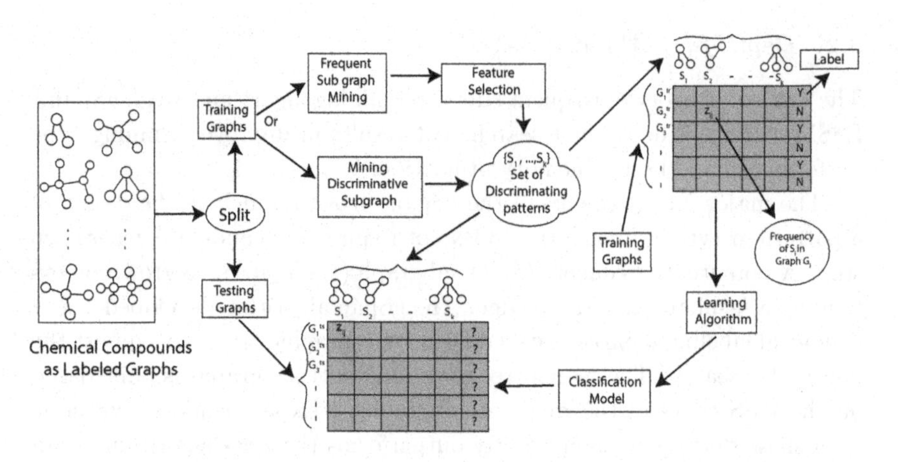

Fig. 3. Chemical compound classification using signature substructures.

compounds that can be synthesized by combinatorial chemistry is far more than what existing screening techniques can handle. Therefore, computational techniques which can classify chemical compounds into a set of classes have been developed. The main idea of the first category of approaches is to represent each chemical compound by a set of descriptors derived from the compound's physicochemical, geometric, and topological properties[12] (bonding pattern, encoding shape, etc.). Once chemical compounds are represented as vectors in the descriptors space, traditional data mining and machine learning algorithms can be applied for clustering, similarity search, and classification.

A second category of approaches build a kernel-based classification framework.[13] In this framework, a chemical compound is represented as a graph with labeled nodes representing atoms and edges representing bonds. The first step in this framework is the development of a kernel function to assign a similarity between every pair of compounds in the dataset. Once the kernel similarity matrix is computed, a kernel-based classifier (e.g., Support Vector Machines (SVM)) can be built to learn a classification model. Ravaliola *et al.*[13] introduced the idea of kernel functions for chemical compound classification and proposed three kernel functions (Tanimoto, MinMax, Hybrid) which showed superior performance when applied on three publicly available datasets. Several graph kernel functions have been proposed including kernels that calculate the global similarity between two graphs (e.g., product graph, random walk based kernels, shortest path based kernels) and other kernels which are based on the local substructures-based similarity between graphs (e.g., subgraph kernels).[25]

The third category of approaches for chemical compound classification is based on mining discriminative substructures from the graphs representing the chemical compounds.[3] A chemical compound is represented as a graph with labeled nodes representing atoms, edges representing the covalent bonds, and edge labels correspond to the bond order. Figure 3 shows a schematic description of the steps in the discriminative substructure-based chemical compound classification. The first step in this approach is to mine a set of discriminative subgraphs from a subset of the graphs (training dataset). Once these discriminative substructures are discovered, each chemical compound is represented as a vector in which the ith entry equals the number of times (frequency) the ith substructure appears in the graph representation of the chemical compound. Once each compound is represented as a vector in the substructure space, traditional classification techniques can be applied for the classification task.

Several algorithms have been proposed for discovering discriminative subgraphs. Deshpande *et al.*[3] proposed a model for classifying chemical compounds by first mining the frequent substructures from each class of chemical compounds. The number of frequent substructures could be very large which can have drastic implications on the learning and classification performance. The algorithm thus employs a feature selection algorithm to select a subset of the frequent substructures that have high discriminatory power. Once the discrimination features (subgraphs) are discovered, each chemical compound is represented by a frequency vector. Due to the high-dimensionality and the sparsity of the dataset, the authors recommend to employ support vector machines (SVM) for building the classification model because SVM has been shown effective in learning from high dimensional and sparse training datasets. The performance of the classification power of models built on the frequent subgraphs was evaluated on eight classification problems. These eight problems were constructed from three main data sets: 1) Predictive Toxicology Evaluation Challenge, 2) National Cancer Institute DTP AIDS Antiviral Screen program, and 3) The Center of Computational Drug Discovery's anthrax project.

The authors observed that mining frequent substructures with low support yields better classification performance in most cases, albeit at the cost of an increased running time. However, the authors noted that building classifiers on extremely large number of features with some features having low support might lead to overfitting the training dataset.

A major disadvantage of this two-phase approach (mine then filter) is that mining all frequent subgraph patterns is practically very slow; this is specially true when mining frequent patterns with low support. Several algorithms have been proposed for mining only discriminative subgraphs from a set of graphs with class labels. The LEAP (Descending Leap Mine) algorithm[26] introduced the concept of "pruning by structural proximity" for pruning futile search branches. A branch is pruned if the upper bound of the interestingness score of the current node and its descendants is not promising. Moreover, sibling branches can be pruned since the best scores between neighbor branches in the search tree are likely similar. Experiments on 11 graph datasets of small molecules, from the PubChem website (`http://pubchem.ncbi.nlm.nih.gov`), show that the classification models built using subgraph patterns mined with the LEAP algorithm outperform other graph classification methods in terms of the accuracy of classification and the running time of the algorithm.

The CORK algorithm[27] employs a greedy approach for feature selection on frequent subgraphs. The feature selection approach is integrated in the gSpan mining algorithm to prune the search space for discriminative subgraphs. Experiments on Anti-cancer screen datasets (NCI) and protein graphs datasets show that classifiers built on patterns mined by the CORK algorithm are competitive to those built on subgraphs selected by state-of-the-art feature selection approaches including rankers using Pearsons Correlation Coefficient, Information Gain, and the Sequential Cover method.

To alleviate the large running time of discovering significant subgraphs, other algorithms have followed a novel approach for mining significant subgraphs by mining feature vectors representing the graphs. The GraphSig algorithm[14,28] uses the significantly overrepresented substructures with *p-values* below a user-specified threshold to build a classification model. Once the graph dataset is converted to a database of histograms, significant sub-histograms are mined for each of the classes in the database using the GraphRank algorithm.[29] In GraphSig, each graph representing a chemical module is converted into a binary feature vector whose entries correspond to the presence or absence of the mined histogram in the module. Support vector machines classifiers are used to build a classification model. Experimental evaluation on several anticancer screen datasets demonstrate that the GraphSig algorithm outperforms representative algorithms from frequent-pattern-based (LEAP), graph kernel-based (OA), and structure keys-based (Joelib) approaches in terms of accuracy. Moreover, the GraphSig's classification accuracy is competitive to that of the Daylight algorithm which is a fingerprints-based approach. In terms of running time, the

GraphSig algorithm has significant performance advantage over frequent-pattern-based algorithms.

5. Discovering Protein Structural Motifs

Over the past decades, the number of known protein structures has been increasing at an unprecedented rate, due to advancements in protein structure determination techniques such as nuclear magnetic resonance (NMR) spectroscopy, X-ray crystallography, and low-temperature electron microscopy (cryo-EM). As of October 30, 2012, there are 85,848 protein structures in the Protein Data Bank (PDB)[a].[30] Determining the structures of these proteins plays a crucial role in identifying the functions of these proteins. Since it is a fundamental axiom of biology that there is a strong relationship between three-dimensional structure of a protein and its structure.[15] Despite having the structural information about so many proteins, the functions of a lot of these proteins are still unknown. Structure comparison enable scientists to assess the similarity between proteins and assign a newly-discovered protein to a protein fold or family. Several protein structural alignment algorithms have been proposed (e.g, DALI[31] , CE[32]). Most of these algorithms report a global alignment between a pair of proteins which define a superposition of one protein onto the other.

Proteins in the same family have significantly many common structural motifs which are structural fragments of various sizes with fixed three dimensional arrangements of residues.[16] These structural motifs have direct applications in protein classification, function prediction, and folding. In the structural motif discovery problem, the task is to discover protein structural motifs that appear frequently in a set of proteins. Moreover, given several families of proteins, structural motifs that appear frequently in one family of proteins and rarely in the others can also be discovered. These structural motifs can be further investigated to study their roles in protein functions and stability.

5.1. *Representing Proteins as Graphs*

The first step in protein structural motif discovery is to represent the protein's tertiary structure (the 3D structure) as a graph. A protein P, with n amino acids, is a collection of its elements. More formally, $P = \{P_1, P_2, \cdots, P_n\}$, where P_i is i^{th} element. We use the element-based

[a] www.pdb.org

representation of protein structure where each element is a residue that is represented by the 3D coordinates of the $C\alpha$-atom, i.e. P_i is the 3D coordinates of $C\alpha_i$. Thus, a protein is represented by a set of points in three-dimensional space representing the coordinates of the $C\alpha$-atoms on the protein backbone.

Huan *et al.*[16] surveyed graph representations of protein structures for the purpose of mining structural motifs. A protein is represented as a graph whose vertices represent the amino acids and each vertex is labeled by the corresponding residue type. There are two types of edges in this graph representation. First, there is the *bond* edge between adjacent amino acid residues on the primary sequence of the protein. The second type of edges is the *proximity* edge which is an edge between a residue and its nearest neighbors, non-consecutive amino acid residues. The nearest neighbor residues are computed based on various distance functions. Several approaches have been proposed for computing the proximity edges for a given protein structure.[33]

(1) **Distance edges:** In this approach, a proximity edge is added between two vertices ($C\alpha$-atoms) if the distance between the vertices is less than a user-defined threshold, δ. The value of δ is chosen from the range $6.5 - 9.5$ Å. Since the constructed graph depends on the threshold employed, the performance and quality of the mined structural motifs are highly dependent on the distance threshold used to construct the contact graph.
(2) **Delaunay edges:** The Delaunay tessellation of a set of points in 3D is a type of triangulation such that no point is inside the circumcircle of any triangle. Finding the Delaunay tessellation of a set of points in d-dimensions can be converted to finding the convex hull of the points in $(d+1)$-dimension.[34] There is a proximity edge between two points ($C\alpha$-atoms) if and only if there is an empty sphere such that the two points are on the sphere's boundary.[33] A post-processing is applied after all the proximity edges are added to remove any edge that connects two points with distance greater than a threshold δ.
(3) **almost-Delaunay edges:** Since there is imprecision in the atomic coordinates of the atoms in protein structure determination, the almost-Delaunay approach adds an edge between two points x and y if by perturbing all the points by ϵ, the two points can lie on an empty sphere. It is clear that the Delaunay edges are a subset of almost-Delaunay edges because with $\epsilon = 0$, we get all the Delaunay edges.[33]

5.2. *Mining Family-Specific Subgraphs*

Given a set of n proteins, belonging to two families, represented as n graphs ($\mathcal{G} = \{G_1, G_2, \cdots, G_n\}$) and a minimum support threshold, $\sigma \in [0, 1]$, Huan *et al.*[33] used a subgraph mining algorithm (FSSM[9]) to mine all subgraphs that occurr in at least $\sigma \times n$ graphs. The number of frequent subgraphs is typically very large, specially for low minimum support values. Most of these subgraph patterns can be frequent in both families and thus they are not family-dependent. For these patterns to be useful in subsequent tasks in the knowledge discovery process such as the task of classifying proteins into families, only subgraphs with discrimination power should be retained.

5.2.1. *Mining k-coherent subgraphs*

A graph pattern is *k-coherent* if the mutual information between the graph and every k-subgraph pattern (with k nodes) is greater than a user-defined threshold t.

The anti-monotone *k-coherent* property implies that for a *k-coherent* graph, every subgraph of size at least k is also k-coherent. Anti-monotone interestingness properties can be seamlessly integrated with existing pattern-growth graph mining algorithms (e.g., gSpan, FSSM). The property can be used to effectively prune the search space since a subgraph pattern that is not *k-coherent* cannot be extended to discover *k-coherent* subgraphs.

For classification purposes, each protein (graph) is represented by a feature vector where the features represent the coherent subgraphs and the ith entry in the vector corresponds to the number of occurrences of the ith subgraph in the graph representation of the protein. Once all the proteins are represented in the coherent subgraph space, traditional classification techniques (e.g., SVM) are employed to learn a classification model which is used to assign proteins to families. Extensive experiments on four structural and functional families in the SCOP database showed that mining frequent subgraph extracts family-specific protein structural motifs. Moreover, classifiers built with these structural motifs as features have high accuracy (above 90%).[35,36] The first dataset used in these studies contains proteins from two families: the nuclear receptor ligand-binding domain proteins (NB) and the prokaryotic serine protease (PSP). The second dataset contains proteins from the eukaryotic serine protease (ESP) and the prokaryotic serine protease (PSP) families.

The results showed that structural motifs mined from graphs constructed with distance edges and delaunay edges have biological meaning. Moreover, the results showed that the number of frequent substructures depends on the graph representation and that the distance-based graph representation produced the largest number of structural motifs.

5.2.2. *Maximal frequent patterns as fingerprints*

For the purpose of mining coherent subgraphs, the definition of the occurrence of a pattern in a graph is too restrictive. The restrictive definition is based on induced subgraph isomorphism which requires a graph G' to be isomorphic to an induced subgraph in G to declare that the subgraph G' occurs in G. Moreover, all coherent subgraph patterns were mined which resulted in a large number of patterns with significant overlap between patterns because all the subgraphs of a coherent subgraph are also coherent.

The definition of coherent subgraph is strict and does not tolerate missing edges that could occur due to local connectivity variations in proteins. The authors in[16] used simple subgraph isomorphism for defining the occurrence of a subgraph pattern in a graph. Moreover, to address the problem of the significant overlap between patterns, the authors proposed to mine only maximal frequent subgraphs.

For mining family-specific motifs, the proposed approach only mines subgraphs that are frequent (support is at least σ) in a group of proteins in one family and not frequent (support is at most δ) in the background database. Throughout the experiment, σ was set to 90% and δ to 5%. A non-redundant set of 4,600 protein structures with no more than 60% pairwise sequence similarity was used as the background database. Two datasets from the SCOP database were used in the experiments. The first dataset includes 43 proteins from the "trypsin-like serine proteases" SCOP superfamily and the other dataset has 29 proteins from the SCOP family "protein kinases, catalytic subunit." Experiments on these two protein datasets with the SCOP background database shows that the proposed approach is effective in discovering family-specific patterns which include known active sites and biologically meaningful motifs.

6. Challenges and Future Research

There are two significant challenges for frequent subgraph mining algorithms. First, the set of the frequent patterns is often too large. This is important since, in most applications, discovering the entire set of frequent patterns is only the first phase in the knowledge discovery process. Therefore, these frequent patterns need to be filtered and refined to be useful in subsequent analysis steps, e.g., classification, link prediction, and graph indexing. For example, in chemical compound classification, only discriminative graph patterns are of interest. Several Algorithms have been proposed to mine only discriminative subgraphs from a set of graphs with class labels (e.g., LEAP,[26] CORK[27]). Several algorithms have been introduced to integrate connectivity constraints with frequency to mine frequent dense subgraph patterns. The CLAN algorithm[19] mines closed cliques. A subgraph is a closed clique if it is a complete graph that occurs in at least *minsup* input graphs and that does not have a super-graph that is a clique that occurs in the same input graphs. Since cliques have a strict connectivity constraint, the Cocain algorithm[20] mines for coherent closed quasi-cliques.

The second challenge for frequent subgraph mining algorithms is the exponential search space of the frequent subgraph patterns which renders existing graph mining algorithms practically infeasible when applied on large graph datasets. To overcome the combinatorial explosion in the number of frequent patterns, several approaches have been proposed to report only succinct non-redundant set of subgraph patterns. Two examples in this direction are the CloseGraph algorithm[37] which mines closed patterns and the SPIN[38] algorithm which mines maximal patterns. Another research direction is to mine only representative set of the patterns which is a sample of the set of frequent patterns. The ORIGAMI approach[39] reports a set of representative patterns such that the patterns in the set are different from each other and each unreported frequent pattern has at least one pattern in the representative set to which it is similar. A sampling approach that is based on the Metropolis-Hastings algorithm was proposed to sample the output space of frequent subgraphs.[40] Another related work is the Summarize-Mine algorithm that generates randomized summaries in order to reduce the size of the dataset.[41] A summary of each graph is built and the mining process is applied on the compressed graph database.

For graph database in which the labels are unique (these graphs are often referred to as relation graphs), several efficient algorithms have been proposed. Mining unique labeled graphs does not require the costly operation of subgraph isomorphisim that dominates the computation cost in general frequent subgraph mining. The Crochet algorithm[21] mines cross-all-graph quasi-cliques which are groups of vertices that induce quasi-cliques in each of the input graphs. Since the occurrence requirement of the cross-all-graphs quasi-clique is too strict in the Crochet algorithm, the same authors proposed the Crochet+ algorithm[22] to mine frequent cross-graph quasi-cliques. The MULE algorithm proposed an efficient enumeration approach for mining frequent subgraphs without any connectivity constraints from protein protein interaction networks for several species.[42]

Another active graph mining research area focuses on mining dense subgraphs (sometimes referred to as dense subnetworks, modules, or communities) from a single graph. Lee *et al.*[43] has an excellent review on the topic. A comprehensive survey of the computational approaches for module discovery in protein interaction networks can be found in Li *et al.*[44]

Considering that graphs are increasingly being adopted for representing biological data, the overarching research area of graph mining, including mining significant subgraph patterns from massively large graph datasets, becomes of utmost importance and surely will gain more attention from the research community. We expect some efforts will be devoted to developing parallel and sampling algorithms for mining significant subgraphs. Mining significant subgraphs from dynamic graphs will also attract some attention.

Acknowledgement

This publication was made possible by grants from the National Center for Research Resources (P20RR016471) and the National Institute of General Medical Sciences (P20 GM103442) from the National Institutes of Health.

References

1. R. Milo, S. Shen-Orr, S. Itzkovitz, N. Kashtan, D. Chklovskii, and U. Alon. Network motifs: Simple building blocks of complex networks. In *Science*, pp. 824–827 (2002).
2. M. Koyuturk, A. Grama, and W. Szpankowski. An efficient algorithm for detecting frequent subgraphs in biological networks. In *Bioinformatics*, pp. 200–207 (2004).
3. M. Deshpande, M. Kuramochi, and G. Karypis. Frequent sub-structure-based approaches for classifying chemical compounds. In *ICDM*, pp. 35–42 (2003).
4. Y. Huang, H. Li, H. Hu, X. Yan, M. S. Waterman, H. Huang, and X. J. Zhou, Systematic discovery of functional modules and context-specific functional annotation of human genome, *Bioinformatics*. **23**(13), i222–i229 (2007).
5. X. Yan, P. S. Yu, and J. Han. Graph indexing: A frequent structure-based approach (2004).
6. A. Inokuchi, T. Washio, and H. Motoda. An apriori-based algorithm for mining frequent substructures from graph data. In *PKDD*, pp. 13–23 (2000).

7. M. Kuramochi and G. Karypis. Frequent subgraph discovery. In *ICDM*, pp. 313–320 (2001).
8. X. Yan and J. Han. gspan: Graph-based substructure pattern mining. In *ICDM*, pp. 721–724 (2002).
9. J. Huan, W. Wang, and J. Prins. Efficient mining of frequent subgraphs in the presence of isomorphism. In *ICDM*, pp. 549–552 (2003).
10. S. Nijssen and J. N. Kok. A quickstart in frequent structure mining can make a difference. In *KDD*, pp. 647–652 (2004).
11. J. S. Handen, The industrialization of drug discovery, *Drug Discovery Today.* **7**(2), 83–85 (2002).
12. P. Willett, Chemical similarity searching, *J. Chemical Information and Computer Science.* **38**(6), 983–996 (1998).
13. L. Ravaliola, S. Swamidass, H. Saigo, and P. Baldi, Graph kernels for chemical informatics, *Neural Networks.* **18**(8), 1093–1110 (2005).
14. S. Ranu and A. K. Singh, Mining statistically significant molecular substructures for efficient molecular classification, *Journal of Chemical Information and Modeling.* **49**, 2537–2550 (2009).
15. G. A. Petsko and D. Ringe, *Protein Structure and Function.* New Science Press Ltd., Oxford, UK (2002).
16. J. Huan, D. Bandyopadhyay, W. Wang, J. Snoeyink, J. Prins, and A. Tropsha, Comparing graph representations of protein structure for mining family-specific residue-based packing motifs, *Journal of Computational Biology (JCB).* **12**(6), 657–671 (2005).
17. A. P. Gasch and M. B. Eisen, Exploring the conditional coregulation of yeast gene expression through fuzzy k-means clustering, *Genome Biology.* **3**(11), research0059.1–0059.22 (2002).
18. H. K. Lee, A. K. Hsu, J. Sajdak, J. Qin, and P. Pavlidis, Coexpression analysis of human genes across many microarray data sets, *Genome Res.* **14** (6), 1085–1094 (2004).
19. J. Wang, Z. Zeng, and L. Zhou. Clan: An algorithm for mining closed cliques from large dense graph databases. In *Proceedings of the 22nd International Conference on Data Engineering*, ICDE '06, p. 73 (2006).
20. Z. Zeng, J. Wang, L. Zhou, and G. Karypis. Coherent closed quasi-clique discovery from large dense graph databases. In *Proceedings of the 12th ACM SIGKDD international conference on Knowledge discovery and data mining*, KDD '06, pp. 797–802 (2006).
21. J. Pei, D. Jiang, and A. Zhang. On mining cross-graph quasi-cliques. In *Proceedings of the eleventh ACM SIGKDD international conference on Knowledge discovery in data mining*, KDD '05, pp. 228–238 (2005).
22. D. Jiang and J. Pei, Mining frequent cross-graph quasi-cliques, *ACM Trans. Knowl. Discov. Data.* **2**(4), 16:1–16:42 (jan, 2009).
23. A. Inokuchi, T. Washio, and H. Motoda, Complete mining of frequent patterns from graphs: Mining graph data, *Mach. Learn.* **50**(3), 321–354 (2003).
24. R. Agrawal and R. Srikant. Fast algorithms for mining association rules in large databases. In *Proceedings of the 20th International Conference on Very Large Data Bases*, VLDB '94, pp. 487–499 (1994).

25. A. M. Smalter, J. Huan, and G. H. Lushington. Gpm: A graph pattern matching kernel with diffusion for chemical compound classification. In *BIBE '08*, pp. 1–6 (2008).
26. X. Yan, H. Cheng, J. Han, and P. S. Yu. Mining significant graph patterns by leap search. In *SIGMOD* (2008).
27. M. Thoma, H. Cheng, A. Gretton, J. Han, H.-P. Kriegel, A. J. Smola, L. Song, P. S. Yu, X. Yan, and K. M. Borgwardt. Near-optimal supervised feature selection among frequent subgraphs. In *SIAM Int. Conf. on Data Mining*, pp. 1075–1086 (2009).
28. S. Ranu and A. K. Singh. Graphsig: A scalable approach to mining significant subgraphs in large graph databases. In *Proceedings of the 2009 IEEE International Conference on Data Engineering*, ICDE '09, pp. 844–855 (2009).
29. H. He and A. K. Singh. Graphrank: Statistical modeling and mining of significant subgraphs in the feature space. In *ICDM*, pp. 885–890 (2006).
30. H. M. Berman, J. Westbrook, Z. Feng, G. Gilliland, T. N. Bhat, H. Weissig, I. N. Shindyalov, and P. E. Bourne, The protein data bank, *Nucleic Acids Research*. **28**, 235–242 (2000).
31. L. Holm and C. Sander, Protein structure comparison by alignment of distance matrices, *J. Mol. Biol.* **233**(1), 123–138 (1993).
32. I. N. Shindyalov and P. E. Bourne, Protein structure alignment by incremental combinatorial extension (CE) of the optimal path, *Protein Eng.* **11**, 739–747 (1998).
33. J. Huan, D. Bandyopadhyay, J. Prins, J. Snoeyink, A. Tropsha, and W. Wang. Distance-based identification of structure motifs in proteins using constrained frequent subgraph mining. In *in Proceedings of the IEEE Computational Systems Bioinformatics (CSB)*, pp. 227–238 (2006).
34. R. K. Singh, A. Tropsha, and I. I. Vaisman, Delaunay tessellation of proteins: Four body nearest neighbor propensities of amino acid residues, *Journal of Computational Biology (JCB)*. **3**(2), 213–222 (1996).
35. J. Huan, W. Wang, A. Washington, J. Prins, R. Shah, and A. Tropsha. Accurate classification of protein structural families using coherent subgraph analysis. In *in Proceedings of the Pacific Symposium on Biocomputing (PSB)*, pp. 411–422 (2004).
36. J. Huan, W. Wang, D. Bandyopadhyay, J. Snoeyink, J. Prins, and A. Tropsha. Mining protein family specific residue packing patterns from protein structure graphs. In *RECOMB'04* (2004).
37. X. Yan and J. Han. Closegraph: Mining closed frequent graph patterns. In *KDD*, pp. 286–295 (2003).
38. J. Huan, W. Wang, J. Prins, and J. Yang. Spin: mining maximal frequent subgraphs from graph databases. In *KDD*, pp. 581–586 (2004).
39. V. Chaoji, M. A. Hasan, S. Salem, J. Besson, and M. J. Zaki, Origami: A novel and efficient approach for mining representative orthogonal graph patterns, *Journal of Statistical Analysis and Data Mining*. **1**(2), 67–84 (2008).
40. M. A. Hasan and M. J. Zaki, Output space sampling for graph patterns, *Proc. VLDB Endow.* **2**(1), 730–741 (2009).
41. C. Chen, C. X. Lin, M. Fredrikson, M. Christodorescu, Yan, and J. Han. Mining graph patterns efficiently via randomized summaries. In *VLDB* (2009).

42. M. Koyuturk, Y. Kim, S. Subramaniam, W. Szpankowski, and A. Grama, Detecting conserved interaction patterns in biological networks, *Journal of Computational Biology.* **13**(7), 1299–1322 (2006).
43. V. E. Lee, N. Ruan, R. Jin, and C. Aggarwal. A survey of algorithms for dense subgraph discovery. In eds. C. C. Aggarwal and H. Wang, *Managing and Mining Graph Data*, vol. 40, *Advances in Database Systems*, pp. 303–336. Springer US (2010).
44. X. Li, M. Wu, C.-K. Kwoh, and S.-K. Ng, Computational approaches for detecting protein complexes from protein interaction networks: a survey, *BMC Genomics.* **11**(Suppl 1), S3 (2010).

Chapter 7

On the Integration of Prior Knowledge in the Inference of Regulatory Networks

Catharina Olsen[1], Benjamin Haibe-Kains[2], John Quackenbush[3] and Gianluca Bontempi[1]

[1] *Machine Learning Group (MLG), Computer Science Department, Université Libre de Bruxelles (ULB), Brussels, Belgium*

[2] *Bioinformatics and Computational Genomics Laboratory, Institut de recherches cliniques de Montréal, Montreal, QC, Canada*

[3] *Department of Biostatistics and Computational Biology, Dana-Farber Cancer Institute and Department of Biostatistics, Harvard School of Public Health, Boston, MA, USA*

Scientific literature in biology and medicine contains a wide amount of knowledge which is unfortunately difficult to exploit and integrate with data generated with high-throughput technologies. This is particularly relevant when we address system-level problems characterized by large complexity and dimensionality. This chapter aims to focus on recent results on the integration of prior knowledge and genomic data for the inference of regulatory networks. First we present existing tools to retrieve prior knowledge from different sources such as PubMed and structured biological databases. Then we focus on the current state-of-the-art of network inference methods combining genomic data and prior knowledge. In the second half of the chapter we present a case study based on two publicly available cancer data sets using the tool *predictionet* showing the usefulness of prior knowledge for network inference.

1. Introduction

The advent of high-throughput technologies such as microarrays and RNA sequencing paved the way to the unraveling and the modeling of gene regulatory networks. However, the fact that the collection of experimental measures advances at a rapid pace should not lead to ignore the large amount of prior knowledge contained in scientific literature and/ or genomic repositories. The reason is not simply that we should avoid reinventing the wheel but is essentially related to the nature of data issued by new technologies. Typical data sets used for modeling regulatory networks include gene expressions from microarray experiments and more recently from RNA-seq experiments. Though technology evolved from hybridization to sequencing with an expected improvement in accuracy, a major issue remains unsolved: high-throughput technologies measure at the same time a number of variables which is enormously larger than the number of samples with detrimental impact on the quality and robustness of the models[1,2] . For this reason the integration of prior knowledge (e.g., about the existence of interactions) in the modeling process might play a crucial role by constraining the model selection and by regularizing the estimation process.

The effective integration of prior knowledge in a modeling procedure demands the availability of three components: the first should be able to retrieve (possibly in an automated manner) knowledge by harvesting existing repositories, the second should code the retrieved knowledge in a compact and reusable way and the third should take advantage of the effective coding to improve the accuracy of the inference process.

This is the reason why we structure the overview of the state-of-the-art in three sections. In the first one, we present an overview of existing tools to gather prior information. In the second section, we describe common techniques to code existing prior knowledge in a quantitative and reusable manner. In the third section, we discuss the most common methods for integrating prior knowledge in the inference process. In particular we will discuss Bayesian network inference techniques which are definitely the most known methodology to integrate prior knowledge with measured observations. In spite of their historical role, the use of Bayesian networks for large scale inference tasks is computationally restricted by the number of genes that can be efficiently investigated. This led in recent years to the development of several alternatives which aim to adopt feature selection strategies to model locally the interconnectivity between nodes[3] . Whenever these techniques appear to be competitive in terms of accuracy and computational efficiency, they disregard the use of prior knowledge.

In order to address this issue the authors of this chapter recently proposed *predictionet*, a network inference method capable of dealing with large number of genes at once and taking into account an a priori score on the existence of gene interactions. This method, which will be detailed in Section 4.2.1, uses the maximum relevance minimum redundancy feature selection criterion to select the most interactors for each gene and a criterion based on interaction information for subsequent orientation. Prior knowledge is integrated in both steps using a linear constrained combination of the network inferred using genomic data and the network based on prior knowledge.

The last part of the chapter will present an example of prior integration in the inference process. We will use the *predictionet* tool to infer a gene regulatory network on the basis of two publicly available genomic data sets and prior knowledge retrieved from PubMed abstracts and biological structured databases.

2. Retrieval of prior information

In this section, we present three existing tools for prior retrieval by focusing on the interface and the format of output returned to the user.

iHOP/GIM[4,5] The online service *information Hyperlinked Over Proteins* (iHOP) is designed to allow the user to retrieve biomedical literature associated to a gene of interest. The lack of an automated retrieval for multiple genes led to the development of the *Gene Interaction Miner* (GIM) which provides an interface to upload a list of genes and retrieve all associated publications, see Fig. 1. The output of a GIM query is a list of undirected gene-gene interactions each annotated with related references and an output graph in which the edges' weights are the number of found citations.

GeneMANIA[6] A functional association network is made of nodes (e.g., genes or proteins) and undirected edges which represent co-functionality of the associated nodes. The weight of an edge corresponds to the confidence in the co-functionality as extracted from the given data source. *GeneMANIA* builds functional association networks using different data sources and subsequently fuses them into a single final network.

The definition of an edge connecting two nodes can be done according to several criteria [a], the most common types are presented in the following.

- Co-expression: two genes are connected when Pearson correlated according to a data set linked to a publication.
- Pathway: a link is added if a reaction within a pathway is shared, according to various databases.

[a]http://pages.genemania.org/help/

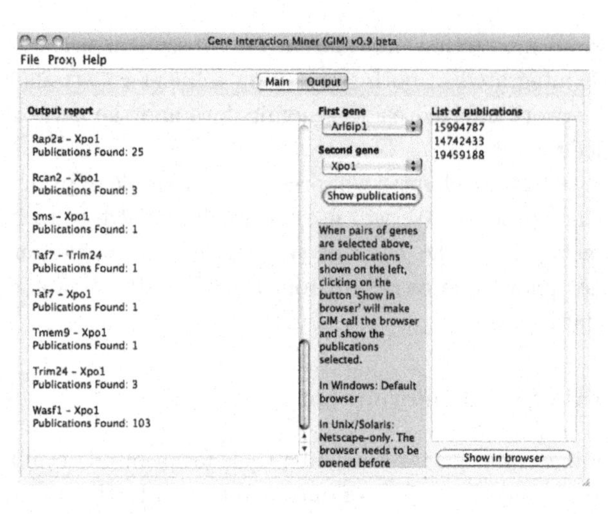

Fig. 1. Screenshot of the output returned by iHOP when using an example list provided by developers.

- Physical interactions: two gene products are linked when found to interact in a protein-protein interaction study.
- Predicted: a link is added for each retrieved functional relationship from another organism via orthology.
- Shared protein domains: a link is added when the same protein domain is shared, collected from domain databases.

After uploading a gene list and selecting the desired sources, the web server presents a network in which edges are color coded by source, see Fig. 2. Different exporting options are available for both the network graph, the connections themselves and the search parameters.

Predictive Networks[7] The *Predictive Networks* (PN) web application, see Fig. 3 was developed by the Dana-Farber Cancer Institute in collaboration with Entagen. It is made of two main modules. The first mines for relevant biological knowledge by analyzing PubMed abstracts, full-text articles in PubMed Central and structured biological databases like Pathway commons[b]. The second performs network inference by integrating the retrieved knowledge and uploaded expression data and will be detailed in Section 4.2.1.

[b]pathwaycommons.org

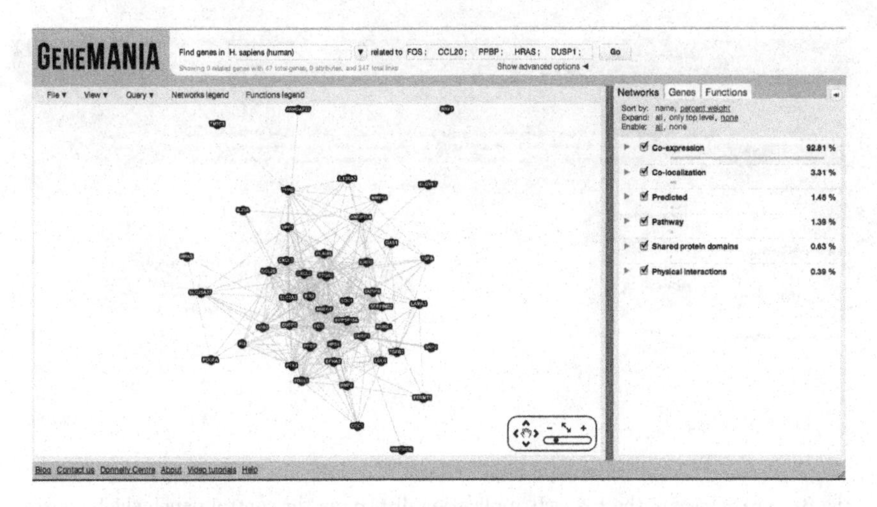

Fig. 2. Screenshot of GeneMANIA web-application using the genes from the experimental study presented in Section 5.

The text mining algorithm works by looking for triples having the structure *subject-predicate-object* where the subject and the object are gene names and the predicate includes terms such as 'regulates' and 'is inhibited by'. The search terms entered in the web application are contained in a user-uploaded gene list for which the web application returns a directed network containing all detected connections. The retrieved prior knowledge can then be downloaded in different formats, exporting each retrieved connection together with its source.

3. Coding of prior representation

Though several tools exist to retrieve prior knowledge in a given medical and/or biological context, it is less straight-forward how to translate this prior knowledge in a format usable by methods like Bayesian networks or gene prioritization tools. This problem is considered as one of the most difficult tasks in Bayesian learning.[8] We start this section by presenting different techniques developed for Bayesian network inference, followed by the prior representation used in *predictionet* and we conclude by presenting a way to transform different data sources, including prior knowledge, for subsequent gene prioritization.

From prior knowledge to prior over graph structures The general idea of Bayesian network learning is to represent probabilistic relationships

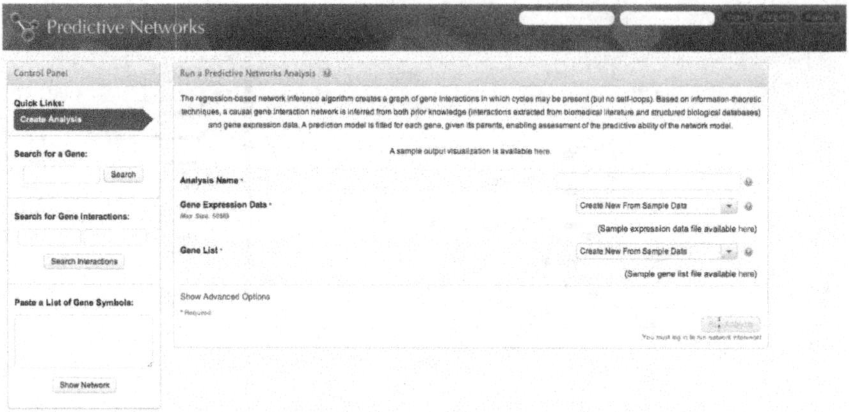

Fig. 3. Front page of the PN web application displaying the control panel which allows to specify one gene, a gene interaction or a list of genes. The *Predictive Networks Analysis* panel allows the user to provide data and carry out a network inference procedure using *predictionet*.

between variables using a graphical model.[9] Originally, this technique was used to encode uncertain expert knowledge[10] , that is experts would give a certain probability for variable X_i influencing another variable X_j. The quality of an inferred network is assessed via its posterior probability which is proportional to the product of the model's prior probability and the likelihood of the model being generated by the given data[11]

$$posterior(model|data) \propto prior(model)\ likelihood(model, data). \quad (1)$$

That is, a network will obtain a high posterior whenever both the likelihood of a model given the data and the prior probability for this model yield high values. The prior probability of a model will be denoted by $p(\mathcal{G})$.

The main problem in including prior knowledge in Bayesian networks is that the prior has to be defined over an entire directed acyclic graph structure whereas the 'usual' prior knowledge is given edge by edge. Different techniques have been proposed in literature to address the issue and will be discussed in what follows.

Penalizing[8] Given n nodes (e.g., genes or proteins) prior knowledge is transformed into an $n \times n$ adjacency matrix, denoted *prior network*, such that each known relationship between two variables X_i and X_j implies an edge between X_i and X_j in the prior network. In this approach[8] , the prior probability of any graph $\mathcal{G}$ is computed by first counting the number of edges in which this graph $\mathcal{G}$ differs from the prior network, denoted by δ. A prior probability for the graph $\mathcal{G}$ is obtained by using δ as exponent to a constant penalizing factor $0 < \kappa \leq 1$. Whenever the number of differences is high, the exponentiation will result in a small value whereas a small number of differences will yield a higher exponentiation value

$$p(\mathcal{G}) = c\kappa^{\delta}, \tag{2}$$

where c is a normalization constant. This procedure ensures that between two graphs, one with high and the second with low similarity to the prior network, the first will yield a higher probability than the second.

Completing partial knowledge[11] This method consists of two steps. The first completes the prior knowledge by assigning probabilities to all edges which have not yet been considered while the second codes the prior knowledge as a prior probability over network structures.

In the first step, it is assumed that for each edge between variables X_i and X_j three options are possible: $X_i \rightarrow X_j$, $X_i \leftarrow X_j$ and $X_i \not\leftrightarrow X_j$. The edges for which the user did not provide any information will be updated using the following rules.

- If none of the three options has an associated prior probability, they will be set to be equally probable

$$p(X_i \rightarrow X_j) = p(X_i \leftarrow X_j) = p(X_i \not\leftrightarrow X_j) = \frac{1}{3}$$

- If one option has a probability of p, then the other two options will be assigned the probability $(1-p)/2$.
- If two options have probabilities p' and p'' respectively, the third option will have a probability of $1 - p' - p''$.

Subsequently, this prior over links is formalized to a prior over a Bayesian network structure using the concept of oriented graphs. Oriented graphs are a superset of directed acyclic graphs (DAGs) as they include graphs with cycles of size greater than two. Contrary to DAGs, these oriented graphs can be decomposed into pairs of variables[11] . Assuming that the prior knowledge for one pair is independent from that of another pair, the prior over a structure is then the product of prior probabilities over all pairs of variables

$$p(\mathcal{G}) = \prod_{i,j \in 1,\ldots,n, i \neq j} p(X_i - X_j), \tag{3}$$

where $X_i - X_j \in \{X_i \to X_j, X_i \leftarrow X_j, X_i \not\leftrightarrow X_j\}$ specifies the type of interaction between X_i and X_j in $\mathcal{G}$. The set of oriented graphs is larger than the set of DAGs, therefore the probability $p(\mathcal{G})$ is not yet a prior distribution over the set of Bayesian networks. In order to obtain such a distribution, the authors[11] propose two methods to take these additional structures into account: a uniform correction

$$p_{uniform}(\mathcal{G}) = c_1 + p(\mathcal{G}) \tag{4}$$

and a proportional correction

$$p_{proportional}(\mathcal{G}) = c_2 \cdot p(\mathcal{G}), \tag{5}$$

with c_1 and c_2 are appropriately chosen normalization factors.

Gibbs distribution In this framework, the prior knowledge is represented by a prior network and then a score is computed such that the lower the score the higher the amount of prior knowledge. This score can then be used to compute a Gibbs distribution which forms the prior over the network structure. This score is known as the *energy function*[12] . The prior knowledge is transformed into an $n \times n$ matrix U such that $u_{ij} = \zeta_1$ if there is a known relationship from gene X_i to gene X_j. It is assigned a value ζ_2 if there is no known relationship, $0 < \zeta_1 < \zeta_2$. The energy of the prior network can then be computed as

$$E(\mathcal{G}) = \sum_{\{i,j\} \in \mathcal{G}} u_{ij}. \tag{6}$$

The prior probability $p(\mathcal{G})$ of a network $\mathcal{G}$ is then modeled by a Gibbs distribution

$$p(\mathcal{G}) = \frac{1}{Z} \exp\{-\zeta E(\mathcal{G})\}, \tag{7}$$

where $\zeta > 0$ is a hyperparameter and $Z = \sum_{\mathcal{G} \in \mathcal{G}} \exp\{-\zeta E(\mathcal{G})\}$ a normalization factor.

Strength of connection[13] Whenever, an information retrieval tool returns not only the prior information concerning the existence of an edge but also the number of times this connection has been cited, this additional information can be interpreted as confidence this interaction has obtained. That is, the higher the citation count $counts_{ij}$ for an interaction between X_i and X_j the higher should be the confidence in this interaction. Since certain regulatory mechanisms have been investigated more than others, the number of counts for each pair of variables X_i and X_j will be bounded superiorly by a value $\max_c$. Then, a score between zero and one is computed by

$$p_{ij} = \frac{counts^*_{ij}}{\max_{ij}\{counts^*_{ij}\}}. \tag{8}$$

where $counts^*_{ij} = \min\{counts_{ij}, \max_c\}$ before rescaling.

Each entry p_{ij} in this matrix $P = (p_{ij})$ can be interpreted as prior belief about the existence of an interaction between a pair of variables X_i and X_j; it takes values in the interval $[0, 1]$.

Kernels[14] The kernel representation of prior knowledge about n nodes consists in creating an $n \times n$ matrix where the (i, j)th entry accounts for the similarity between the nodes X_i and X_j, obtained on the basis of the data set (e.g., co-expression in microarray data) or prior knowledge.[15,16] A common choice for representing graphs is the diffusion kernel[17,18]

$$K = \exp(-\tau L), \tag{9}$$

where $\tau > 0$ and the *Laplacian* $L = D - A$ such that D is the diagonal matrix of vertex degrees and A the adjacency matrix of prior knowledge. However, to evaluate this kernel the Laplacian has to be diagonalized[18] which has a complexity of $O(n^3)$ and returns a dense matrix and is therefore difficult to use when thousands of genes are investigated[16] .

4. Integration of prior knowledge in the inference process

We will discuss here the most important computational techniques used to integrate existing prior knowledge in the inference process.

4.1. *Bayesian networks*

The inference of a Bayesian network from data relies on two steps: the structure learning and the parameter learning. Since the parameter learning part is quite conventional for a given structure,[19] we will restrict the discussion here to the structure learning. Methods for structural learning belong to two main classes: constraint-based methods and score-based methods. The former class relies on independence tests to determine the structure of the network and is typically completely data-driven, whereas the latter aims to select the structure which optimizes a global fitting score with the possibility of taking the prior knowledge into account. As we will discuss hereafter, there have been several suggestions on how to integrate prior knowledge during the structure learning phase using score-based approaches.

4.1.1. *Score-based algorithms*

Once a fitting score is associated to a network structure, structural learning boils down to an optimization problem in the space of structures. This is the rationale of score-based algorithms which aim at identifying the structure with the highest associated score. However, since finding the optimal structure is NP-hard[20] heuristic search techniques (e.g., greedy hill-climbing, beam search and simulated annealing) are used in practice though no guarantee exists about the convergence to a global optimum[21] .

Prior knowledge has been included in score-based algorithms in both steps of the inference process that is in the search algorithm and in the scoring function itself.

search algorithm: this technique consists in using the prior knowledge to seed the network search[22] , such that the search starts out from a network in which each connection is an interaction that has been found to exist in previous research. On the other hand prior knowledge can restrict the search space[23] such that only those structures in which known interactions are present are tested whereas those structures without any known interactions are avoided.

scoring function: this technique consists in creating a specific Bayesian metric s taking into account prior knowledge using the following combination of prior probability $p(\mathcal{G})$ of the graph $\mathcal{G}$ and the posterior probability $p(\mathbf{D}|\mathcal{G})$ of the data $\mathbf{D}$ given the structure $\mathcal{G}$

$$s = \log p(\mathcal{G}) + \log p(\mathbf{D}|\mathcal{G}) + c, \tag{10}$$

where c is a constant. Proposed metrics following this structure include the **BDe** metric[8] and the **BNCR** metric.[24]

The main drawbacks of Bayesian methods are 1) the difficulty to transform prior knowledge into a prior over network structures and 2) the restriction to data sets with not more than a few hundred variables. Especially the latter problem is clearly hindering the application of Bayesian network inference methods to expression data sets given that those usually contain ten thousand variables and more.

4.2. *Feature selection methods*

Bayesian network inference suffers from the high computational cost associated with the algorithms. They are rarely applicable when the number of variables is around hundreds while thousands of variables are commonly present in genomic data sets. An alternative is provided by adopting feature selection techniques to address the structural identification problem. The idea consists in using feature selection techniques to identify the neighborhood of each node and use this local information to infer the global structure. Since the feature selection task remains of large dimensionality, filter selection techniques based on information-theory provide a possible solution. In the basic variant, known as Relevance Networks[25] , mutual-information is used as pairwise score between variables and a threshold to eliminate edges with too little confidence. More complex methods use trivariate scores based on mutual information together with a threshold on this score to infer the network[26–28] . For a detailed review on these methods we refer the reader to[3,29,30] . Though effective in real tasks, these methods are not designed to include prior knowledge. This is the reason why the authors of this chapter proposed in[7] an original method named *predictionet* aiming to combine the idea of inferring networks based on information-theory with a scheme to integrate prior knowledge.

4.2.1. *Predictionet*

Predictionet is the inference engine of the web application *Predictive Networks* discussed in Section 2. The prior knowledge retrieved by the text mining module is transformed following an adaption of the 'strength of connection' approach (see Section 3) to include negative evidence identified by a predicate such as "does not regulate". This implies obtaining an adjacency matrix $P = (p_{ij})$ with values in $[-1, 1]$, where a high number of positive counts between two genes X_i and X_j will correspond to a high prior score $p_{ij} > 0$. A large number of counts for the absence of an edge will result in a low negative value for p_{ij}.

Inferring a network integrating prior knowledge In a first step, *predictionet* builds an undirected network based on the minimum redundancy maximum relevance (mRMR) criterion[28,31] . This network, represented by a weighted adjacency matrix $M = (m_{ij})$ with $m_{ij} \in [0, 1]$, is combined with the prior knowledge matrix P by linear constrained combination[32]

$$w \cdot P + (1 - w) \cdot M, \tag{11}$$

where w is a weighting factor taking values in the interval $[0, 1]$.

After the combination is computed, the top scoring parents are kept for each gene, the number of parents is the user-specified value *maxparents*. For this network, the set of triplets such that $X_i - X_j - X_k$ with $X_i \not- X_k$ is determined. For each triplet the interaction information

$$I(X_i, X_k) - I(X_i, X_k|X_j) \tag{12}$$

is computed with the goal of identifying as many v-structures (triplets $X_i \rightarrow X_j \leftarrow X_k$) as possible: if the interaction for a triplet is negative then a v-structure can be deduced[33–36] . For each pair $X_i - X_j$, the causality score c_{ij} is defined as the maximum negative interaction information over the set X_K of variables such that $X_i - X_j - X_k$ with $X_i \not- X_k$ for all $X_k \in X_K$. That is, considering all triplets in which X_j is the center node and X_i and X_k are adjacencies of X_j, the score between X_i and X_j is defined by the triplet with the maximum negative interaction information. The rescaled adjacency matrix $C = (c_{ij}) / \max_{i,j}\{|c_{ij}|\}$ can then be combined with prior knowledge as before

$$w \cdot P + (1 - w) \cdot C. \tag{13}$$

By pruning the network in such a way that only those edges with a positive score are kept, we obtain the final network.

Cross-validation assessment The goal of network inference is to reverse engineer from data a stable graphical model able to explain gene interactions and at same time predict the impact of specific manipulations. This is the case when we want to elucidate complex phenotypes such as disease susceptibility[37] or predict expression levels after experimentally perturbing the expression of one gene or a set of genes by knock-out or knock-down experiments.

A common problem with network inference from expression data is variability due to the large dimensionality of the modeling task, the small sample size and the high level of noise in the measurements. This means that by removing or adding few samples, the network topology could change significantly.

In order to address this issue, *predictionet* adopts simple linear dependencies between nodes and makes use of a cross-validation scheme to annotate the inferred model with a measure of accuracy and stability.

For each node $X_i, i = 1, \ldots, n$ a linear model

$$X_i = \beta_0 + \sum_{X_j \in \mathbf{pa}(X_i)} \beta_j X_j, \tag{14}$$

is fitted where $\mathbf{pa}(X_i)$ denotes the set of parents of X_i returned by the structural step. The fitting of the model parameters β_j is obtained using the expression values of the corresponding variables $\mathbf{pa}(X_i)$.

Then a k fold cross-validation scheme is implemented such that for k times a network is inferred using $k-1$ folds of the training set and tested on the remaining fold. Cross-validation returns two important measures of quality for each edge: a measure of stability obtained by counting the appearance of the edge in each run and a measure of accuracy assessing how the procedure is useful in predicting the effect of manipulations or perturbations.

4.3. *Gene prioritization methods*

The previously presented methods are mainly aiming at predicting i) gene interactions and ii) gene expression values based on a modeling assumption such as a linear relation between parent genes and the target gene as in equation (14). These methods use prior information during the inference process to obtain better networks compared to those inferred using only genomic data. A different approach to account for prior knowledge is employed by the family of 'gene prioritization' methods. These methods rely on the so-called 'guilt-by-association' principle[38,39] . The underlying idea is not to infer a network but to identify a set of genes which are close to known disease-related genes according to some metric using different sources of prior knowledge. The thus predicted candidate genes are then a good starting point for further experiments.

Different approaches to gene prioritization have been proposed in the literature: methods based on the computation of topological criteria for known protein-protein networks[40] , methods based on random walks or the related diffusion kernel to rank candidate genes for protein-protein interaction data[41] and methods that ranking candidate genes by combining different data sources via kernel representations[42–44] . Here, we will focus our attention on a publicly available web tool, namely ENDEAVOUR[42–44] . This tools allows the integration of multiple data sources such as prior knowledge, different expression data sets, transcriptional motifs, sequence similarity to name a few. This tool[42] starts by building profiles for each of the different data sources based on the set of known genes. Then, the test genes are ranked based on their similarity to the training genes. This is done for each data source separately and finally a combination of these rankings provides the final ranking of the test genes. An improvement of the results based on the statistics used in[42] has been achieved by kernel representation of the data sources and the subsequent combination of kernels[43] . ENDEAVOUR has been applied to identify candidate genes for different diseases such as obesity and diabetes[45] amongst others[42,46] .

5. Integration of data and prior knowledge: a case study

In order to better illustrate the role and the impact of prior knowledge on the process of inferring a regulatory network from expression data, we present the R/Bioconductor package *predictionet* implementing the inference algorithm described in Section 4.2.1. This package contains a set of scripts whose functionalities can be roughly regrouped into three categories.

(1) *Network inference based on genomic data and prior knowledge:* The user can choose the weighting factor between information from data and from prior knowledge and the maximum number of parents for each gene.
(2) *Validation based on stability of the inferred network and on prediction scores based on a linear regression model using cross-validation:* The inference function has been extended to a cross-validation scheme which automatically computes the stability and the prediction scores.
(3) *Visualization by exporting to *.gml file format:* Each network can be exported together with edge and node properties for subsequent visualization using tools such as Cytoscape[47] .

All of these functionalities will be described in more detail in the following section.

For the case study, we use two recent gene expression data sets[48, c] of primary colon tumors from human patients. It is now well-established that it is not individual genes, but rather biological pathways and networks, that drive carcinogenesis. In order to fully understand the way in which these networks interact (or fail to do so) in disease states, we must learn both the structure of the underlying networks and the rules that govern their behavior. Therefore we decided here to model the gene interactions in the RAS pathway, well-known to be a key player in the development of colon cancer. Such approach could analogously be applied to normal colon tissue samples to subsequently identify gene interactions that are gained or lost in the tumors; however, such an analysis is beyond the scope of our case study.

The first data set, hereafter denoted by `jorissen`, consists of 290 colon tumors hybridized on the Affymetrix GeneChip (HG-U133PLUS2) composed of 54,675 probesets and the second, called `expO`, consists of 292 colon tumors hybridized on the same Affymetrix GeneChip. We restrict the analysis to a subset of genes related to the RAS signaling pathway. In particular, we consider the set of genes differentially expressed between colorectal cancer cell lines carrying the RAS mutation and those with the wild-type RAS gene identified in[49] . Prior knowledge is given by the known interactions downloaded with the *Predictive Networks* web application and stored in a prior adjacency matrix. These three objects are available in the *predictionet*[13] package and can be loaded in an R session with the following commands.

```
data(expO.colon.ras)
data(jorissen.colon.ras)
```

The first one loads the four objects `annot.ras`, `data.ras`, `demo.ras` and `priors.ras`, whereas the second command loads four objects named `*2.ras`. We will focus on the data sets `data.ras` and `data2.ras` as well as on the `priors.ras`. Note that `priors.ras` and `priors2.ras` are equivalent as we selected the same 259 variables for both data sets. Some details on the data and prior objects:

[c]http://www.intgen.org/expo/

- **data*.ras**: matrix of gene expression data, tumor samples in rows, probes/variables in columns.
- **priors*.ras**: matrix of prior information, each entry is the number of times an interaction appeared in a *Predictive Networks* source; parents/sources in rows, children: targets in columns.

5.1. *Inferring the networks*

In order to infer a network using the available data and prior knowledge, the maximum number of parents for each gene has to be specified as well as the prior strength. Here, we first select the 50 most variant genes and infer a network with maximally four parents for each gene. This maximum number of parents ensures that on one hand enough triplets of genes are available for the subsequent orientation while on the other hand avoiding fully connected triplets that cannot be inferred using *predictionet*. As an example, a network using a prior weight of 0.5 can be inferred using the following R commands:

```
goi = dimnames(annot.ras)[[1]][order(abs(log2(annot.ras[
    ,"fold.change"])), decreasing=T)][1:50]

mynet.exp0 = netinf2(data=data.ras[,goi],
    priors=priors.ras[goi,goi], priors.count=T, priors.weight=0.5,
    maxparents=4, method="regrnet")

mynet.jorissen = netinf2(data=data2.ras[,goi],
    priors=priors.ras[goi,goi], priors.count=T, priors.weight=0.5,
    maxparents=4, method="regrnet")
```

As discussed in Section 4.2.1, *predictionet* makes available a cross-validation scheme to compute the stability of edges in the network and to evaluate its predictive ability. The function `netinf.cv` is essentially a wrapper invoking several times the basic inference function. In order to assess the impact of prior knowledge on the final outcome we infer networks for three different prior weights values $\{0, 0.25, 0.5, 0.75\}$ and we perform a 10-fold cross-validation. The choice of the prior weight value depends heavily on the data's and the prior's quality. If there is no strong evidence for preferring one over the other, a prior weight of 0.5 is a good starting point as this gives equal importance to both data and priors.

```
mynet.exp0.cv.pw0 <- netinf.cv(data=data.ras[,goi],
    priors=priors.ras[goi,goi], priors.count=TRUE, priors.weight=0,
    maxparents=mymaxparents, method="regrnet",
    seed=54321,nfold=10,categories=3)
mynet.exp0.cv.pw25 <- netinf.cv(data=data.ras[,goi],
    priors=priors.ras[goi,goi], priors.count=TRUE,
    priors.weight=0.25, maxparents=mymaxparents, method="regrnet",
    seed=54321,nfold=10,categories=3)
mynet.exp0.cv.pw50 <- netinf.cv(data=data.ras[,goi],
    priors=priors.ras[goi,goi], priors.count=TRUE,
    priors.weight=0.5, maxparents=mymaxparents, method="regrnet",
    seed=54321,nfold=10,categories=3)
mynet.exp0.cv.pw75 <- netinf.cv(data=data.ras[,goi],
    priors=priors.ras[goi,goi], priors.count=TRUE,
    priors.weight=0.75, maxparents=mymaxparents, method="regrnet",
    seed=54321,nfold=10,categories=3)

mynet.jorissen.cv.pw0 <- netinf.cv(data=data2.ras[,goi],
    priors=priors2.ras[goi,goi], priors.count=TRUE,
    priors.weight=0, maxparents=mymaxparents, method="regrnet",
    seed=54321,nfold=10,categories=3)
mynet.jorissen.cv.pw25 <- netinf.cv(data=data2.ras[,goi],
    priors=priors2.ras[goi,goi], priors.count=TRUE,
    priors.weight=0.25, maxparents=mymaxparents, method="regrnet",
    seed=54321,nfold=10,categories=3)
mynet.jorissen.cv.pw50 <- netinf.cv(data=data2.ras[,goi],
    priors=priors2.ras[goi,goi], priors.count=TRUE,
    priors.weight=0.5, maxparents=mymaxparents, method="regrnet",
    seed=54321,nfold=10,categories=3)
mynet.jorissen.cv.pw75 <- netinf.cv(data=data2.ras[,goi],
    priors=priors2.ras[goi,goi], priors.count=TRUE,
    priors.weight=0.75, maxparents=mymaxparents, method="regrnet",
    seed=54321,nfold=10,categories=3)
```

5.2. *Validation*

The validation of the inference procedure will be carried out in two steps. The first one is based on the definition of a "true network" using GeneMANIA and allows us to compare the performance of *predictionet* with state-of-the-art methods. The second part of the validation is independent of such gold standard definition and solely relies on the outcome of the cross-validation procedure and concerns three aspects: the stability of the inferred network, the quality of cross-validated predictions and a comparison between the two inferred networks.

5.2.1. *Comparison with state-of-the-art*

As presented in Section 2, GeneMANIA allows to download a set of interactions. We will use these interactions as gold standard network in this part of the chapter. This "true" network allows us to compute a performance score for each of the inferred networks and ultimately to show that *predictionet* performs similarly to state-of-the-art methods when applied to data only. When integrating priors, we can show that the performance increases however caution has to be applied as the interactions from GeneMANIA and *Predictive Networks* are stemming from similar research and therefore favoring networks with these interactions. In order to avoid this bias as much as possible, we excluded the main sources of *Predictive Networks* when querying GeneMANIA, that is structured data bases and interactions stated in published studies. Instead, we only downloaded those interactions which have been derived from computing Pearson correlation between variables in publicly available data sets, a source of prior knowledge not available to *Predictive Networks.*

State-of-the-art methods available as R package In the R package *bnlearn*[50] several algorithms for Bayesian network inference have been implemented. For the comparison, we will use one score-based and one constraint-based algorithm, namely hill-climbing with scoring metrics BIC or AIC and the Grow-Shrink algorithm. To the best of our knowledge, there are currently no implementations which allow the easy integration of prior knowledge to Bayesian inference. Therefore, we will compare the performance of the different networks without prior knowledge and the investigate whether its inclusion yields better results.

Performance score All inferred networks will be compared to the "true" network retrieved using GeneMANIA as described before. The networks inferred using *predictionet* and state-of-the-art methods are structurally very different in the sense that the level of sparsity depends on the specific inference algorithm. Therefore, we will restrict all inferred networks to the top scoring edges, the number of edges depending on the algorithm which infers the sparsest network. For each of these networks we will compute the F-score (using *predictionet*'s `eval.network` function) using true positives (edges present in the inferred network and in the true network), false positives (inferred but not present in the true network) and false negatives (not inferred but in the true network)

$$F = \frac{2TP}{2TP + FP + FN} \tag{15}$$

which is the harmonic average of precision and recall. The F-score takes values in the interval $[0, 1]$ where zero corresponds to no true positives and one to a perfect network.

Results In Table 1, we present the F-scores corresponding to the inferred networks using *predictionet* with four different weights, a score-based and a constraint-based algorithm applied to the two different data sets. The main observation is that all methods taking only data into account during the inference obtain F-scores below 0.1. The F-score values are improved

Table 1. F-scores for the different methods on both data sets using interactions retrieved from GeneMANIA as gold standard. Before evaluation, all networks are pruned to keep only the top scoring edges. The number of edges corresponds to that of the sparsest network (34 for the `exp0` data set and 53 for the `jorissen` data set).

method	`exp0` data			`jorissen` data		
	# edges	F-score	p-val	# edges	F-score	p-val
predictionet, w=0	34	0.0754	0.055	53	0.0989	0.113
predictionet, w=0.25	51	0.1072	0	71	**0.1372**	0
predictionet, w=0.5	58	0.1304	0	70	0.1315	0.002
predictionet, w=0.75	59	**0.1360**	0	70	0.1315	0.002
score-based	258	0.0305	0.881	246	0.04142	0.97
constraint-based	77	0.0488	0.479	80	0.0859	0.269

when taking into account the prior knowledge. This holds for both data sets. The quality of the networks with respect to the choice of the prior weight depends on the data set. That is, for the `exp0` data set a higher weight needs to be put on prior knowledge whereas for the `jorissen` data choosing $w = 0.25$ gives the best result.

Another way to assess the performance of a network is to compare its performance to that of randomly inferred networks. Here, we generated 1000 random networks with the same number of edges as the inferred networks, that is 34 for the `exp0` data set and 53 for the `jorissen` data set. The second constraint on the random topologies is the maximal number of parents for each variable which we set to four as with the networks inferred with *predictionet*. The p-values are then computed by the number of random networks with a better performance divided by the number of random networks. The p-values are presented in Table 1 together with the corresponding F-scores. It can be observed that when using no prior knowledge only *predictionet* for the expO data set yields an almost significant result. Solely the networks combining data and prior knowledge are significantly better than random networks for both data sets.

5.2.2. *Stability*

The stability of each edge in the network is computed by counting how frequently it is inferred during cross-validation. In other terms, an edge that was inferred in six out of the 10 cross-validation runs has the stability of 0.6. In Fig. 4, we present the stability values of each inferred edge for the four chosen prior weight values. It can be observed that the inclusion of prior knowledge leads to higher stability for both data sets.

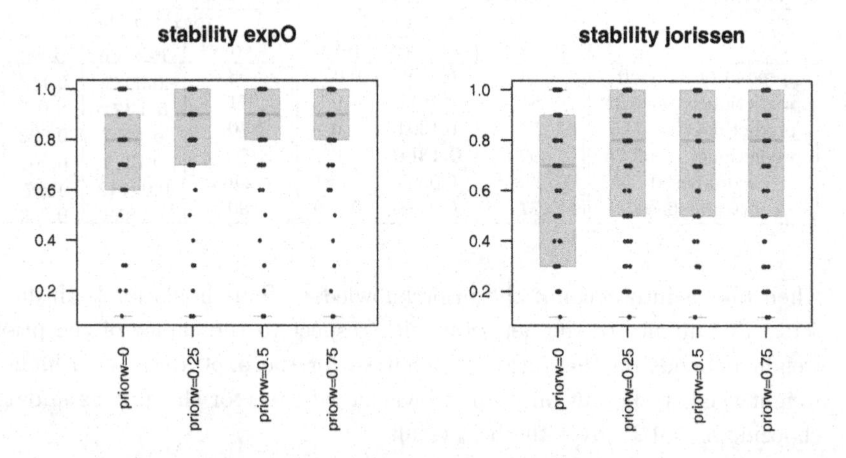

Fig. 4. Stability of edges for four different prior weights computed by 10-fold cross-validation. The x-axis specifies the chosen prior weight and the y-axis corresponds to the stability: an edge that is inferred for all 10 cross-validation runs has a stability of one. The lowest value is 0.1 for an edge that has been inferred in only one cross-validation run. the stability of each inferred edge corresponds to a point in the boxplot. Each box specifies the median and the first and third quartile.

5.2.3. *Prediction*

The quality of the prediction is returned by the R^2 value of the cross-validated predictions for each gene

$$R_i^2 = 1 - \frac{\sum_s (X_{is} - \hat{X_{is}})^2}{\sum_s (X_{is} - \bar{X}_s)^2}, i = 1, \ldots, n \tag{16}$$

where X_{is} is the real value of the sth sample of the ith gene and $\hat{X}_{is}$ is the prediction returned by the linear model on the basis of the observed value of the parents. Note that a perfect prediction corresponds to R^2 equal to 1. In Fig. 5, we present the distribution of the R^2 values for each of the 50 genes in the network. We observe an increase of predictive power with increasing prior strength for both data sets. It is worthy then to remark that the increase of stability due to the integration of prior knowledge does not occur at the detriment of accuracy.

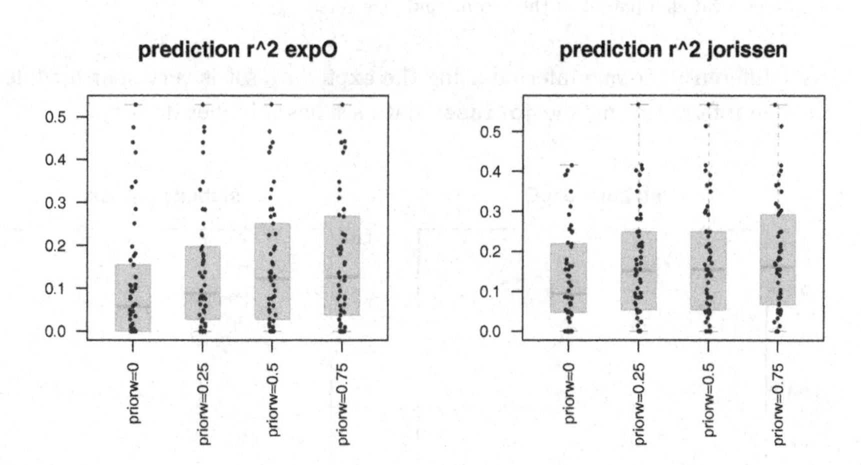

Fig. 5. The x-axis specifies the chosen prior weight and the y-axis corresponds to the obtained R^2 prediction accuracy. Each point is the averaged R^2 prediction accuracy of one gene in the network over the 10 cross-validation folds.

5.2.4. *Comparing networks for the two data sets*

By inferring networks with prior weight equal to zero for both data sets, we can determine how strong the influence of the training set is for the inference procedure. We can observe in Fig. 6 that the two networks are

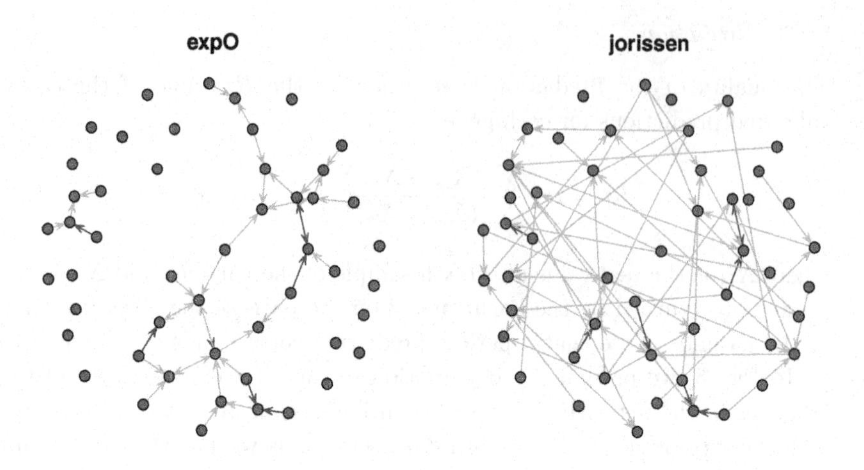

Fig. 6. Networks inferred using only data. In dark gray: edges that have been inferred in both networks. A bidirectional edge between two nodes means that the first one has been identified as a parent of the second and vice versa.

very different: the one inferred using the `expO` data set is very sparse while the one inferred using the `jorissen` data set has a higher density.

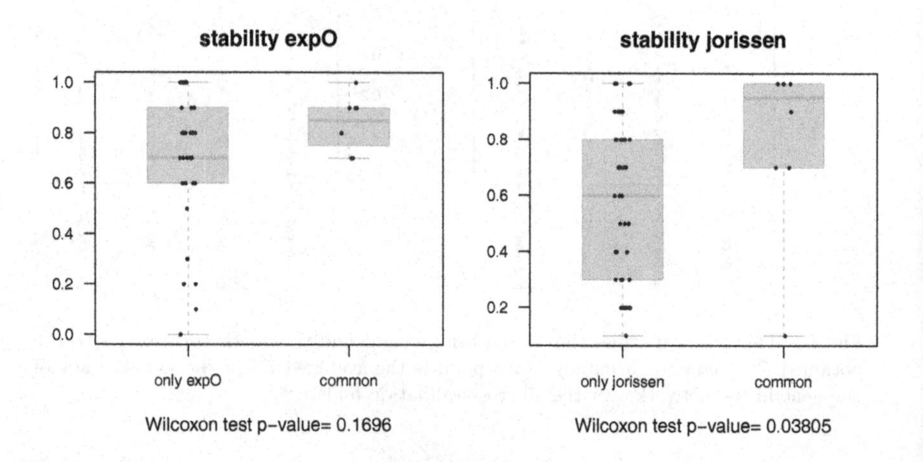

Fig. 7. Stability analysis by taking apart edges that are common to both networks; prior weight equal to zero.

In the following, we will try to determine whether the set of edges common to both networks is also more stable in cross-validation. Therefore,

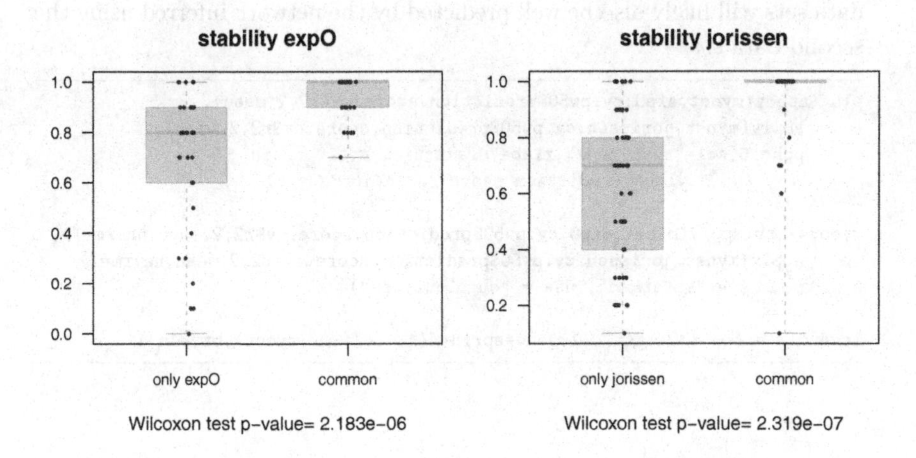

Fig. 8. Stability analysis by taking apart edges that are common to both networks; prior weight equal to 0.5.

we compute the stability of the common edges separately from the non-common edge first for networks inferred using only data (Fig. 7). The difference in stability is not significant for the `exp0` data set and significant with a p-value of ~ 0.03 for the `jorissen` data set using a Wilcoxon test. Considering the networks that are inferred using both data and priors with a prior weight of 0.5, the stability of common edges is significantly higher (p-values of 2.1e-06 and 2.3e-07, respectively; Fig. 8) for both data sets.

```
stab.expO<-list(
        "specific"=mynet.expO.cv$edge.stability[which(mynet.overlap==1)],
        "common"=mynet.expO.cv$edge.stability[which(mynet.overlap==3)])
wt.expO <- wilcox.test(x=stab.expO$specific, y=stab.expO$common)

        stab.jorissen<-list(
        "specific"=mynet.jorissen.cv$edge.stability[which(mynet.overlap==2)],
        "common"=mynet.jorissen.cv$edge.stability[which(mynet.overlap==3)])
        wt.jorissen <- wilcox.test(x=stab.jorissen$specific,
            y=stab.jorissen$common)
```

When computing the correlation between the cross-validated R^2 scores of the two data sets, we can observe a significant correlation between them, compare Fig. 9. This implies that the prediction performances are similar for both data sets in the sense that genes with high R^2 scores in one of the data sets will likely also be well predicted by the network inferred using the second data set.

```
plot(apply(mynet.exp0.cv.pw50$prediction.score.cv$r2,2,mean),
     apply(mynet.jorissen.cv.pw50$prediction.score.cv$r2,2,mean),
     pch=16,col="royalblue",xlab="Prediction score r2 in
          exp0",ylab="Prediction score r2 in jorissen")

mycor<-cor(apply(mynet.exp0.cv.pw50$prediction.score.cv$r2,2,mean,na.rm=T),
     apply(mynet.jorissen.cv.pw50$prediction.score.cv$r2,2,mean,na.rm=T),
     method = "spearman", use = "complete.obs")

legend(x = "bottomright", legend=sprintf("cor=%.2g",mycor),bty="n")
```

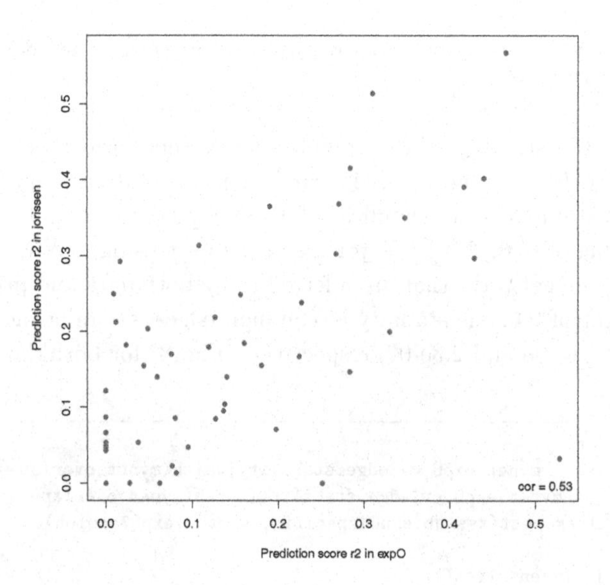

Fig. 9. Prediction scores estimated in `exp0` and `jorissen` and their corresponding Spearman correlation coefficient having inferred both networks using a prior weight of 0.5.

5.2.5. *The network: edges in accordance with the prior vs new edges*

In Fig. 10, we present the network inferred using prior weight equal to 0.5. The nodes have been colored according to the associated R^2 score, where red corresponds to high R^2 scores and blue to low scores. Each edge's color corresponds to whether or not it was part of the prior knowledge, red and green respectively. The network has been visualized using Cytoscape[47]. The R-package *predictionet* offers an export option for object obtained by running `netinf2` or `netinf.cv`.

```
netinf2gml(mynet.jorissen.cv.pw50)
```

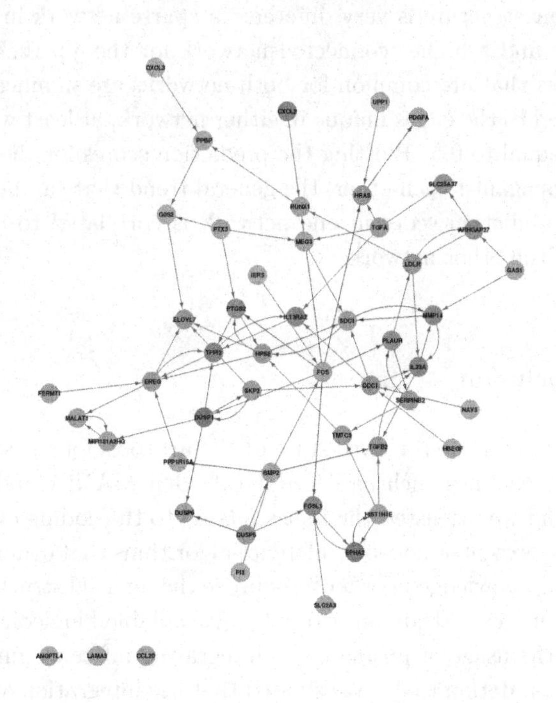

Fig. 10. Inferred network for `jorissen` data using a maximum number of four parents and prior weight equal to 0.5. The graph was created using Cytoscape[47]. The nodes have been color coded corresponding to their prediction score R^2: the higher the value, the darker the node. The edges are color coded by their presence in the prior knowledge: edges present in the prior are dark gray and new edges are light gray.

5.3. *Discussion*

In this case study, we have shown that *predictionet* performs comparably to state-of-the-art methods when applied to genomic data only and furthermore is able to take advantage of both genomic data and prior sources of information. We used a cross-validation scheme to compute the stability of each edge and were able to obtain higher stability for networks inferred with a combination of genomic data and prior knowledge compared to the stability of edges inferred using only data. This improvement through integration of data and prior could also be observed when computing the prediction performance. These observations hold true for both tumor data sets. When however comparing the two inferred networks it can be observed that their structure is very different: a sparse network in case of the `exp0` data set and a higher connected network for the `jorissen` data set. The few edges that are common for both networks are significantly more stable compared to the edges unique in either network, at least when using a prior weight equal to 0.5. Plotting the prediction scores for the genes in the two networks against each other, the general trend that can be observed is that higher prediction score in one network is correlated to a high prediction score in the other network.

6. Conclusion

In this chapter, we reviewed state-of-the-art tools for gathering prior knowledge in genomics such as iHOP/GIM, GeneMANIA and Predictive Networks and we discussed the issues related to the coding of such knowledge. Then we presented network inference algorithms that combine genomic data and prior knowledge in order to improve the inferred structure. The chapter ends with a case study on two publicly available biomedical data sets illustrating the usage of *predictionet*: integrating inference functionalities with a set of validation tools. We showed that the integration of prior knowledge is beneficial for network inference both in terms of stability and prediction accuracy. Our current work is focusing on confirming these preliminary assessments by defining a validation framework based on knock-down experiments.

Acknowledgement

Funding: GB and CO were supported by the Belgian French Community ARC (Action de Recherche Concertée) funding.

References

1. M.-L. T. Lee, F. C. Kuo, G. A. Whitmore, and J. Sklar, Importance of replication in microarray gene expression studies: Statistical methods and evidence from repetitive cdna hybridizations, *Proceedings of the National Academy of Sciences.* **97**(18), 9834–9839 (2000). doi: 10.1073/pnas.97.18.9834. URL `http://www.pnas.org/content/97/18/9834.abstract`.
2. P. P. Łabaj, G. G. Leparc, B. E. Linggi, L. M. Markillie, H. S. Wiley, and D. P. Kreil, Characterization and improvement of RNA-Seq precision in quantitative transcript expression profiling., *Bioinformatics (Oxford, England).* **27**(13), i383–91 (July, 2011). doi: 10.1093/bioinformatics/btr247. URL `http://www.pubmedcentral.nih.gov/articlerender.fcgi?artid=3117338\&tool=pmcentrez\&rendertype=abstract`.
3. P. E. Meyer, C. Olsen, and G. Bontempi, *Transcriptional Network Inference Based on Information Theory.* Wiley-VCH Verlag GmbH & Co. KGaA (2011).
4. R. Hoffmann and A. Valencia, Implementing the ihop concept for navigation of biomedical literature, *Bioinformatics.* **21**(suppl 2), ii252–ii258 (2005). doi: 10.1093/bioinformatics/bti1142. URL `http://bioinformatics.oxfordjournals.org/content/21/suppl_2/ii252.abstract`.
5. A. Ikin, C. Riveros, P. Moscato, and A. Mendes, The gene interaction miner: a new tool for data mining contextual information for protein-protein interaction analysis., *Bioinformatics (Oxford, England).* **26**(2), 283–284 (Jan., 2010). ISSN 1367-4811. doi: 10.1093/bioinformatics/btp652. URL `http://dx.doi.org/10.1093/bioinformatics/btp652`.
6. S. Mostafavi, D. Ray, D. Warde-Farley, C. Grouios, and Q. Morris, Genemania: a real-time multiple association network integration algorithm for predicting gene function, *Genome Biology.* **9**(Suppl 1), S4 (2008). ISSN 1465-6906. doi: 10.1186/gb-2008-9-s1-s4. URL `http://genomebiology.com/2008/9/S1/S4`.
7. B. Haibe-Kains, C. Olsen, A. Djebbari, G. Bontempi, M. Correll, C. Bouton, and J. Quackenbush, Predictive networks: a flexible, open source, web application for integration and analysis of human gene networks, *Nucleic Acids Research.* **40**(D1), D866–D875 (2012). doi: 10.1093/nar/gkr1050. URL `http://nar.oxfordjournals.org/content/40/D1/D866.abstract`.
8. D. Heckerman, D. Geiger, and D. Chickering, Learning Bayesian networks: The combination of knowledge and statistical data, *Machine Learning.* **20**, 197–243 (1995).
9. D. Heckerman. A tutorial on learning with bayesian networks. Technical report, Microsoft Research (1995).
10. D. Heckerman. A bayesian approach to learning causal networks. In *In Uncertainty in AI: Proceedings of the Eleventh Conference*, pp. 285–295 (1995).
11. R. Castelo and A. Siebes, Priors on network structures. Biasing the search for Bayesian networks, *Int. J. Approx. Reasoning.* **24**(1), 39–57 (2000).
12. S. Imoto, T. Higuchi, T. Goto, K. Tashiro, S. Kuhara, and S. Miyano. Combining microarrays and biological knowledge for estimating gene networks via

Bayesian networks. In *In Proceedings of the IEEE Computer Society Bioinformatics Conference (CSB 03*, pp. 104–113, IEEE (2003).

13. B. Haibe-Kains, C. Olsen, G. Bontempi, and J. Quackenbush. *predictionet: Inference for predictive networks designed for (but not limited to) genomic data* (2012). URL `http://compbio.dfci.harvard.edu,http://www.ulb.ac.be/di/mlg`. R package version 1.1.5.
14. P. Pavlidis, J. Weston, J. Cai, and W. N. Grundy. Gene functional classification from heterogeneous data. In *Proceedings of the fifth annual international conference on Computational biology*, RECOMB '01, pp. 249–255, ACM, New York, NY, USA (2001). ISBN 1-58113-353-7. doi: 10.1145/369133.369228. URL `http://doi.acm.org/10.1145/369133.369228`.
15. G. R. Lanckriet, M. Deng, N. Cristianini, M. I. Jordan, and W. S. Noble, Kernel-based data fusion and its application to protein function prediction in yeast., *Pacific Symposium on Biocomputing. Pacific Symposium on Biocomputing*. pp. 300–311 (2004). ISSN 1793-5091. URL `http://view.ncbi.nlm.nih.gov/pubmed/14992512`.
16. K. Tsuda, H. Shin, and B. Schölkopf, Fast protein classification with multiple networks, *Bioinformatics*. **21**(2), 59–65 (Jan., 2005). ISSN 1367-4803. doi: 10.1093/bioinformatics/bti1110. URL `http://dx.doi.org/10.1093/bioinformatics/bti1110`.
17. J. Vert, Graph-driven features extraction from microarray data using diffusion kernels and kernel cca, *Advances in Neural Information Processing Systems* (2003). URL `http://ci.nii.ac.jp/naid/10021201067/en/`.
18. J. Shawe-Taylor and N. Christianini, *Kernel Methods for Pattern Analysis*. Cambridge University Press (2004).
19. G. Cooper and E. Herskovits, A bayesian method for the induction of probabilistic networks from data, *Machine Learning*. **09**(4), 309–347 (October, 1992). ISSN 0885-6125. URL `http://www.ingentaconnect.com/content/klu/mach/1992/00000009/00000004/00422779`.
20. D. Chickering, D. Geiger, and D. Heckerman. Learning bayesian networks is NP-hard. Technical report (1994).
21. D. Koller, N. Friedman, L. Getoor, and B. Taskar. Graphical models in a nutshell. In eds. L. Getoor and B. Taskar, *An Introduction to Statistical Relational Learning*. MIT Press (2007).
22. A. Djebbari and J. Quackenbush, Seeded bayesian networks: Constructing genetic networks from microarray data, *BMC Systems Biology*. **2**(1), 57 (2008). ISSN 1752-0509. doi: 10.1186/1752-0509-2-57. URL `http://www.biomedcentral.com/1752-0509/2/57`.
23. E. Almasri, P. Larsen, G. Chen, and Y. Dai. Incorporating literature knowledge in bayesian network for inferring gene networks with gene expression data. In *Proceedings of the 4th international conference on Bioinformatics research and applications*, ISBRA'08, pp. 184–195, Springer-Verlag, Berlin, Heidelberg (2008). ISBN 3-540-79449-2, 978-3-540-79449-3. URL `http://dl.acm.org/citation.cfm?id=1791494.1791512`.
24. S. Imoto, K. Sunyong, T. Goto, S. Aburatani, K. Tashiro, S. Kuhara, and S. Miyano, Bayesian network and nonparametric heteroscedastic regression

for nonlinear modeling of genetic network, *Proc. 1st IEEE Computer Society Bioinformatics Conference.* **1**, 219–227 (2002).
25. A. Butte and I. Kohane, Mutual information relevance networks: functional genomic clustering using pairwise entropy measurements., *Pac Symp Biocomput.* pp. 418–429 (2000). ISSN 1793-5091. URL `http://view.ncbi.nlm.nih.gov/pubmed/10902190`.
26. A. Margolin, I. Nemenman, and K. B. et al, Aracne: an algorithm for the reconstruction of gene regulatory networks in a mammalian cellular context, *BMC Bioinformatics.* **7** (2006).
27. J. Faith, B. Hayete, and J. T. et al, Large-scale mapping and validation of escherichia coli transcriptional regulation from a compendium of expression profiles, *PLoS Biology.* **5** (2007).
28. P. Meyer, K. Kontos, F. Lafitte, and G. Bontempi, Information-theoretic inference of large transcriptional regulatory networks, *EURASIP Journal on Bioinformatics and Systems Biology* (2007).
29. P. Meyer. *Information-Theoretic Variable Selection and Network Inference from Microarray Data.* PhD thesis, Université Libre de Bruxelles, Belgium (2008).
30. F. Emmert-Streib, G. Glazko, A. Gokmen, and R. De Matos Simoes, Statistical inference and reverse engineering of gene regulatory networks from observational expression data., *Frontiers in Genetics.* **3**(8) (2012). ISSN 1664-8021. doi: 10.3389/fgene.2012.00008. URL `http://www.frontiersin.org/bioinformatics_and_computational_biology/10.3389/fgene.2012.00008/abstract`.
31. C. Ding and H. Peng, Minimum redundancy feature selection from microarray gene expression data, *Journal of Bioinformatics and Computational Biology.* **3**, 185–205 (2005).
32. R. Meir. Bias, variance and the combination of estimators; the case of linear least squares. In *In Advances in Neural Information Processing Systems 7*, Morgan Kaufmann (1995).
33. W. Luo, K. D. Hankenson, and P. J. Woolf, Learning transcriptional regulatory networks from high throughput gene expression data using continuous three-way mutual information., *BMC bioinformatics.* **9**(1), 467+ (Nov., 2008). ISSN 1471-2105. doi: 10.1186/1471-2105-9-467. URL `http://dx.doi.org/10.1186/1471-2105-9-467`.
34. J. Watkinson, K.-c. Liang, X. Wang, T. Zheng, and D. Anastassiou, Inference of regulatory gene interactions from expression data using Three-Way mutual information, *Annals of the New York Academy of Sciences.* **1158**(1), 302–313 (Mar., 2009). ISSN 1749-6632. doi: 10.1111/j.1749-6632.2008.03757.x. URL `http://dx.doi.org/10.1111/j.1749-6632.2008.03757.x`.
35. G. Bontempi and P. E. Meyer. Causal filter selection in microarray data. In *ICML*, pp. 95–102 (2010).
36. G. Bontempi, B. H. Kains, C. Desmedt, C. Sotiriou, and J. Quackenbush, Multiple-input multiple-output causal strategies for gene selection, *BMC Bioinformatics.* **12**(1), 458+ (Nov., 2011). ISSN 1471-2105. doi: 10.1186/1471-2105-12-458. URL `http://dx.doi.org/10.1186/1471-2105-12-458`.

37. P.-R. Loh, G. Tucker, and B. Berger, Phenotype prediction using regularized regression on genetic data in the dream5 systems genetics b challenge, *PLoS ONE*. **6**(12), e29095 (12, 2011). doi: 10.1371/journal.pone.0029095. URL `http://dx.doi.org/10.1371%2Fjournal.pone.0029095`.
38. D. Nitsch, L.-C. Tranchevent, B. Thienpont, L. Thorrez, H. Van Esch, K. Devriendt, and Y. Moreau, Network analysis of differential expression for the identification of Disease-Causing genes, *PLoS ONE*. **4**(5), e5526+ (May, 2009). doi: 10.1371/journal.pone.0005526. URL `http://dx.doi.org/10.1371/journal.pone.0005526`.
39. L.-C. Tranchevent, F. Capdevila, N. Nitsch, B. De Moor, P. De Causmaecker, and Y. Moreau, A guide to web tools to prioritize candidate genes, *Briefings in Bioinformatics*. **12**(1), 22–32 (Jan., 2010). URL `https://lirias.kuleuven.be/handle/123456789/260982`.
40. J. Xu and Y. Li, Discovering disease-genes by topological features in human protein-protein interaction network, *Bioinformatics*. **22**(22), 2800–2805 (Nov., 2006). ISSN 1367-4811. doi: 10.1093/bioinformatics/btl467. URL `http://dx.doi.org/10.1093/bioinformatics/btl467`.
41. S. Köhler, S. Bauer, D. Horn, and P. N. Robinson, Walking the interactome for prioritization of candidate disease genes., *American journal of human genetics*. **82**(4), 949–958 (Apr., 2008). ISSN 1537-6605. doi: 10.1016/j.ajhg.2008.02.013. URL `http://dx.doi.org/10.1016/j.ajhg.2008.02.013`.
42. S. Aerts, D. Lambrechts, S. Maity, P. Van Loo, B. Coessens, F. De Smet, L.-C. Tranchevent, B. De Moor, P. Marynen, B. Hassan, P. Carmeliet, and Y. Moreau, Gene prioritization through genomic data fusion, *Nat Biotech*. **24**(5), 537–544 (May, 2006). ISSN 1087-0156. doi: 10.1038/nbt1203. URL `http://dx.doi.org/10.1038/nbt1203`.
43. T. De Bie, L.-C. Tranchevent, L. M. M. van Oeffelen, and Y. Moreau, Kernel-based data fusion for gene prioritization, *Bioinformatics*. **23**(13), i125–i132 (July, 2007). ISSN 1460-2059. doi: 10.1093/bioinformatics/btm187. URL `http://dx.doi.org/10.1093/bioinformatics/btm187`.
44. L.-C. Tranchevent, R. Barriot, S. Yu, S. Van Vooren, P. Van Loo, B. Coessens, B. De Moor, S. Aerts, and Y. Moreau, Endeavour update: a web resource for gene prioritization in multiple species, *Nucleic Acids Research*. **36**(suppl 2), W377–W384 (2008). doi: 10.1093/nar/gkn325. URL `http://nar.oxfordjournals.org/content/36/suppl_2/W377.abstract`.
45. C. C. Elbers, C. N. Onland-Moret, L. Franke, A. G. Niehoff, Y. T. van der Schouw, and C. Wijmenga, A strategy to search for common obesity and type 2 diabetes genes, *Trends in Endocrinology & Metabolism*. **18**(1), 19–26 (2007). doi: 10.1016/j.tem.2006.11.003. URL `http://dx.doi.org/10.1016/j.tem.2006.11.003`.
46. I. Ebermann, H. Scholl, P. Charbel Issa, E. Becirovic, J. Lamprecht, B. Jurklies, J. M. Millán, E. Aller, D. Mitter, and H. Bolz, A novel gene for usher syndrome type 2: mutations in the long isoform of whirlin are associated with retinitis pigmentosa and sensorineural hearing loss, *Human Genetics*. **121**, 203–211 (2007). ISSN 0340-6717. doi: 10.1007/s00439-006-0304-0. URL `http://dx.doi.org/10.1007/s00439-006-0304-0`.

47. M. E. Smoot, K. Ono, J. Ruscheinski, P.-L. L. Wang, and T. Ideker, Cytoscape 2.8: new features for data integration and network visualization., *Bioinformatics (Oxford, England)*. **27**(3), 431–432 (Feb., 2011). ISSN 1367-4811. doi: 10.1093/bioinformatics/btq675. URL `http://dx.doi.org/10.1093/bioinformatics/btq675`.
48. R. N. Jorissen, P. Gibbs, M. Christie, S. Prakash, L. Lipton, J. Desai, D. Kerr, L. A. Aaltonen, D. Arango, M. Kruhffer, T. F. Orntoft, C. L. Andersen, M. Gruidl, V. P. Kamath, S. Eschrich, T. J. Yeatman, and O. M. Sieber, Metastasis-associated gene expression changes predict poor outcomes in patients with dukes stage b and c colorectal cancer, *Clin Cancer Res* (2010).
49. A. H. Bild, G. Yao, J. T. Chang, Q. Wang, A. Potti, D. Chasse, M.-B. Joshi, D. Harpole, J. M. Lancaster, A. Berchuck, J. A. Olson, J. R. Marks, H. K. Dressman, M. West, and J. R. Nevins, Oncogenic pathway signatures in human cancers as a guide to targeted therapies., *Nature*. **439**(7074), 353–7 (Jan., 2006). doi: 10.1038/nature04296. URL `http://www.ncbi.nlm.nih.gov/pubmed/16273092`.
50. M. Scutari and K. Strimmer. Introduction to graphical modelling. URL `http://arxiv.org/abs/1005.1036` (June, 2011).

3. Classification, Trend Analysis and 3D Medical Images

Chapter 8

Classification and its Application to Drug-Target Interaction Prediction

Jian-Ping Mei[1], Chee-Keong Kwoh[1], Peng Yang[1,2] and Xiao-Li Li[2]

[1]*School of Computer Engineering, Nanyang Technological University 50 Nanyang Avenue, Singapore 639798*

[2]*Institute for Infocomm Research, Agency for Science, Technology & Research (A*STAR) 1 Fusionopolis Way, Singapore 138632*

Classification is one of the most popular and widely used supervised learning tasks, which categorizes objects into predefined classes based on known knowledge. Classification has been an important research topic in machine learning and data mining. Different classification methods have been proposed and applied to deal with various real-world problems. Unlike unsupervised learning such as clustering, a classifier is typically trained with labeled data before being used to make prediction, and usually achieves higher accuracy than unsupervised one.

In this chapter, we first define classification and then review several representative methods. After that, we study in details the application of classification to drug-target interaction prediction, which is a critical challenging problem in drug discovery, i.e., drug-target prediction, due to the challenges in predicting possible interactions between drugs and targets.

1. Classification

Classification is the process of finding a model or function that describes and distinguishes data classes or concepts.[1] It is one of the most important tasks that supervised learning is applied to. Supervised learning is an important machine learning method which learns a model or a function with the help of supervision. Other than classification, supervised learning is also used for regression analysis. The goal of classification analysis is simply to know the class labels while regression analysis is to predict values for certain variables or model the relationships between a dependent variable and one or more independent variable.

The rapid development of technologies, such as microarrays, high-throughput sequencing, genotyping arrays, mass spectrometry, and automated high-resolution image acquisition techniques, have led to a dramatic increase in availability of biomedical data.[2] Facing large amount of data, computational methods, which are cheaper and more efficient, arise to be useful complement to support traditional experimental methods in many biomedical researches and applications. As an important data analysis tool, classification has been applied for handling many important tasks in bioinformatics, including Sequence annotation,[3–5] Protein function prediction,[6] Protein structure prediction,[7] Gene regulatory network inference,[8,9] Protein-protein interaction prediction,[10] disease gene identification[11,45] and drug-target interaction prediction.[12–14] Many of these tasks are to search the answer of a question with "positive" or "negative". For example, to predict whether two proteins interact or not, a protein is enzyme or non-enzyme, a piece of sequence is coding or non-coding, and so on. This type of prediction can be directly handled with binary classification, where "positive" and "negative" are treated as two class labels. It also can be solved through regression methods. Instead of directly answering "positive" or "negative", binomial regression methods produce the likelihood or the degree of being "positive" or "negative", based on which the final result can be easily obtained by cutting with a certain threshold, i.e., "positive" if the likelihood is larger than the threshold, and "negative" if the likelihood value is below the given threshold. Next we give more detailed introduction on several representative classification methods, which are most widely used in computational biology.

In the following subsections, we will introduce a number of representative supervised learning methods. We represent the training data that consist of n labeled examples or data objects as $D = \{\mathbf{x}_i, y_i\}_{i=1}^n$, where each $\mathbf{x}_i$ is a p-dimensional vector, i.e., $\mathbf{x}_i = (x_{i1} \ldots x_{ip})^T$, and y_i is its associated class label.

1.1. *k-Nearest Neighbor (k-NN)*

k-Nearest Neighbor (k-NN)[15] is instance-based classification. In k-NN, an unlabeled object is assigned to the most common class among its k most nearest neighbors in the training set. In order to decide the k nearest neighbors of the given object, the distances between this object and all the labeled objects need to be calculated. The number of neighbors k is an important parameter in k-NN. Setting k to different values, k-NN may produce different results.

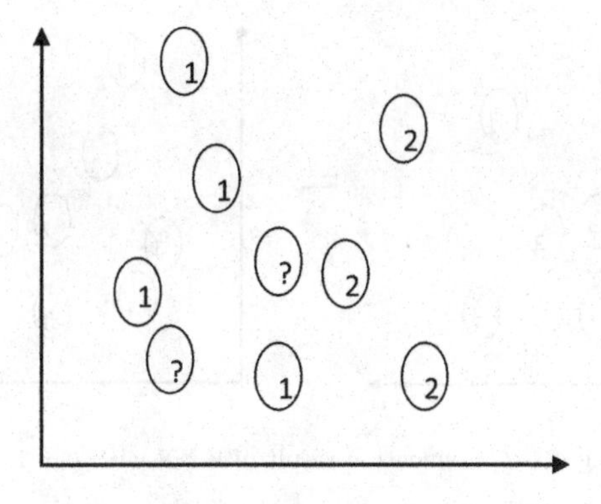

Fig. 1. A simple two-dimensional dataset.

Now we use a simple example to illustrate how labeled data are used in k-NN to predict the class labels of those unlabeled test objects. Fig. 1 shows a simple two-dimensional dataset. This dataset consists of seven labeled objects belonging to two classes "1" or "2" and two unlabeled objects marked with "?" First, we set $k = 1$. In this case, each unlabeled object is assigned to the same class as its nearest neighbor. Fig. 3 shows the classification result of the two unlabeled objects with $k = 1$. Since the nearest neighbor of the first unlabeled object, i.e., the one that is located at the left lower corner, is labeled as class 1, this object is also labeled as 1. Similarly, since the nearest neighbor of the other unlabeled object (located near the center) is labeled as class 2, the class label is also predicted as class 2 for this object. When $k > 1$, the neighbors of an unlabeled object possibly have different class labels, and in such a case, the unlabeled object is typically assigned to the most common class among its neighbors. Fig. 3 shows the classification result of k-NN with $k = 3$. It is seen that the object in the left lower corner is still labeled as class 1 as all its three nearest neighbors are in this class. However, the other object is now labeled as class 1 as two of its nearest neighbors belong to class 1, although its most nearest neighbor belongs to class 2. Here, once k is decided, all the neighbors are considered to be equally important in deciding the class of the unlabeled object. Another way is to assign different weights to the neighbors (e.g. based on their distances to the unlabeled object) so that the k neighbors have different levels of significance of their votes.

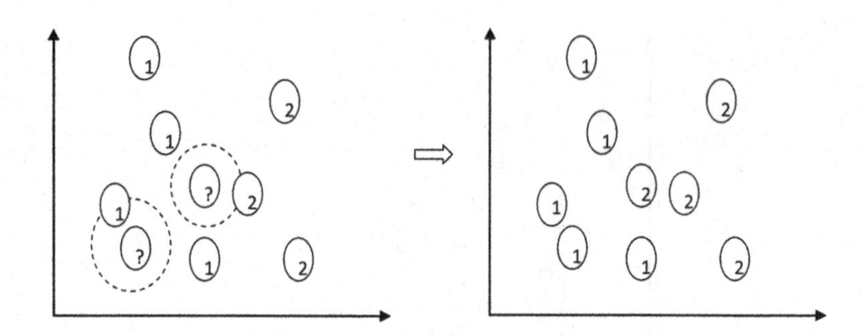

Fig. 2. Classification result of k-NN with $k = 1$.

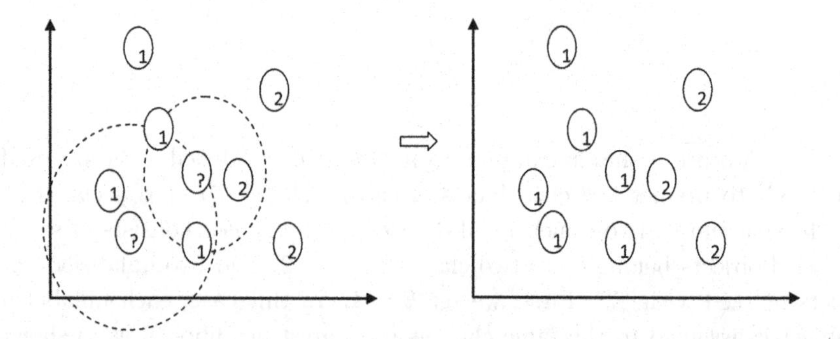

Fig. 3. Classification result of k-NN with $k = 3$.

1.2. *Support Vector Machine*

The classic Support Vector Machine (SVM)[16] is a linear binary classifier. Given a p-dimensional dataset where the training samples belong to two classes, the goal of a linear classifier is to find a $p-1$ dimensional hyperplane which separates the samples in two classes as illustrated in Fig. 4. Among many of such kind of hyperplanes, the one maximizes the separation or margin of the two classes is of the most interest, and the corresponding classifier is called the maximum margin classifier. In SVM, the margin is the distance from the hyperplane to the nearest samples in each of the classes. Samples located on the boundary of each class are called support vectors.

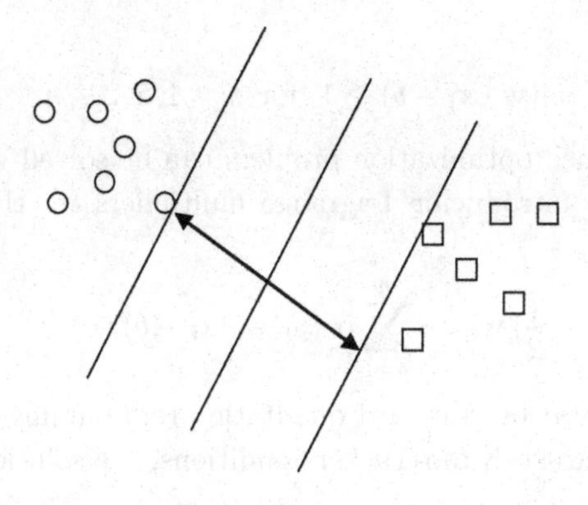

Fig. 4. Example of linearly separable dataset in a two dimensional space. An optimal hyperplane is the one that maximizes the distance between two classes.

1.2.1. *Linear SVM*

Now we formally define the linear SVM. For a set of n training samples, where each object with label -1 or 1 is a p-dimensional vector, we may represent it as $D = \{(\mathbf{x}_i, y_i) | \mathbf{x}_i \in R^p, y_i \in \{-1, 1\}\}_{i=1}^{n}$, where x_i represents the data and y_i represents the label information. Assume that the dataset is linearly separable, then there exists a $\mathbf{w}$ and $\mathbf{b}$ such that the following inequalities are valid for all $\mathbf{x}_i \in D$:

$$\mathbf{w} \cdot \mathbf{x}_i - b \geq 1 \ \text{ if } \ y_i = 1 \tag{1}$$

$$\mathbf{w} \cdot \mathbf{x}_i - b \leq -1 \ \text{ if } \ y_i = -1 \tag{2}$$

The above two inequalities can be written into one as below

$$y_i(\mathbf{w} \cdot \mathbf{x}_i - b) \geq 1 \tag{3}$$

Among the training samples, vectors $\mathbf{x}_i$ for which

$$y_i(\mathbf{w} \cdot \mathbf{x}_i - b) = 1 \tag{4}$$

are called support vectors, and they define the boundary of the two classes.

The distance or margin between the two classes is $\frac{2}{\|\mathbf{w}\|}$. The goal is to find the optimal hyperplane or to decide $\mathbf{w}$ and $\mathbf{b}$ to maximize this margin subject to (3), which requires all the training samples to be correctly classified. Since maximizing $\frac{2}{\|\mathbf{w}\|}$ is equivalent to minimizing $\frac{1}{2}\|\mathbf{w}\|^2$, we can solve the above maximization problem by solving the equivalent minimization problem as below

$$\min \frac{1}{2}\|\mathbf{w}\|^2 \tag{5}$$

subject to

$$y_i(\mathbf{w} \cdot \mathbf{x_i} - b) \geq 1 \quad \text{for} \quad i = 1, 2, \ldots, n. \tag{6}$$

This constrained optimization problem can be solved with the method of Lagrange. By introducing Lagrange multipliers α_i, the Lagrangian is constructed as

$$\frac{1}{2}\|\mathbf{w}\|^2 - \sum_{i=1}^{n} \alpha_i (y_i(\mathbf{w} \cdot \mathbf{x_i} - b) - 1) \tag{7}$$

which can be solved by standard quadratic programming techniques. According to the Karush-Kuhn-Tucker conditions, the solution of $\mathbf{w}$ is in the form as below:

$$\mathbf{w} = \sum_{i=1}^{n} \alpha_i y_i \mathbf{x}_i \tag{8}$$

The above formula shows that $\mathbf{w}$ is a linear combination of the training samples. When $y_i(\mathbf{w} \cdot \mathbf{x}_i - b) = 1$, $\alpha_i > 0$; for other cases, $\alpha_i = 0$. This means that $\mathbf{w}$ is only defined by a small number of support vectors, i.e., the training samples located at the boundary of the classes, rather than all the training samples.

In the above formulation, we assume that the dataset is linearly separable, or there exists a hyperplane that can divide the samples according to their class labels without any classification error. In cases that such kind of hyperplane does not exist, we may want to find a hyperplane that correctly divides the samples into two different classes as many as possible. This is called the soft margin method. Slack variables $\xi_i \geq 0$ are introduced to formulate this idea. The constraints are now become

$$y_i(\mathbf{w} \cdot \mathbf{x}_i - b) \geq 1 - \xi_i \tag{9}$$

Since a larger ξ_i corresponds to a larger error in the classification of x_i, we want to penalize large ξ_i through minimizing the objective function as below

$$\min \frac{1}{2}\|\mathbf{w}\|^2 + C \sum_{i=1}^{n} \xi_i \tag{10}$$

where C is the weight parameter of the penalty term. With Lagrange multipliers $\alpha_i \geq 0$ and $\beta_i \geq 0$, the problem to be solved is written as

$$\min \frac{1}{2}\|\mathbf{w}\|^2 + C \sum_{i=1}^{n} \xi_i - \sum_{i=1}^{n} \alpha_i (y_i(\mathbf{w} \cdot \mathbf{x_i} - b) - 1 + \xi_i) - \sum_{i=1}^{n} \beta_i \xi_i \tag{11}$$

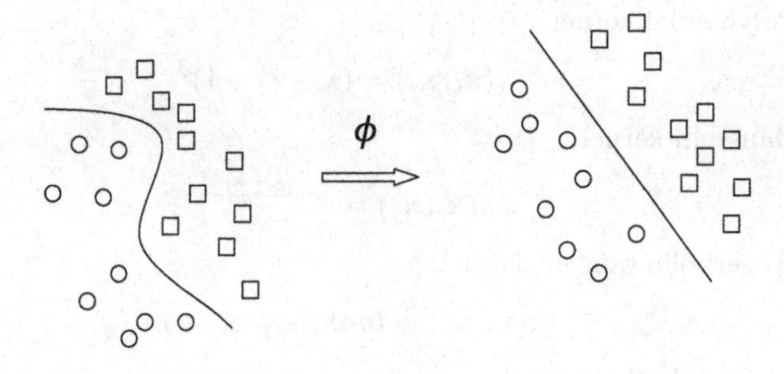

Fig. 5. A non-linearly separable dataset becomes linearly separable after mapping ϕ.

1.2.2. *Kernel SVM*

In many cases, the training data are not linearly separable. To solve the problem, Kernel-based approaches are applied to map the original space into a high or infinity dimensional feature space, i.e., $\mathbf{x} \to \phi(\mathbf{x})$, which makes the data easier to be separated in Fig. 5. A kernel function κ is used to calculate the inner product of the vectors in the high dimensional space in terms of the vectors in the original space:

$$\kappa(x_i, x_j) = \phi(\mathbf{x}_i) \cdot \phi(\mathbf{x}_j) = \phi(\mathbf{x}_i)^T \phi(\mathbf{x}_j) \tag{12}$$

As $\|\mathbf{w}\| = \mathbf{w}^T\mathbf{w}$, substituting (8) into (7), the dual of SVM is the following optimization problem:

$$\max_{\alpha_i} = \sum_{i=1}^{n} \alpha_i - \frac{1}{2} \sum_{i=1}^{n} \sum_{j=1}^{n} \alpha_i \alpha_j y_i y_j x_i^T x_j \tag{13}$$

subject to

$$\alpha_i \geq 0 \text{ for } i = 1, 2, \ldots, n. \tag{14}$$

$$\sum_{i=1}^{n} \alpha_i y_i = 0 \tag{15}$$

By substituting $\mathbf{x}_i$ with $\phi(\mathbf{x_i})$ in the above formula, we get the objective function in the mapped space, and with the kernel function given in (12), we have the following form without defining the mapping explicitly:

$$\max_{\alpha_i} = \sum_{i=1}^{n} \alpha_i - \frac{1}{2} \sum_{i=1}^{n} \sum_{j=1}^{n} \alpha_i \alpha_j y_i y_j \kappa(\mathbf{x}_i, \mathbf{x}_j) \tag{16}$$

Below are the three commonly used kernels:

- Polynomial kernel

$$\kappa(\mathbf{x}_i, \mathbf{x}_j) = (\mathbf{x}_i \cdot \mathbf{x}_j + 1)^d \tag{17}$$

- Gaussian kernel

$$\kappa(\mathbf{x}_i, \mathbf{x}_j) = e^{-\frac{\|\mathbf{x}_i - \mathbf{x}_j\|^2}{\beta}} \tag{18}$$

- Hyperbolic tangent kernel

$$\kappa(\mathbf{x}_i, \mathbf{x}_j) = tanh(h\mathbf{x}_i \cdot \mathbf{x}_j + c) \tag{19}$$

where h is the scale factor and c is the offset.

Since any positive-definite matrix could be treated as a kernel matrix, kernel SVM can be used to make prediction based on a similarity matrix, which records pairwise similarities between objects. To make sure kernel SVM performs stably, some preprocess is needed if the similarity matrix given is not positive-definite.

1.3. *Bayesian classification*

Bayesian classifiers[17] are statistical classifiers based on Bayes theorem. A Bayesian classifier generates the probability or membership of an object with respect to each of the classes. Assume X is an object that is to be classified or labeled and Y is the hypothesis that X belongs to some class, then $P(Y = c/X)$ is the probability that X belongs to the cth class. According to the Bayes theorem, this probability of $Y = c$ conditioned on X can be calculated with posterior probability $P(X/Y = c)$, and prior probabilities $P(X)$ and $P(Y)$:

$$P(Y = c/X) = \frac{P(X/Y = c)P(Y = c)}{P(X)} \tag{20}$$

In the above formula, $P(X)$ is constant for any c. If $P(Y = c)$ is unknown, it is usually assumed that all classes have equal probability or it is estimated by $\frac{n_c}{n}$, the ratio of the number of objects in class c. The left remaining problem is how to calculate $P(X/Y = c)$. To simplify computation, the values of attributes are assumed to be conditionally independent to each other, i.e., given the class label of an object, there are no dependence relationships among the attributes. Assume x_j is the value of the jth feature, and there are p features in total, then based on this assumption,

$$P(X/Y = c) = \prod_{j=1}^{p} P(x_j/Y = c) \tag{21}$$

and the classifier is called the Naive Bayes Classifier.

If the kth attribute is categorical, then

$$P(x_j/Y = c) = \frac{n_{jc}}{n_c} \tag{22}$$

where n_c is the number of objects in class c, and n_{jc} is the number of objects in class c that have the value of the kth attribute equal to x_j.

If the kth attribute is continuous-values with a probability distribution g, e.g., the Gaussian distribution with mean μ_c and standard deviation δ_c, then

$$P(x_j/Y = c) = g(x_j) = \frac{1}{\sqrt{2\pi}\delta_c} e^{-\frac{(x_j-\mu_c)^2}{2\delta_c^2}} \tag{23}$$

Once the probabilities $P(X/Y = c)$ for all $c = 1, 2, \ldots, k$ are calculated, X is assigned to the class with the largest posterior probability, i.e., X is labeled as class f, where $f = \arg\max_c P(Y = c/X)$.

1.4. *Decision trees*

A decision tree is a tree structure where each internal node denotes a test on an attribute, each branch denotes an outcome of the test, and each leaf node represents a class. Once a decision tree has been constructed with training data, a new unlabeled sample is tested against the decision tree from the top node to the leaf node which corresponds to the predicted class of the new sample.

Given a set of training objects, a decision tree is built in a top-down recursive divide and conquer manner. A critical problem needs to be considered in construction of the tree is how to select the attributes for testing. Entropy or equivalently information gain and Gini index are commonly used for attribute selection. The entropy measures the purity of the partitions: the smaller the entropy or the larger the information gain, the purer the partitions are. Thus, the attribute with the minimum entropy or highest information gain is chosen as the test attribute for the current node. Assume the training data consists of n labeled objects that are distributed in k classes, and each class c and contains n_c objects, then the expected information needed to classify a given sample is

$$E = -\sum_{c=1}^{k} p_c \log_2 p_c \tag{24}$$

where p_c is the probability that an arbitrary object belongs to class c. It is estimated by $\frac{n_c}{n}$. For a feature a, which has h distinct values, the entropy

Table 1. Drug-Target Interaction Data

id	$Chemical^d$	$Sequence^t$	$Category^d$	PPI^t	Interact
1	Average	Small	Different	Interact	No
2	Average	Small	Different	Non-interact	No
3	Large	Small	Different	Interact	Yes
4	Small	Large	Different	Interact	Yes
5	Small	Average	Same	Interact	Yes
6	Small	Average	Same	Non-interact	No
7	Large	Average	Same	Non-interact	Yes
8	Average	Large	Different	Interact	No
9	Average	Average	Same	Interact	Yes
10	Small	Large	Same	Interact	Yes
11	Average	Large	Same	Non-interact	Yes
12	Large	Large	Different	Non-interact	Yes
13	Large	Small	Same	Interact	Yes
14	Small	Large	Different	Non-interact	No

based on the partitioning into k subsets by a is calculated by

$$E(a) = \sum_{j=1}^{h} P(j)E(j) \tag{25}$$

where $P(j) = \frac{n_j}{n}$, n_j is the number of objects of which the value of feature a is equal to j, and

$$E(j) = -\sum_{c=1}^{k} p_{cj} \log_2 p_{cj} \tag{26}$$

is the entropy of the jth value of the ath attribute, where $p_{cj} = \frac{n_{cj}}{n_j}$

$$Gain(a) = E - E(a) \tag{27}$$

Now we use the Drug-Target Interaction data in Table 1 as an example to show how to calculate the Entropy of each attribute. This data consists of fourteen samples that belong to two classes: *Interact* or Not *Interact* (corresponding to the last column in Table 1). Each sample is a drug-target pair that is described by four attributes, namely $Chemical^d$, $Sequence^t$, $Category^d$ and PPI^t, where the superscript d denotes a drug-based feature and t denotes a target-based feature. Assume that $\{d_i, t_j\} \in \mathcal{I}$ is a set of drug-target pairs which are known to interact, and $\mathcal{I}$ involves n_d drugs and n_t targets. The two drug-based features are defined as the sum of similarities between the query drug and these n_d drugs, i.e., $Chemical^d$ is chemical structure similarity, and $Category^d$ is the category similarity. Similarly, the two target-based features are the sum of similarities between the query target and these n_t targets, where $Sequence^t$ represents the sequence similarity and PPI^t denotes the protein-protein interactions. Typically, these features are numerical, which means they take continuous values. For the convenience of illustration, here we use categorical values for each feature, which may be obtained by setting thresholds from the continuous values. Take the first sample for example. It is observed that the drug and the target do not interact if the chemical structure of this drug is not very similar with other drugs, the drug is from different category with other drugs, the sequence of the target is not similar with other targets, and the target protein interacts with other proteins.

It is shown that $P(Interact = Yes) = \frac{9}{14}$, $P(Interact = No) = \frac{5}{14}$, So the Entropy of *Interact* or the expected information needed to classify a sample is

$$E = -\left(\frac{9}{14}\log_2\frac{9}{14} + \frac{5}{14}\log_2\frac{5}{14}\right) = 0.940 \tag{28}$$

Now we calculate the Entropy of the attribute $Chemical^d$. This attribute has three values *Average*, *Large*, and *Small*, which occured 5, 4, and 5 times, respectively, i.e., $P(Average) = \frac{5}{14}$, $P(Large) = \frac{4}{14}$ and $P(Small) = \frac{5}{14}$. Among the five samples of which $Chemical^d$ is *Average*, two are $Interact = Yes$, three are $Interact = No$, thus the Entropy of *Average* is

$$E(Average) = -\left(\frac{2}{5}\log_2\frac{2}{5} + \frac{3}{5}\log_2\frac{3}{5}\right) = 0.971 \tag{29}$$

Similarly

$$E(Large) = -\left(\frac{4}{4}\log_2\frac{4}{4} + 0\log_2 0\right) = 0 \tag{30}$$

$$E(Small) = -\left(\frac{3}{5}\log_2\frac{3}{5} + \frac{2}{5}\log_2\frac{2}{5}\right) = 0.971 \tag{31}$$

So the Entropy of $Chemical^d$ is

$$E(Chemical^d) \tag{32}$$

$$= P(Average)E(Average) + P(Large)E(Large) + P(Small)E(Small) \tag{33}$$

$$= \frac{5}{14}0.971 + \frac{4}{14}0 + \frac{5}{14}0.971 = 0.694 \tag{34}$$

and the Information Gain of $Chemical^d$ is

$$Gain(Chemical^d) = E - E(Chemical^d) = 0.940 - 0.694 = 0.246 \tag{35}$$

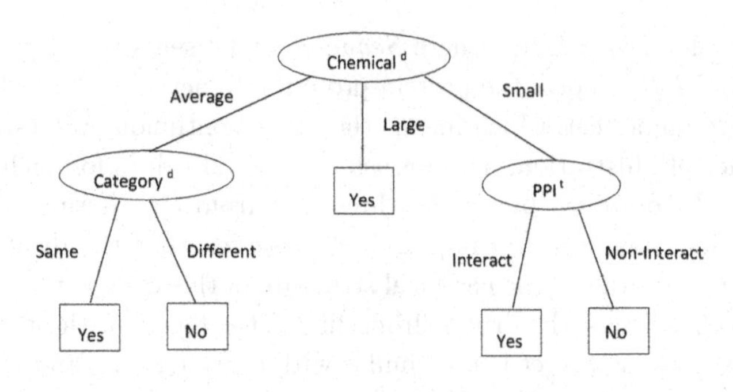

Fig. 6. The decision tree of the Drug-Target Interaction data.

With the same steps, we can calculate the Gain of the other three attributes: $Gain(Sequence^t) = 0.029$, $Gain(Category^d) = 0.152$, and $Gain(PPI^t) = 0.048$. Since $Chemical^d$ has the largest Gain, it is the best attribute of the current stage that should be selected for testing.

Once the best attribute is decided and represented as an intermediate node of the tree, branches below this node are added where each branch corresponds to a possible value this attribute takes. For each value, the subset of samples having this value of the current attribute are taken as the input of the next iteration for further splitting. This process continues until all samples under consideration have the same class label. A complete decision tree of this dataset is shown in Fig. 6.

The tree constructed to correctly classify all the training samples may be over-fitting. Pruning handles the over-fitting problem by removing least reliable branches. Other than higher classification accuracy, pruning also results in a simplified tree which makes the test process faster. Pruning performed during the construction of the tree is called Prepruning. It stops the construction early with less purity. Pruning can also be performed by removing branches from a fully grown tree. This type is called post-pruning. Fig. 7 shows the unpruned and pruned decision three of the Fisher's iris data. This dataset consists of 50 samples from each of three species of Iris (setosa, virginica and versicolor). Four features were measured from each sample: sepal length (SL), sepal width (SW), petal length (PL), and petal width (PW).

ID3 is a popular decision tree algorithm proposed by Quinlan,[18] and C4.5[19] is an extension of ID3 with improved computing efficiency as well as other more functions, including dealing with continuous values, handling attributes with missing values, and avoiding over fitting. Another

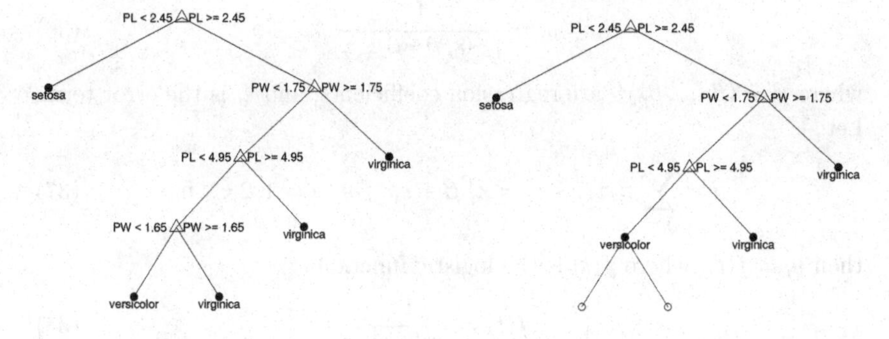

Fig. 7. The decision tree of the Iris data. (a) The unpruned three, (b) The tree with pruning.

algorithm called Classification and regression trees (CART) proposed by Breiman[20] produces either classification or regression binary trees, depending on whether the dependent variable is categorical or numeric. Chen *et al.*[2] reviewed tree-based classification approaches and their applications in bioinformatics.

1.5. *Regression models for classification*

Other than these previously reviewed supervised learning methods which are widely used for classification, regression models may also be used for classification analysis. Regression methods model the relationship between a dependent variable and one or more independent variables. Specifically, regression is to analyse how the value of the dependent variable changes when any one of the independent variable varies while other independent variables are fixed. The dependent variable is the output variable or response variable, and the independent variables are input variables or explanatory variables. Next we discuss two regression models namely the Logistic Regression and the Regularized Least Squares, which are frequently used for classification purpose.

1.5.1. *Logistic Regression*

Logistic Regression is a type of binomial regression that predicts the probability of the outcome of a "yes or no" type trial using logistic function. Formally, the Logistic Regression models the relationship between dependent variable y_i and independent variables $\mathbf{x}_i = (x_{i1} \dots x_{ip})^T$ by

$$y_i = \frac{1}{e^{-(\mathbf{x}_i^T \boldsymbol{\beta} + \epsilon_i)} + 1} \tag{36}$$

where $\boldsymbol{\beta} = (\beta_1 \dots \beta_p)^T$ are regression coefficients, and ϵ_i is the error term. Let

$$t = \sum_{j=1}^{p} \beta_j x_{ij} + \epsilon_i = \mathbf{x}_i^T \boldsymbol{\beta} + \epsilon_i \quad for \quad i = 1, 2, \dots n \tag{37}$$

then $y_i = f(t)$, where $f(t)$ is the logistic function

$$f(t) = \frac{1}{e^{-t} + 1} \tag{38}$$

A property of the logistic function is like distribution functions, its output is between 0 and 1 for any input in the full range from negative infinity to positive infinity, i.e., $f(t) \in [0, 1]$ for $t \in (-\infty, \infty)$. The coefficients are usually estimated with maximum likelihood estimation with iterative algorithms such as Newton's Method. Once the coefficients are learned, the Logistic Regression can be used for binary classification where the predicted value $\hat{y}_i$ is the probability of being "yes".

1.5.2. *Regularized Least Squares*

Unlike many other regression models, such as the logic regression, the Regularized Least Squares (RLS) method does not require the examples to be represented as feature vectors explicitly as it learns the model and makes prediction with a kernel matrix $\mathbf{K}$, where each entry $k_{ij} \in K = \kappa(\mathbf{x}_i, \mathbf{x}_j)$ is defined by a certain kernel function, e.g., Gaussian kernel in (18). For a dataset with labels $\mathbf{y} = (y_1, y_2, \dots, y_n)^T$, and kernel matrix $\mathbf{K}$, the Regularized Least Squares (RLS) is to find coefficients $\mathbf{c} = (c_1, c_2, \dots, c_n)^T$ to minimize the following value

$$\frac{1}{2} \|\mathbf{y} - \mathbf{K}\mathbf{c}\|_2^2 + \frac{\delta}{2} \mathbf{c}^T \mathbf{K} \mathbf{c} \tag{39}$$

where the first term is the least squares term and the second term is the regularization term with weight δ. The solution of $\mathbf{c}$ that minimizes the above value has a simple closed form as below

$$\mathbf{c} = (\mathbf{K} + \delta I)^{-1} \mathbf{y} \tag{40}$$

Once $\mathbf{c}$ is obtained, we can use it to predict the label $\hat{y}$ of a new data object $\hat{\mathbf{x}}$ by

$$\hat{y} = \hat{\mathbf{k}}_c^T = \hat{\mathbf{k}}^T (k + \delta \mathbf{I})^{-1} \mathbf{y} \tag{41}$$

$\hat{\mathbf{k}}$ is an n-dimensional vector where each dimension $\hat{k}_i$ is the value of the kernel function between this object and a training example, i.e., $\hat{k}_i = \kappa(\hat{\mathbf{x}}, \mathbf{x_i})$.

In real applications, the similarity matrix recording a certain type of similarity between each pair of examples may be treated as a kernel matrix. Since kernel matrix is positive definite, some preprocessing may be needed to transform the given similarity matrix into a positive definite matrix.

1.6. *Ensemble classifier*

An ensemble classifier is not a specific type of classifier as those introduced earlier. Instead, it is a classifier ensemble, which combines or aggregates the predictions of several individually trained classifiers called base classifiers to produce a final result. A simple ensemble classifier is illustrated in Figure 8. Through aggregating, the prediction of an ensemble classifier is usually more accurate than any of the individual classifiers. An important problem is how to train each of the base classifiers. Since ensemble makes sense only if the outputs of the base classifiers are different. To generate disagreements in the prediction, base classifiers may be trained with different ways, e.g. initial weights, different parameters, different subsets of features, and different portions of the training set. The two well-known ensemble methods *Bagging*[21] and *Boosting*[22–24] mainly focus on the last way to train the base classifiers, and another popular method *Random Forest*[25] makes use of the last two strategies.

In the *Bagging* method, each classifier is trained on a random sample of the training set. More specifically, a set of samples to be used for training a base classifier is generated by randomly drawing with replacement from the training samples. Although each individual classifier could result in higher test-set error when trained with a subset of training samples, the combination of them could produce lower test-set error than using a single classifier trained with all the training samples. Breiman[21] showed that *Bagging* is effective on "unstable" learning algorithms, such as decision tree and neural network, where small changes in the training set result in large changes in predictions. Unlike *Bagging*, where the generation of training set for one classifier is independent on other classifiers, in *Boosting*, the training set used for each base classifier is chosen based on the performance of the earlier classifiers. Examples that were incorrectly predicted by previous classifiers are selected more often than those were correctly predicted. By that way, Boosting attempts to make subsequent classifiers be able to more accurately predict examples for which the current ensemble's

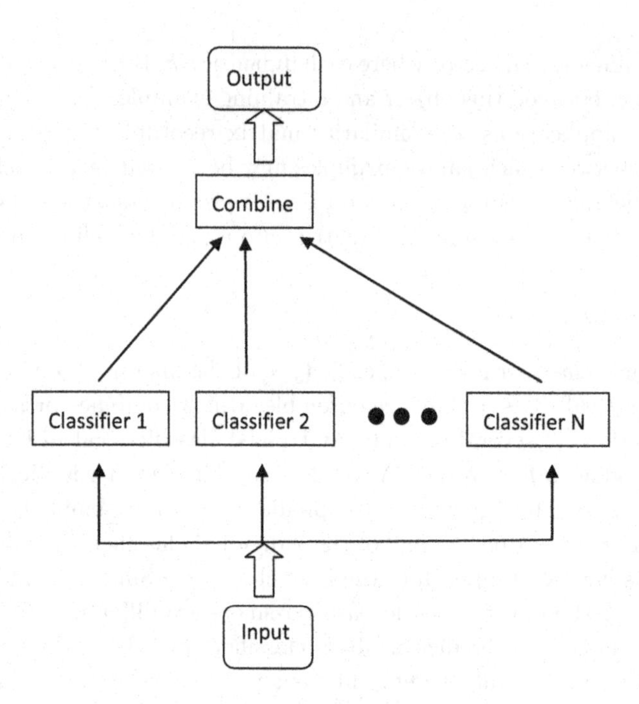

Fig. 8. An ensemble of classifiers.

performance is poor. The *Random Forest* combines the *Bagging* idea to select training samples and features. The selection of a random subset of features is an example of the random subspace method as discussed by Ho,[26] which is especially useful for handling high-dimensional data, e.g., gene expression data. It projects the original high dimensional space into different low subspaces so that the problems caused by high-dimensionality are avoid. Although decision tree is often used as base classifiers in these ensemble methods, other types of classifiers may also be used to produce base predictions in an ensemble.

Once all the base classifiers are trained, they generate predictions for new samples to be classified. Voting strategy is a commonly used way to combine these predictions to give the final class label for the input. Assuming that the majority of the classifiers would make the correct prediction, voting labels the sample as the class that is predicted by most of the base classifiers. Instead of equally weighting the classifiers, the aggregating weight for each base classifier may also be adapted according to their performance.[24]

2. Drug-target interaction prediction

Empowered by advances in genomic research together with efforts made in chemical and pharmacological research, large-scale public databases of different types of information are now available to be used for drug discovery with system level analysis.

In this section, we take the drug-target interaction prediction as an example to present detailed discussion on the use of classification technique for handling this specific task in computational biology. Some background knowledge of drug-target interaction prediction is first given. After that, recent studies on drug-target interaction prediction with classification are discussed. Finally, experimental studies on benchmark datasets are given to evaluate the performance of several different classification approaches in drug-target interaction prediction.

2.1. *Background*

A classical way of drug discovery is to screen drug candidates, i.e., ingredients from approved drugs, to identify those that have a desirable therapeutic effect. Advances in biotechnology in recent years open the way to study genome-wide responses of chemical compounds with high throughput screening of large compound libraries against molecular targets. These targets are the products of genes with a desirable phenotype after perturbation. Identification of drug-target interaction is an important part of the drug discovery pipeline. The great advances in molecular medicine and the human genome project provide more opportunities to discover unknown associations in the drug-target interaction network. These new interactions may lead to the discovery of new drugs and also are useful for helping understand the causes of side effects of existing drugs. However, using experimental way to determine drug-target interactions is costly and time-consuming. The increasing accessibility of large and diverse databases makes it possible to discover new knowledge from the existing data with statistics and computer-based data analysis algorithms. Such kind of method, called *in silico* method, becomes a potential complementary way to experimental method.

In *silico* approaches for prediction of drug-target interactions are generally categorized into drug-based approaches and target-based approaches. Drug-based approaches screen candidate drugs, compounds or ligands to predict whether they interact with a given target based on the assumption that similar drugs share the same target. The similarity of drugs is measured in different ways with respect to different aspects. Martin *et al.*[27] compared drugs according to their chemical structures, Campillos *et al.*[28] proposed to use side-effect to measure the similarity between drugs. Assuming that similar targets bind to the same ligand, target-based approaches, on the other hand, compare proteins to predict whether they bind to the given ligand, or whether they are the targets of the given drug or compound. More specifically, for a given drug, new targets are identified by comparing candidate proteins to the known targets of this drug with respect to certain descriptors such as amino acid sequence, binding sites, or ligands that bind to them. Haupt and Schroeder[29] reviewed computational methods to find new targets for already approved drugs for the treatment of new diseases based on the structural similarity of their binding sites. Candidate targets are compared by the chemical similarity of ligands that bind to them.[30] Different from these classic drug-based or target-based approaches, chemogenomics approaches have been proposed to consider the interactions between drugs and a protein family rather than a single target.[31–34]

Recently, machine learning approaches have been applied to this task to explore the whole interaction space. In the supervised bipartite graph learning approach,[12] the chemical space and the geometric space are mapped into a unified space so that those interacting drugs and targets are close to each other while those non-interacting drugs and targets are far away from each other. After the mapping function to such a unified space is learned, the query pair of drug and target are mapped in the same way to that unified space, and the probability of interaction between them is their closeness in the mapped space. It has been shown that the combination of supervised learning based on drug and target independently performs very well.[13] This approach is called the Bipartite Local Model (BLM). For a query pair of drug and target, a model of the query drug is learned with a certain classifier based on the information of its known targets. Then the probability of interaction between this drug and the query target is predicted with this model. The same procedure is applied to obtain the probability of interaction between them from the target side. Finally, an overall probability of interaction for the query pair is calculated by combing these two probabilities. It has been reported that the result based the knowledge of both directions, i.e., from the drug side and from the target side, is much better than those based on each single one. The same idea is adopted by another two following works.[35,36] Semi-supervised approach is used instead of supervised approach to learn the local model by Xia *et al.*[35] Laarhoven[36] found that using the kernel based on the topology of the known interaction network alone is able to obtain a very good performance, although together with other types of similarities can further improve the results. Other than using one type of drug-drug similarity and one type of target-target similarity, Perlman *et al.*[37] used multiple types of drug-drug similarities and target-target similarities and combined them as features to describe each drug-target pair for model learning. Next, we present the details of how the drug-target prediction task is handled by three types of classification problems.

2.2. *A binary classification problem*

A relatively straightforward way to predict whether a given pair of drug-target interact is to model it as a binary classification problem.[37] The key problem is how to extract a set of features based on different biological data sources to characterize or represent each drug-target pair. Perlman *et al.*[37] proposed to do this in three steps. First, five drug-drug similarities and three gene-gene similarities are calculated based on different biological and chemical sources. Then, the drug and gene similarities are combined as features to describe each drug-target pair. Feature selection is performed to select important features. Finally, the classifier, e.g., Logistic Regression, is trained with the labeled samples which are described by those selected features.

The whole process is shown in Fig. 9.[37] The drug-drug similarities were computed using chemical structure, Ligand, drug side effects, drug response gene expression profiles, and the Anatomical, Therapeutic and Chemical (ATC) classification system code. The gene-gene similarity measures used are based on protein-protein interactions, sequence, and Gene Ontology (GO). Once all these drug-drug similarities and target-target similarities are obtained, each feature is constructed based on one type of drug-drug similarity and one type of target-target similarity. Specifically, each feature is calculated by combining the drug-drug similarities between the query drug and other drugs and the gene-gene similarities between the query gene and other target genes across all true drug-target associations. Therefore, fifteen features are constructed in such a way. After feature selection, ten features are finally selected. Table 2 shows the prediction results in terms of AUC (area under ROC curve) and AUPR (area under precision-recall curve) with all the features, all the selected features and each single selected feature. Here AUC and AUPR are two performance evaluation

Table 2. Comparison of AUC and AUPR for the four datasets

	All features	0.905	0.935
	selected features	0.908	0.935
Ligand	Sequence similarity	0.851	0.867
Ligand	GO semantic similarity	0.845	0.867
Predicted Side Effect	GO semantic similarity	0.832	0.863
ATC similarity	GO semantic similarity	0.81	0.858
Ligand	PPI closeness	0.809	0.844
Chemical	GO semantic similarity	0.805	0.84
ATC similarity	PPI closeness	0.762	0.809
Chemical	Sequence similarity	0.749	0.763
Predicted Side Effect	PPI closeness	0.729	0.759
Co-expression	Sequence similarity	0.724	0.748

measures. It is shown that using ten selected features gives a comparable result with using all the fifteen features, which is much better than using any of a single feature. It is also shown that when used individually, the combination of Ligand and sequence similarity gives the best feature. Once each drug-target pair is represented as a vector of these features, the prediction problem of whether a query pair interacts simply becomes a binary classification problem that can be solved by many existing classification algorithms, e.g., the Logistic Regression as used in this paper. Other than developing a good data presentation through aggregation of multiple data sources, some other studies focus more on design of new learning algorithms. Next we introduce two recently proposed learning algorithms, namely the Bipartite Graph Learning and the Bipartite Local Model.

2.3. *Bipartite graph model (BGM)*

We assume that the problem under consideration is to predict new interactions between n_d drugs and n_t targets. An $n_d \times n_t$ matrix $\mathbf{A}$ is used to record these known interactions, i.e., $a_{ij} \in \mathbf{A} = 1$ if the ith drug denoted as d_i, is known to interact with the jth target denoted as t_j. All other entries of $\mathbf{A}$ are 0. Assume that n_i interactions in total involve m_d drugs and m_t targets and $m_d < n_d$ and $m_t < n_t$. This means there are some new drug and target candidates of which the corresponding rows and columns of $\mathbf{A}$ are all 0. Other than the interaction network, we also have the chemical similarity matrix of drug and the sequence similarity matrix of target, denoted as Sd and St, respectively.

The bipartite graph learning method learns the correlation between the chemical/genomic space and the interaction space, which is called the 'pharmacological space'. As illustrated in Fig. 10,[12] first, the compounds and

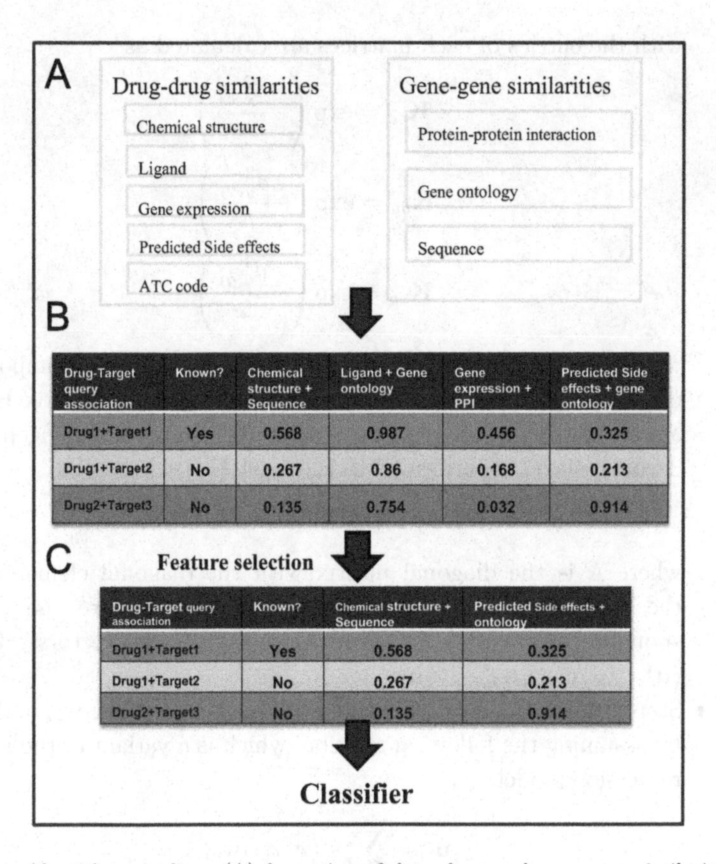

Fig. 9. Algorithm pipeline. (A) formation of drug-drug and gene-gene similarity matrices, (B) integration of the similarities to classification features, (C) classification with feature selection.

proteins are embedded into a unified space called ‘pharmacological space’. The mapping function or model between the chemical/genomic space and the pharmacological space is learned. With this model, any query pair of compound and protein are mapped onto the same pharmacological space. The compound-protein pairs under testing are predicted to be interacting if the two are closer than a threshold in the pharmacological space. The whole process consists of the following steps:[12]

- Step 1: construct a graph-based similarity matrix

$$\mathbf{K} = \begin{pmatrix} \mathbf{K}_{cc} & \mathbf{K}_{cg} \\ \mathbf{K}_{cg}^T & \mathbf{K}_{gg} \end{pmatrix} \tag{42}$$

with the entries of each matrices are calculated as

$$\mathbf{K}_{cc} = \exp\left(-\frac{d_{c_i c_j}^2}{h^2}\right) \tag{43}$$

$$\mathbf{K}_{gg} = \exp\left(-\frac{d_{g_i g_j}^2}{h^2}\right) \tag{44}$$

$$\mathbf{K}_{cg} = \exp\left(-\frac{d_{c_i g_j}^2}{h^2}\right) \tag{45}$$

where d is the shortest distance between two objects (compounds or proteins) on the bipartite graph. The symmetric matrix $\mathbf{K}$ has a scale of $(n_c + n_d) \times (n_c + n_d)$. After $\mathbf{K}$ is constructed, eigenvalue decomposition is performed to $\mathbf{K}$ to get $\mathbf{U}$:

$$\mathbf{K} = \Gamma\Lambda^{1/2}\Lambda^{1/2}\Gamma^T = \mathbf{U}\mathbf{U}^T \tag{46}$$

where Λ is the diagonal matrix with the diagonal elements are the eigenvalues and the columns of matrix Γ are the corresponding eigenvectors. Write $\mathbf{U}$ with its row vectors: $\mathbf{U} = (\mathbf{u}_{c_1}, \ldots, \mathbf{u}_{cn_c}, \mathbf{u}_{g_1}, \ldots, \mathbf{u}_{gn_g})^T$.

- Step 2: For $i = \{1, \ldots, n_c\}$ and $j = \{1, \ldots, n_g\}$, learn $\mathbf{w}_{ci}$ and $\mathbf{w}_{gj}$ by assuming the following relation, which is a variant of the kernel regression model:

$$\mathbf{u}_{ci} = \sum_{i=1}^{n_c} s_c(x, x_{ci})\mathbf{w}_{ci} + \epsilon \tag{47}$$

$$\mathbf{u}_{gj} = \sum_{j=1}^{n_g} s_g(x, x_{gj})\mathbf{w}_{gj} + \epsilon \tag{48}$$

- Step 3: map the query compound c_q and protein g_q with learned $\mathbf{W}_c$, and $\mathbf{W}_g$:

$$\mathbf{u}_{cq} = \sum_{i=1}^{n_c} s_c(c_q, c_i)\mathbf{w}_{ci} \tag{49}$$

$$\mathbf{u}_{gq} = \sum_{j=1}^{n_g} s_g(g_q, g_j)\mathbf{w}_{gj} \tag{50}$$

- Step 4: The score of interaction between c_q and g_q denoted as p_{c_q, g_q} is calculated as the inner product of the feature vectors in the mapped space

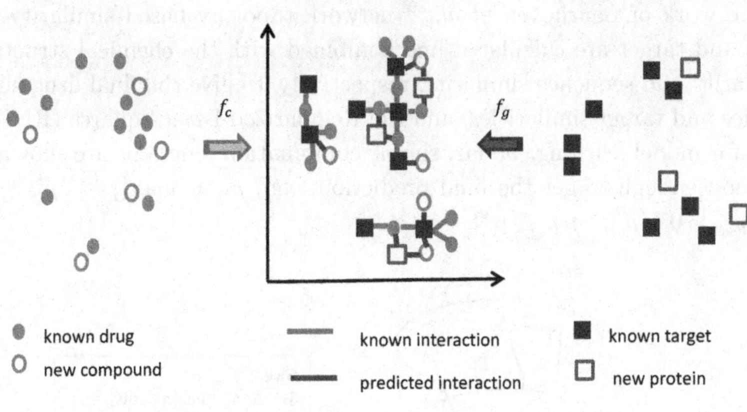

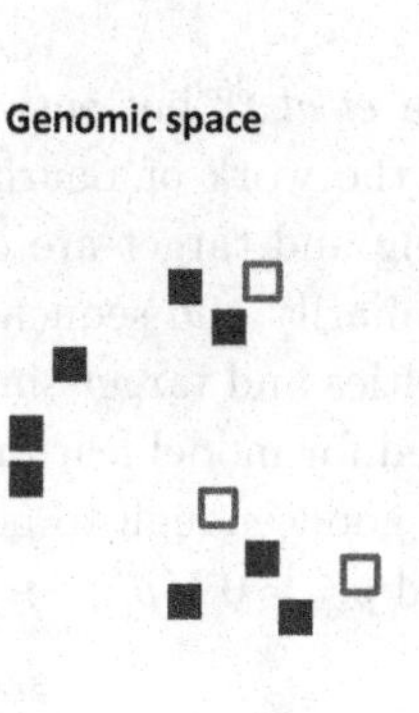

Fig. 10. Bipartite graph model.

$$p_{c_q,g_q} = < \mathbf{u}_{cq}, \mathbf{u}_{gq} > \tag{51}$$

2.4. *Bipartite local model (BLM)*

To predict p_{ij}, the probability that a drug d_i and a target t_j interacts, the basic bipartite local model is described as follows. A local model of d_i is first learned based on the known targets of this drug and the similarities between these targets. This model is then used to predict $p_{ij}^{d\to t}$ the probability of interaction between this drug to the tested protein. The model learning and prediction process is performed independently from the query target side to get $p_{ij}^{t\to d}$. After that, they are combined with some function f to get the final result $p_{ij} = f(p_{ij}^{d\to t}, p_{ij}^{t\to d})$. Fig. 11 illustrates the idea of drug-target interaction prediction with learning from the drug and target independently.

This framework was first proposed by Bleakley and Yamanishi,[13] and then was further extended in Ref.[35] and Ref.[36] Under the same BLM framework, different results may be produced due to the differences in drug-drug similarity $\mathbf{S}_d$ and target-target similarity $\mathbf{S}_t$, the classifier, and the way how $p_{ij}^{d\to t}$ and $p_{ij}^{t\to d}$ is combined, i.e., the function f. For example, Bleakley and Yamanishi[13] used Support Vector Machine (SVM) as the classifier with chemical structure similarity for drug and sequence similarity for protein targets, respectively. The same similarities were used by

Xia *et al.*,[35] but with a semi-supervised approach for local model learning. In the work of Laarhoven *et al.*,[36] network topology based similarity for drug and target are calculated and combined with the chemical structure similarity and sequence similarity, respectively, to give the final drug similarities and target similarities, and the Regularized Least Squares (RLS) is used for model learning. So far, simple combination functions are shown to be good enough to get the final prediction, e.g., $p_{ij} = \max\{p_{ij}^{d\rightarrow t}, p_{ij}^{t\rightarrow d}\}$,[13] and $p_{ij} = 0.5(p_{ij}^{d\rightarrow t} + p_{ij}^{t\rightarrow d})$.[36]

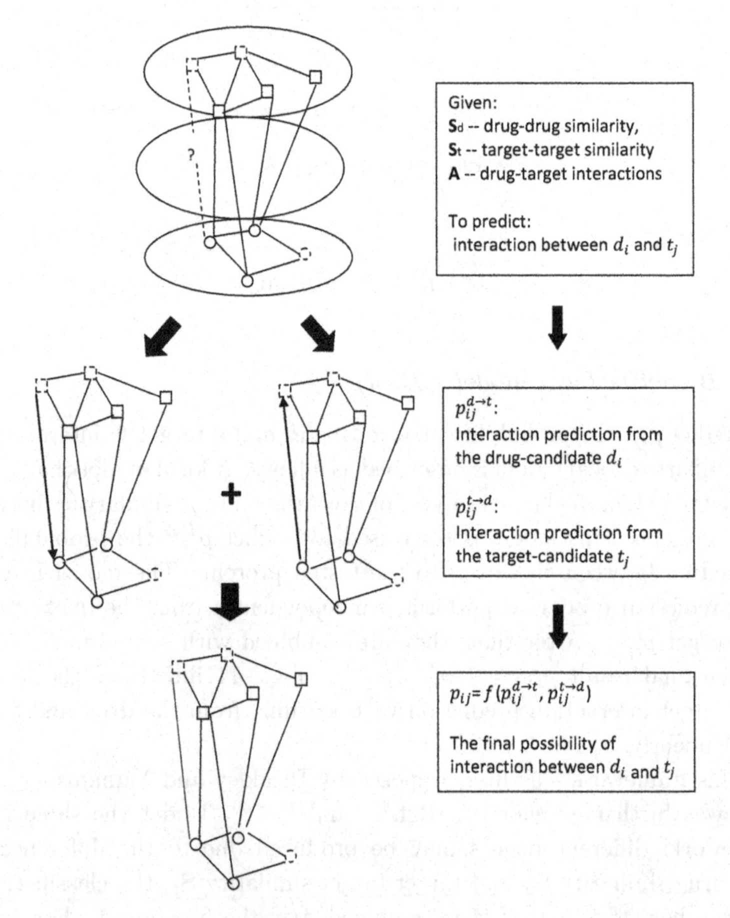

Given:
Sd -- drug-drug similarity,
St -- target-target similarity
A -- drug-target interactions
To predict:
interaction between d_i and t_j
?
+
$p_{ij}^{d\rightarrow t}$:
Interaction prediction from the drug-candidate d_i
$p_{ij}^{t\rightarrow d}$:
Interaction prediction from the target-candidate t_j
$p_{ij}=f(p_{ij}^{d\rightarrow t}, p_{ij}^{t\rightarrow d})$
The final possibility of interaction between d_i and t_j

Fig. 11. Drug-target interaction prediction with learning from the drug and target independently.

2.5. *Enhanced BLM with training data inferring for new drug/target candidates*

Generally, supervised learning performs better than unsupervised learning. However, a good performance of supervised learning is largely dependent on the amount and quality of the labeled training data. When the drug candidate is new, it has no existing targets that can be used as positive labeled training data and the model for this drug thus cannot be learned. Similarly, supervised local model learning does not work for new target candidates. To extend the application domain of BLM to new drug and target candidates, we presented a training data inferring procedure and integrated it into BLM.[14] Based on the assumption that drugs which are similar to each other interact with the same targets, training data for a new drug candidate could be possibly inferred from its neighbors. The neighbors of a new drug candidate generally refer to those drugs that share some similar properties with the new drug candidate, e.g. similar in chemical structure.

For a drug candidate d_i that has no known targets, we infer the weighted interaction profile for d_i with the following formula

$$\mathbf{l}(i) = \mathbf{s}_i^d \mathbf{A} \tag{52}$$

where each dimension

$$l_j(i) = \sum_{h=1}^{n_d} s_{ih}^d a_{hj} \tag{53}$$

Here vector $\mathbf{s}_i^d$ is the ith column of $\mathbf{S}_d$, which records the similarities between d_i and all the other drugs, s_{ih}^d is the similarity between two drugs d_i and d_h, and vector $\mathbf{l}(i)$ is the inferred interaction profile for d_i, where each dimension $l_j(i)$ corresponds to the weight of the interaction between d_i and t_j. The above formula shows that the interaction weight of d_i with respect to the jth target is the sum of interactions between its neighbors and this target weighted by the similarity between this drug and its neighbors. More specifically, this simple formula defines that for a given new drug candidate d_i, its weight of interaction with respect to a target is high if many of its neighbors interact with this target, and the final weight to a target is influenced more by a neighbor with a larger similarity than those with smaller similarities. To allow neighbors with large similarities only to contribute, a threshold may be used to reduce the impact of those non-important neighbors to 0. Alternately, an exponential function with bandwidth β given as below may be applied:

$$\mathbf{l}(i) = e^{(\mathbf{s}_i^d/\beta)}\mathbf{A} \tag{54}$$

To ensure the value of each $l_j(i)$ is in the range of [0, 1], linear scale is performed subsequently. The procedure of inferring training data for new target candidates is not discussed in details here as it is similar to the procedure of inferring training data for new drug candidates as presented above.

Learning from neighbors allows drugs and targets to obtain training data when themselves do not have any known interactions. This procedure actually introduces some degree of globalization into the original local model to give more chances or an enlarged scope for the learning process. However, too much globalization is not desired as it will decrease the local characteristics and make the models of each drug or target less discriminative. Moreover, the low quality of neighbors may add in noise and cause a negative impact when neighbors' preferences are too much relied upon. In the current study, we only activate the neighbor-based training data inferring for totally new candidates. For other cases, we still train the model locally on its own preference, i.e., the known interactions.

3. Experimental study

Now we give some experimental results to compare the performance of the BGM method, the BLM method and the BLMN method (BLM with neighbor-based training data inferring) for the task of drug-target interaction prediction. From the experimental results, we have the following observations: first, BLM-based approaches outperform BGM; second, with neighbor-based training data inferring, BLMN performs better than the classic BLM; third, network topology based similarity is helpful to improve the prediction.

3.1. *Datasets*

We are the four groups of datasets that have been first analysed by Yamanishi and Araki[12] and then later by several other researchers.[13,35,36,38] These four datasets correspond to drug-target interactions of four important categories of protein targets, namely enzyme, ion channel, G-protein-coupled receptor (GPCR) and nuclear receptor, respectively [a]. Table 3 gives some statistics of each of the datasets.

[a]The datasets were download from http://web.kuicr.kyoto-u.ac.jp/supp/yoshi/drugtarget/

Table 3. Some statistics of the four datasets. n_d: the total number of drugs, n_t: the total number of targets, E: the total number of interactions, $\bar{D}_d$: the average number of targets for each drug, $\bar{D}_t$: the average number of targeting drugs for each target, $D_d = 1$: the percentage of drugs that have only one target, and $D_t = 1$: the percentage of targets that have one targeting drug.

Dataset	Enzyme	Ion Channel	GPCR	Nuclear Receptor
n_d	445	210	223	54
n_t	664	204	95	26
E	2926	1476	635	90
$\bar{D}_d$	6.58	7.03	2.85	1.67
$\bar{D}_t$	4.41	7.24	6.68	3.46
$D_d = 1(\%)$	39.78	38.57	47.53	72.22
$D_t = 1(\%)$	43.37	11.27	35.79	30.77

Each dataset is described by three types of information in the form of three matrices. Together with the drug-target interaction information, the drug-drug similarity, and target-target similarity are also available. Four interaction networks were retrieved from the KEGG BRITE,[39] BRENDA,[40] SuperTarget[41] and DrugBank.[42] The drug-drug similarity is measured based on chemical structures from the DRUG and COMPOUND sections in the KEGG LIGAND database[39] and is calculated with SIMCOMP.[43] The target-target similarity is measured based on the amio acid sequences from the KEGG GENES database[39] and is calculated with a normalized version of Smith-Waterman score.

3.2. *Approaches compared*

We compare the following approaches:

- BGM:[12] Bipartite graph model;
- BY(2009):[13] Bipartite local model;
- Laarhoven *et al.* (2011):[36] Bipartite local model with network-based similarity;
- BLM: Ignoring 'new candidate' in BLMN;
- BLMN: BLM with neighbor-based training data inferring .

Among the above methods, BGM requires eigendecomposition of a $(n_c + n_d) \times (n_c + n_d)$ matrix, which is computational consuming for large datasets. The BY(2009), Laarhoven *et al.* (2011) and BLM are three variants of the classic BLM method, which is not applicable to new candidates. BLMN is the modified BLM method which can be used to predict the interaction between any compounds and proteins.

3.3. *Evaluation*

Leave-one-out cross validation (LOOCV) is performed. In each run of prediction, one drug-target pair is left out by setting the corresponding entry of matrix **A** to 0. Then we try to recover its true value using the remaining data. We measure the quality of the predicted interaction matrix **P** by comparing it to the true interaction matrix **A** in terms of the area under ROC curve or true positive rate (TPR) vs. false positive rate (FPR) curve (AUC) and the area under the precision vs. recall curve (AUPR). TPR is equivalent to recall. Assume that TP, FP, TN, FN represent true positive, false positive, true negative, and false negative, respectively, then

$$TPR/recall = \frac{TP}{TP + FN} \tag{55}$$

$$FPR = \frac{FP}{FP + TN} \tag{56}$$

$$precision = \frac{TP}{TP + FP} \tag{57}$$

Since in the current task, the known interactions are much less than those unknown ones, the precision-recall curve should be a better measurement than the ROC curve here as has been discussed in by Davis and Goadrich.[44]

3.4. *Performance comparison*

Table 4 gives the AUC and AUPR scores of the five approaches on the four datasets. The results of BGM, BY (2009), and Laarhoven *et al.* (2011) are the best ones reported in Ref.[13] and Ref.[36] Both BLMN and BLM are run with three different groups of inputs: *Chem-Seq*, *Network-based*, and *Hybrid*. *Chem-Seq* denotes that chemical similarity is used for drug and sequence similarity is used for target; *Network-based* denotes that the drug-drug similarity and target-target similarity are derived from the existing interaction network; *Hybrid* denotes that the drug-drug similarity and target-target similarity are combinations of the two types of similarities.

It is shown from the table that with a low time complexity, four BLM-based approaches, including three BLM variants and BLMN, produce better results than the BGM method. Among the three BLM variants, the results of BLM and BY(2009) with *Chem-Seq* are similar as the only difference between them is the former use RSL as the classifier while the later use SVM. The results of BLM and Laarhoven *et al.* (2011) with *Network-based* are also close in most of the cases although the later used Kronecker product, which is a more complicated way to combine two types of similarities. In all the cases, BLMN produced better results than the three classic BLM algorithms. This clearly show that neighbor-based training data inferring is very useful for improving the final result when the dataset contains new drug/target candidates.

Despite the consistent improvements of BLMN compared to the other three on all the four datasets, the amounts of improvements differ for different datasets. If we compare the improvements of the proposed approach over the four datasets, it is seen that the improvement with respect to BLM on Nuclear Receptor is the most significant while the improvement on Enzyme and Ion Channel are not so significant. Such kind of difference in performance of the proposed approach is consistent with our expectation according to the difference in the structure of the datasets. Although all the datasets do not contain new drug/target candidates, in our experiment, the real interaction to be predicted is left out. This means drugs and targets with degree equal to 1 turn out to have no positive training data and thus they are simulated to be "new" in the experiments. As shown in Table 3, Nuclear Receptor has a much larger portion of "new" drugs and targets than Ion Channel. Therefore, it has more chances for BLMN to improve the result for Nuclear Receptor where the training data inferring is applied more frequently.

It is also observed that although network-derived similarity alone provides good information, combining biological information can further improves the result especially when the network is sparse, e.g., the results of both BLM and BLMN for Ion Channel with only *Network-based*is very close to those with *Hybrid* while significant improvements are achieved for both approaches on Nuclear Receptor when *Chem-Seq* is further combined with *Network-based* similarity. This shows that combining multiple types of similarities usually gives better results when no single type of similarity is good enough.

4. Summary

Computational method becomes rather important in cases where the data to be analysed are too large to be efficiently handled with traditional experimental method. The data is said to be large if it contains a large number of samples, described by a large number of features, or need to be integrated from several sources. Classification is an important tool that has been studied extensively for data analysis with automatic prediction. Many

Table 4. Comparison of AUC and AUPR for the four datasets

Dataset	Data	Method	AUC	AUPR
Enzyme	*Chem-Seq*	BGM	96.7	83.1
		BY(2009)	97.6	83.3
		BLM	96.1	85.8
		BLMN	98.0	87.3
	Network-based	Laarhoven *et al.* (2011)	98.3	88.5
		BLM	98.2	88.0
		BLMN	99.1	93.1
	Hybrid	Laarhoven *et al.* (2011)	97.8	91.5
		BLM	98.2	91.3
		BLMN	98.8	92.9
Ion Channel	*Chem-Seq*	BGM	96.9	77.8
		BY(2009)	97.3	78.1
		BLM	97.0	81.9
		BLMN	97.8	84.6
	Network-based	Laarhoven *et al.* (2011)	98.6	92.7
		BLM	98.5	92.5
		BLMN	99.0	95.6
	Hybrid	Laarhoven *et al.* (2011)	98.4	94.3
		BLM	98.5	92.7
		BLMN	99.0	95.0
GPCR	*Chem-Seq*	BGM	94.7	66.4
		BY(2009)	95.5	66.7
		BLM	95.1	68.1
		BLMN	98.1	78.8
	Network-based	Laarhoven *et al.* (2011)	94.7	71.3
		BLM	94.4	70.6
		BLMN	97.5	84.6
	Hybrid	Laarhoven *et al.* (2011)	95.4	79.0
		BLM	95.7	76.2
		BLMN	98.4	86.5
Nuclear Receptor	*Chem-Seq*	BGM	86.7	61.0
		BY(2009)	88.1	61.2
		BLM	86.9	58.4
		BLMN	96.9	80.7
	Network-based	Laarhoven *et al.* (2011)	90.6	61.0
		BLM	90.9	62.9
		BLMN	95.7	80.7
	Hybrid	Laarhoven *et al.* (2011)	92.2	68.4
		BLM	94.0	72.4
		BLMN	98.1	86.6

biology problems involve prediction tasks where the technique of classification may be applied. Binary classification problem with 'positive' or 'negative' prediction is especially common in biology data analysis. We have introduced several popular supervised learning methods for classification including popular classification methods, regression models used for classification, and ensemble classification. We give more detailed discussion of how different classification methods can be used for drug-target interaction prediction. Experimental studies are given to compare the performance of different approaches with benchmark datasets.

Since no single classification algorithm is always better than others, how to select a proper one to use for real-world applications is important. A number of factors need to be considered when deciding which classification algorithm to use for a particular problem. Here we only highlight some of the most important ones. The mathematical representation of the data is one important factor to be considered. In some cases, the data is represented as vectors described by a set of features, and in some other cases, it is represented as pairwise similarities or relationships. Many algorithms such as SVM, Logistic Regression, and Decision tree deal with feature vector data. Further concerns need to be taken on the type of features, i.e., numerical or categorical, and the number of features, because each algorithm is usually designed to be applicable to some certain types of features, and some algorithms are only suitable for handling low-dimensional data. In cases that the data are characterized by pairwise relationships, e.g., the PPI data, or the drug-drug / target-target similarities, Kernel SVM should be used instead of SVM. Other than the data representation, the characteristics of the data also directly affect the performance of the algorithm. For example, for drug-target interaction prediction, if the data contain new drug/target candidates, it is better to use algorithms that have the ability to handle these new drug/target candidates, e.g., the BLM with neighbor-based training data inferring. Finally, the chosen algorithm should be with a proper scalability to be applied to real-world problem, where the scale of the data may be very large.

Other than the learning method, the classification result is also highly dependent on the amount and quality of the given training data and the way the data represented, e.g., a set of selected features or well defined similarity measures. Given the same set of training data, a good data representation with a simple classifier may already produces a good result. Nevertheless, with the same data representation, an advanced classification algorithm is able to make use of it more effectively and hence produces a better result.

References

1. J. Han, M. Kamber, and J. Pei, *Data Mining: Concepts and Techniques*, 3 edn. Morgan Kaufmann (2012).
2. X. Chen, M. Wang, and H. Zhang, The use of classification trees for bioinformatics, *Wiley Interdisciplinary Reviews: Data Mining and Knowledge Discovery*. **1**(1), 55–63 (2011).
3. E. C. Uberbacher and R. J. Mural, Locating protein-coding regions in human dna sequences by a multiple sensor-neural network approach, *Proc. Nati. Acad. Sci. USA*. **88**, 11261–11265 (1991).
4. S. Salzberg, Locating protein coding regions in human dna using a decision tree algorithm, *J. Comput. Biol.* **2**, 473–485 (1995).
5. R. Gupta, P. Wikramasinghe, A. Bhattacharyya, F. A. Perez, S. Pal, and R. V. Davuluri, Annotation of gene promoters by integrative data-mining of chip-seq pol-ii enrichment data, *BMC Bioinformatics*. **11(Suppl 1)**, S65 (2010).
6. K. M. Borgwardt, C. S. Ong, S. Schonauer, S. V. N. Vishwanathan, A. J. Smola, and H.-P. Kriegell, Protein function prediction via graph kernels, *Bioinformatics*. **21 (Suppl. 1)**, i47–i56 (2005).
7. W. A. McLaughlin and H. M. Berman, Statistical models for discerning protein structures containing the dnabinding helix-turn-helix motif, *J Mol Biol*. **330**, 43–55 (2003).
8. F. Mordelet and J.-P. Vert, SIRENE: supervised inference of regulatory networks, *Bioinformatics*. **24**(16), i76–i82 (2008).
9. L. Cerulo, C. Elkan, and M. Ceccarelli, Learning gene regulatory networks from only positive and unlabeled data, *BMC Bioinformatics*. **11** (2010).
10. Y. Qi, Z. Bar-Joseph, and J. Klein-Seetharaman, Evaluation of different biological data and computational classification methods for use in protein interaction prediction, *Proteins*. **63**, 490–500 (2006).
11. P. Yang, X.-L. Li, J.-P. Mei, C.-K. Kwoh, and S.-K. Ng, Positive-unlabeled learning for disease gene identification, *Bioinformatics*. **28**(20), 2640–2647 (2012).
12. Y. Yamanishi, M. Araki, A. Gutteridge, W. Honda, and M. Kanehisa, Prediction of drug-target interaction networks from the integration of chemical and genomic spaces, *Bioinformatics*. **24**, i232–i240 (2008).
13. K. Bleakley and Y. Yamanishi, Supervised prediction of drug-target interactions using bipartite local models, *Bioinformatics*. **25**(18), 2397–2403 (2009).
14. J.-P. Mei, C.-K. Kwoh, P. Yang, X.-L. Li, and J. Zheng, Drug-target interaction prediction by learning from local information and neighbors, *Bioinformatics*. **29**(2), 238–245 (2013).
15. Z. L. Zhou and C. K. Kwoh, The nearest feature midpoint - a novel approach for pattern classification, *International Journal of Information Technology*. **11**(1), 1–15 (2011).
16. G. L. Zhang, I. Bozic, C. K. Kwoh, T. August, and V. Brusic, Prediction of supertype-specifc hla class i binding peptides using support vector machines, *Journal of Immunological Methods*. **320**, 143–154 (2007).
17. C. K. Kwoh and D. F. Gillies, Using hidden nodes in bayesian networks, *Artificial Intelligence*. **88**(1-2), 1–38 (1996).

18. J. R. Quinlan, Induction of decision trees, *Mach. Learn.* **1**, 81–106 (1986).
19. J. R. Quinlan, *C4.5: Programs for Machine Learning.* Morgan Kaufmann, San Mateo, CA (1993).
20. L. Breiman, J. Friedman, R. Olshen, and C. Stone, *Classification and Regression Trees.* CRC Press, Boca Raton, FL (1984).
21. L. Breiman, Bagging predictors, *Machine Learning.* **24**(2), 123–140 (1996).
22. R. Schapire, The strength of weak learnability, *Machine Learning.* **5**(2), 197–227 (1990).
23. L. Breiman. Bias, variance, and arcing classifiers. Technical Report 460, UC-Berkeley, Berkeley, CA (1996).
24. Y. Freund and R. E. Schapire, A decision-theoretic generalization of on-line learning and an application to boosting, *Journal of Computer and System Science.* **55**, 119–139 (1997).
25. L. Breiman, Random forests, *Machine Learning.* **45**(1), 5–32 (2001).
26. T. K. Ho, The random subspace method for constructing decision forests, *IEEE Transactions on Pattern Analysis and Machine Intelligence.* **20**(8), 832–844 (1998).
27. Y. C. Martin, J. L. Kofron, and L. M. Traphagen, Do structurally similar molecules have similar biological activity?, *J. Med. Chem.* **45**, 4350–4358 (2002).
28. M. Campillos, M. Kuhn, A.-C. Gavin, L. J. Jensen, and P. Bork, Drug target identification using side-effect similarity, *Science.* **321**(5886), 263–266 (2008).
29. V. J. Haupt and M. Schroeder, Old friends in new guise: repositioning of known drugs with structural bioinformatics, *Breifings in Bioinformatics* (2011).
30. M. J. Keiser, V. Setola, J. J. Irwin, C. Laggner, A. I. Abbas, S. J. Hufeisen, N. H. Jensen, M. B. Kuijer, R. C. Matos, T. B. Tran, R. Whaley, R. A. Glennon, J. Hert, K. L. H. Thomas, D. D. Edwards, B. K. Shoichet, and B. L. Roth, Predicting new molecular targets for known drugs, *Nature.* **462**, 175–181 (2009).
31. P. R. Caron, M. D. Mullican, R. D. Mashal, K. P. Wilson, M. S. Su, and M. A. Murcko, Chemogeominc approaches to drug discovery, *Curr. Opin. Chem. Biol.* **5**, 464–470 (2001).
32. H. Kubinyi and G. Müller, *Chemogenomics in Drug Discovery.* Wiley-VCH, Weinheim (2004).
33. D. Rognan, Chemogenomic approaches to rational drug design, *British Journal of Pharmacology.* **152**, 38–52 (2007).
34. L. Jacob and J.-P. Vert, Protein-ligand interaction prediction: an improved chemogenomics approach, *Bioinformatics.* **24**(19), 2149–2156 (2008).
35. Z. Xia, L.-Y. Wu, X. Zhou, and S. T. Wong, Semi-supervised drug-protein interaction prediction from heterogeneous biological spaces, *BMC Systems Biology.* **4 (Suppl 2)**, S6 (2010).
36. T. V. Laarhoven, S. B. Nabuurs, and E. Marchiori, Gaussian interaction profile kernels for predicting drug-target interaction, *Bioinformatics* (2011).
37. L. Perlman, A. Gottlieb, N. Atias, E. Ruppin, and R. Sharan, Combining durg and gene similarity measures for drug-target elucidation, *Journal of computational biology.* **18**, 133–145 (2011).

38. X. Chen, M.-X. Liu, and G.-Y. Yan, Drug-target interaction prediction by random walk on the heterogeneous network, *Molecular BioSystems* (2012).
39. M. Kanehisa, S. Goto, M. Hattori, K. Aoki-Kinoshita, M. Itoh, S. Kawashima, T. Katayama, M. Araki, and H. M, From genomics to chemical genomics: new developments in kegg, *Nucleic acids res.* **34**(Database), D354–357 (2006).
40. I. Schomburg, A. Chang, C. Ebeling, M. Gremse, C. Heldt, G. Huhn, and D. Schomburg, BRENDA, the enzyme database: updates and major new developments, *Nucleic Asids Res.* **32**(supl-1), D431–433 (2004).
41. S. Günther, M. Kuhn, M. Dunkel, M. Campillos, C. Senger, E. Petsalaki, J. Ahmed, E. Urdiales, A. Gewiess, L. Jensen, R. Schneider, R. Skoblo, R. Russell, P. Bourne, P. Bork, and R. Preissner, SuperTarget and Matador: resources for exploring drug-target relationships, *Nucleic acids res.* **36** (Database issue), D919–D922 (2008).
42. D. S. Wishart, C. Knox, A. Guo, D. Cheng, S. Shrivastava, D. Tzur, B. Gautam, and M. Hassanali, DrugBank: a knowledgebase for drugs, drug actions and drug targets, *Nucleic acids res.* **36(Database issue)**, D901–906 (2008).
43. M. Hattori, Y. Okuno, S. Goto, and M. Kanehisa, Development of a chemical structure comparison method for integrated analysis of chemical and genomic information in the metabolic pathways, *J. Am. Chem Soc.* **125**(39), 11853–11865 (2003).
44. J. Davis and M. Goadrich. The relationship between precision-recall and roc curves. In *Proc. 23rd International Conference on Machine Learning*, pp. 233–240 (2006).
45. P. Yang, X.-L. Li, M. Wu, C.-K. Kwoh, S.-K. Ng (2011) Inferring Gene-Phenotype Associations via Global Protein Complex Network Propagation. PLoS ONE 6(7): e21502.

Chapter 9

Characterization and Prediction of Human Protein-Protein Interactions

Yi Xiong[1,2], Dan Syzmanski[3,1] and Daisuke Kihara[1,2,*]

[1]*Department of Biological Sciences;* [2]*Department of Computer Science;*
[3]*Department of Agronomy, Purdue University, West Lafayette, IN 47907, USA*
[*] *E-mail: dkihara@purdue.edu*

In this chapter, we introduce classification techniques that are used to predict group membership to which a data object belongs. In the first half of the chapter, we summarize several start-of-the-art classification algorithms. In the latter section, we show examples of applications of the classification techniques for prediction of protein-protein interactions in human proteome.

1. Introduction

Classification is a form of data analysis that can be used to construct a model to describe data classes of interest and predict a predefined class to which an input object belongs. For example, a classification model can be built to categorize a human tissue into a normal or cancer class based on its gene expression pattern. A wide variety of classification algorithms have been proposed by numerous researchers in machine learning, expert systems, and statistics. Although the algorithms are based on different theories, their applications follow the same framework. Generally, data classification is performed in two steps on a data set, which is split into a training set and a testing set [1]. In the first step, a model is built on the training set to describe a set of data classes. The model is represented in a variety of forms, for example, a decision tree or

[*] Corresponding author.

a mathematical formula. In the second step, the model is applied to the testing set to evaluate its performance. If the accuracy of the classification model on the testing set is acceptable, the model will be adopted to classify a new data set for which class labels are not known.

Classification techniques are widely applied for biological problems related to human health. For example, they have been applied to the cancer classification based on gene expression profiles by DNA microarrays. In a work by Golub *et al.* [2], classification models were based on a collection of tumor samples for which the cancer types or the eventual outcome were known. They classified tumor samples to known classes, which could reflect current states (such as different types of cancer) or future clinical outcomes (such as drug response or survival). Nayal and Honig [3] used crystallography data to analyze surface cavities of a nonredundant set of 99 drug-target complexes, and exploited 408 physicochemical, structural, and geometric features to construct a random forest model to predict drug binding sites. Additionally, classification methods were used to predict protein-protein interactions (PPIs) in human proteome, since understanding PPI networks can provide insights into mechanisms of human diseases such as cancers [4]. Furthermore, classification methods have been extensively applied to the identification of protein function and function sites in proteins, using a linear discriminant rule [5], Naïve Bayesian [6], support vector machine [7] and random forest [8].

In this chapter, we first introduce several frequently used state-of-the-art classification algorithms, namely, decision tree, neural network, Naïve Bayesian, support vector machine, and random forest. Then, we use the specific case of human PPI prediction as an example to show how classification methods are applied to this problem.

2. Classification methods

2.1. *Framework of classification*

There are two categories of learning in machine learning, unsupervised and supervised learning. In unsupervised learning, classes to which input objects belong to are not defined in advance and it is aimed at finding grouping or hidden structures in the input data. On the other hand, in supervised learning, input data with known class labels are used to construct a mapping from input data space to output class label space. Typically a classification task consists of four major components, namely, a dataset, a feature set and its representation, a learning algorithm, and model performance evaluation. The first part of this chapter focuses on introducing several commonly used learning algorithms in supervised learning.

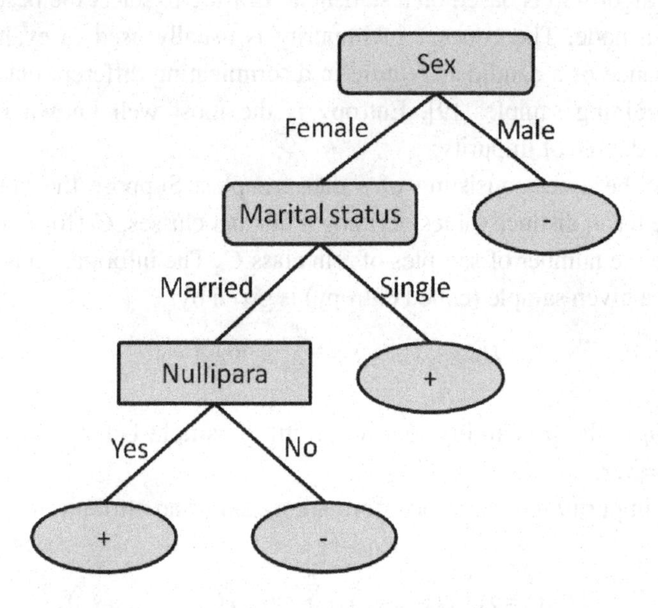

Fig. 1. A decision tree for a person that can indicate whether or not he or she has a breast cancer with high risk. The leaf node with + means a person with a high risk on breast cancer, and - is the class label for a person without high risk on breast cancer. Internal nodes are denoted by rectangles, and leaf nodes are denoted by ovals. The root as a special internal node is denoted by the rounded rectangle.

2.2. *Classification algorithms*

In this section, we introduce several state-of-art classification algorithms, which are decision tree, artificial neural network, Bayesian model, support vector machine, and random forest.

2.2.1. *Decision tree*

The advantage of decision tree-based classification is that it can be easily interpreted and intuitively understandable by human. A decision tree is a flow-chart-like tree structures (Fig. 1), where each internal node denotes a test on an attribute and leaf nodes represent classes or class distribution. The top most node in a tree is the root node, which is the starting point of the classification and where all samples belong to.

The algorithm is based on a statistical measure to select the best feature to split a node. The concept of impurity is usually used to evaluate the performance of a candidate feature in discriminating different class labels in the training samples [9]. Entropy is the most well known index to measure degree of impurity.

Let S be a set consisting of s data samples. Suppose the class label attribute has n distinct values defining n distinct classes, C_i (for i=1,2,..., n). Let s_i be the number of samples of S in class C_i. The information needed to classify a given sample (called entropy) is given by

$$I(s_1, s_2, ..., s_n) = -\sum_{i=1}^{n} p_i \log_2(p_i) \tag{1}$$

where p_i is the probability that an arbitrary sample belongs to class C_i, thus $p_i = s_i/s$.

The impurity reduction or information gain of an attribute A is defined as:

$$\Delta I(A) = I(s_1, s_2, ..., s_n) - \sum_{j=1}^{m} w_j \times I(s_{1j}, s_{2j}, ..., s_{nj}), \tag{2}$$

where attribute A has m distinct values, which divide S into m subsets, and s_{ij} is the number of samples of class C_i in a subset S_j. The term w_j is the weight of the jth subset. The attribute that leads to the maximal reduction of impurity is chosen to split a node among all candidate attributes.

The basic module of the decision tree construction is a greedy algorithm that builds the tree in a top-down heuristic search using the recursive manner. The complete trial and error method for all the possible partitions is intractable, therefore, the top-down heuristic search is employed on the recursive partition process. The search algorithm for the attribute is greedy and it never takes a step back to reconsider its previous choice.

The well-known algorithm ID3 [10] uses information gain (Eq. 2) to select the attribute that will best categorize the samples into individual classes. The ID3 algorithm is then executed recursively on the smaller subsets. However, attributes with continuous values are not allowed in ID3, and the information gain measure is biased on the attributes with many values. As a successor algorithm to ID3, C4.5 [11] creates a threshold to handle continuous attributes and the normalized information gain to avoid the bias.

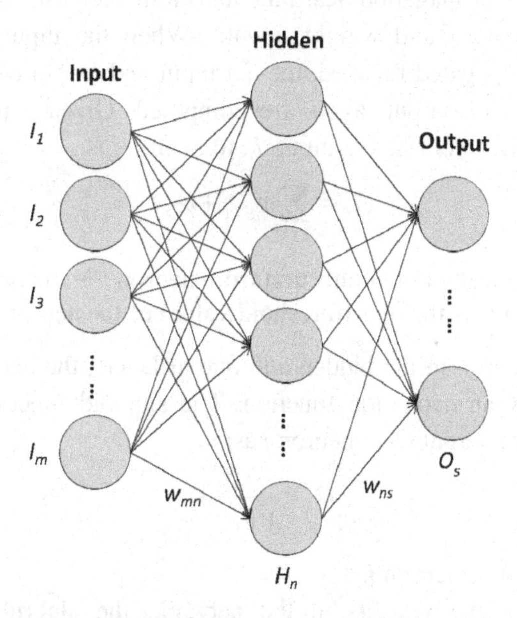

Fig. 2. The interconnected group of neurons in an artificial neural network. Weighted connections exist between consecutive layers.

2.2.2. *Artificial neural network*

Artificial neural network (ANN) is one of the most popular classification techniques, which is in the category of the artificial intelligence field that makes computers or machines simulate the thinking, psychology, evolution, and intelligent behaviors of humans. The theory of artificial neural network is inspired by the structure of human brains, which can be simplified as the network of neurons that transfer information between each other to learn data and make a decision (Fig. 2). The neurons in the input layer correspond to the attribute values of each training sample. The weighted outputs of these nodes are fed into a second layer, which is called a hidden layer. The weighted output of the hidden layer will be input to the next hidden layer. The weighted output of the last hidden layer will be input for neurons in the output layer, which makes the final prediction for given samples. One of the commonly used algorithms for training parameters for neural networks is the back-propagation method [12]. The back-propagation learning algorithm consists of two major phases: propagation and weight update. When the inputs of training samples are propagated forward, the net input and output of each neuron in the hidden and output layers are computed. Given a neuron j in a hidden or output layer, the net input, I_j, to neuron j is

$$I_j = \sum_i w_{ij} O_i + b_j \tag{3}$$

where w_{ij} is the weight of connection from neuron i in the previous layer to neuron j, and b_j is the bias (threshold value) of the neuron.

For each neuron in the hidden and output layers, the net input is then transformed by an activation function. The sigmoid function is usually used to compute output O_j of neuron j as:

$$O_j = \frac{1}{1+e^{-I_j}} \tag{4}$$

where I_j is input of neuron j.

For training the weights in the network, the algorithm compares outputs from the neural network with the actual known class labels. Observed error for each training sample is fed back to update weights so that the error of outputs will be smaller. The process of updating weights is applied to each sample in a training dataset repeatedly until the weights do not change much or until the updates is performed by a pre-determined number of times. Please see a reference [12] for more details about the back-propagation learning algorithm.

2.2.3. *Bayesian model*

Bayesian classification is based on Bayes' theorem. It can predict class membership probabilities, such as the probability that a given sample belongs to a particular class. Each training sample is represented by *n* variables (or attributes; $X_1, X_2, X_3, \ldots, X_n$). Suppose there are *m* classes, $C_1, C_2, C_3, \ldots, C_m$, for the class label variable *Y*. The possibility that a sample belongs to a class C_i is defined by the Bayes theorem as follows:

$$P(C_i \mid X_1, X_2, \ldots, X_n) = \frac{P(X_1, X_2, \ldots, X_n \mid C_i) P(C_i)}{P(X_1, X_2, \ldots, X_n)} \tag{5}$$

The Naïve Bayesian classifier assumes the independent relationship among its input attributes (Fig. 3). Therefore, the Eq. 5 can be rewritten as:

$$P(C_i \mid X_1, X_2, \ldots, X_n) = \frac{P(C_i) \prod_{i=1}^{n} p(X_i \mid C_i)}{P(X_1, X_2, \ldots, X_n)} \tag{6}$$

The class prior possibility $P(C_i)$ is estimated as the fraction of samples belonging to C_i in the whole training set.

A Bayesian model classifies a sample to class C_i if and only if

$$P(C_i \mid X_1, X_2, \ldots, X_n) > P(C_j \mid X_1, X_2, \ldots, X_n) \text{ for } 1 \le j \le m, j \ne i. \tag{7}$$

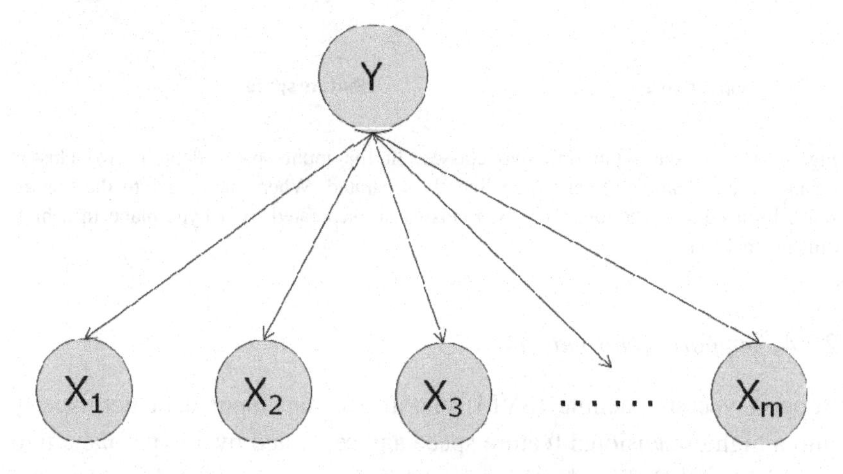

Fig. 3. A graph model of Naive Bayesian. The variable *Y* (class label) is dependent on the input variables (or observed values), which are independent between each other.

In the case that there are only two classes (C_1 and C_2, m=2) (binary classification), a likelihood ratio (*LR*) score usually is used to determine the label of the sample.

$$L(X_1, X_2, ..., X_n) = \frac{p(X_1, X_2, ..., X_n \mid C_1)}{p(X_1, X_2, ..., X_n \mid C_2)} \quad (8)$$

In the naïve Bayesian model, the *LR* can be calculated as the product of the individual likelihood ratios with respect to the features considered separately.

$$L(X_1, X_2, ..., X_n) = \prod_{i=1}^{n} \frac{p(X_i \mid C_1)}{p(X_i \mid C_2)} = \prod_{i=1}^{n} L(X_i) \quad (9)$$

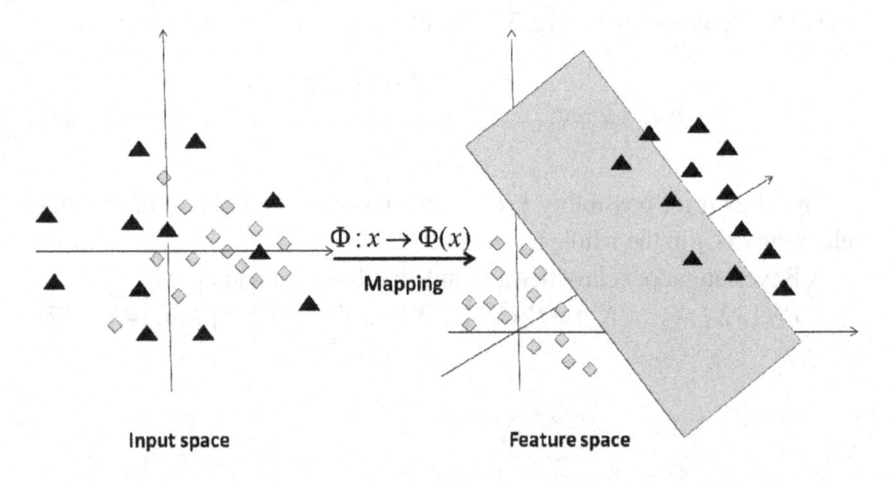

Fig. 4. Hyperplane separating two classes. In the input space, data in two classes (triangles and diamonds) cannot be linearly separated. When mapping into the feature space by a kernel function, these two classes are separated by a hyperplane in a high dimensional space.

2.2.4. *Support vector machine*

Support vector machine (SVM) classifiers map input data nonlinearly into a high dimensional feature space and separated by a hyperplane into two classes [13] (Fig. 4). Given a number of training samples belonging to two classes, a support vector machine constructs a hyperplane or set of hyperplanes in a high dimensional space. Intuitively, a good separation is achieved by selecting the hyperplane that separates the two classes but adopts the maximal distance from any one of the given samples, since the larger the margin is the lower the generalization error of the classifier for unknown samples is. The application of a kernel function allows the algorithm to fit the maximum-margin hyperplane in a transformed feature space. There are mainly three common types of kernel functions. Given the feature vectors (x_i and x_j) of two samples, kernel functions transform the distance between them.

(a) Polynomial function:

$$K(x_i, x_j) = \left[(x_i^T x_j) + 1\right]^n \tag{10}$$

(b) Gaussian radial basis function ($\gamma > 0$ is a parameter):

$$K(x_i, x_j) = e^{\left(-\gamma \|x_i - x_j\|^2\right)} \tag{11}$$

(c) Hyperbolic tangent function (v and c are parameters):

$$K(x_i, x_j) = \tanh(v(x_i^T x_j) + c) \tag{12}$$

Polynomial kernels are well suited for problems where all training data is normalized. The default recommended kernel function would be the radial basis function. The hyperbolic tangent kernel is also known as the sigmoid kernel, which comes from the neural networks field.

2.2.5. *Random forest*

Random forest is an ensemble classifier that consists of a bagging of unpruned decision trees with a randomized selection of features at each split, and outputs the class that is the majority vote of the classes output by individual trees [14]. Generally speaking random forest can improve prediction accuracy over a single decision tree. Random forest consists of two stages of randomization. The first stage is the randomization through bagging or bootstrap aggregation, which creates new training sets by randomly sampling a given set of samples with replacement. The second randomness is to select a random subset of features for each split when building an individual tree.

To construct a random forest, the following steps are employed (Fig. 5). An individual classification tree is developed on a bootstrap sample. At each node in the tree, the split is selected on the randomly chosen subset of features. The tree is grown to full size without pruning. These two step are repeated for n times for n trees to construct a forest. The ensemble classification label is a majority vote of the prediction from all the trees.

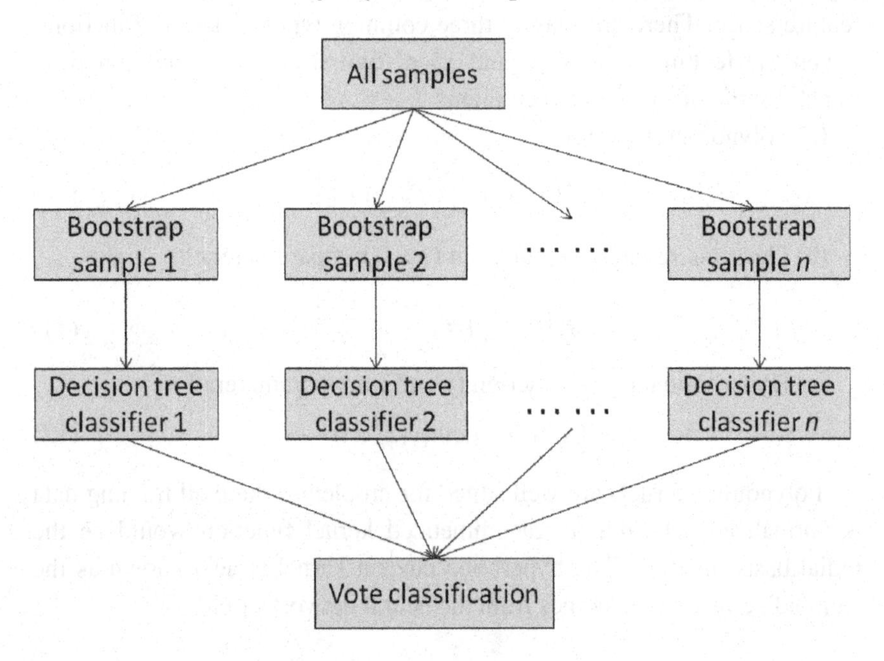

Fig. 5. A flowchart of a random forest model. Each decision tree is trained on a subset of the whole data set, which is chosen by random sampling with replacement. In each tree classifier, a random subset of the features is selected at each split. The final output class of the forest is selected by the majority vote of the classes output by all n trees.

2.3. *Model validation and evaluation*

When a classification model is built using one of the algorithms introduced above, it requires a golden standard dataset to validate and evaluate how well the constructed model works. The golden standard dataset consists of well-labeled samples, which are divided into a training set and testing set. For model validation, there are three methods that are commonly employed to judge the performance of classification models. They are *k*-fold cross-validation, leave-one-out cross-validation, and independent tests. In the *k*-fold cross-validation, a sample set is randomly partitioned into *k* subsets of equal size. Of the *k* subsets, one subset is retained as the validation dataset for testing the model, and the remaining *k*-1 subsets are used for training. The cross-validation process is then repeated *k* times with each of the *k* subsets used exactly once as the validation data. The *k* results from all rounds are finally averaged to generate a single estimation metric. Leave-one-out cross-validation (LOOCV) uses a single instance in the sample set for testing while the remaining instances are used as the training data. This is repeated so that each instance in the sample set is used once as the validation data. This is the same as a *k*-fold cross-validation with *k* being equal to the number of instances in the original sample set. Leave-one-out cross-validation is computationally expensive when the number of samples in the training set is too large. In order to test the model in an unseen sample set, an independent test will be adopted. Independent test is conducted on a dataset which is independent from the training set. Thus, it is mimicking the actual scenario of prediction.

In order to assess the classification performance, various performance criteria are defined:

$$\text{Recall}=\frac{TP}{TP+FN} \tag{13}$$

$$\text{Specificity}=\frac{TN}{TN+FP} \tag{14}$$

$$\text{False positive rate}=\frac{FP}{TN+FP} \tag{15}$$

$$\text{Precision}=\frac{TP}{TP+FP} \tag{16}$$

$$F_1=\frac{2\times \text{Recall}\times \text{Precision}}{\text{Recall}+\text{Precision}} \tag{17}$$

where *TP* is the number of correctly predicted positive samples, *TN* is the number of correctly predicted negative samples, *FP* is the number of negative samples wrongly predicted as positives and *FN* is the number of positive samples wrongly predicted as negatives.

The receiver operating characteristic (ROC) curve is a plot of the sensitivity (also called as recall) versus (1-specificity) for a binary classifier at varying thresholds. The area under the curve (*AUC*) can be used as a threshold-independent measure of classification performance. Alternatively, the precision-recall (PR) curve can be used, which plots recall relative to precision for a binary classifier at varying thresholds.

3. Application to human PPI data

3.1. *Background of human PPI and biological problem statement*

Most of cellular functions are carried out through protein interactions. PPI data identifies interactions of proteins in a cell, which can provide insights into mechanisms that underlie human diseases. PPI data may also lead to new drug development to prevent the diseases. Since the current human PPI map is estimated to be far from complete [4], there is a strong need to increase the coverage of the human interactome by classifications (also called predictions here). In the last decade, the high throughput experimental technologies such as the yeast two hybrid (Y2H) assay and targeted mass spectrometry are employed to investigate protein-protein interaction networks in a whole organism scale [15]. However, it is pointed out by some researchers that these high-throughput experimental methods have high false positive rates, and analysis of the high-throughput datasets has shown that they do not overlap much with each other [16, 17]. Accurate computational methods are therefore necessary to complete the interactome, which will compensate time-consuming and expensive experimental methods for identifying PPIs. Here, using recent works on PPI data as examples, we will see how classification techniques are used in practice. The methods capture observed features of interacting proteins and are applied to predict novel interactions between protein pairs. Note that although PPIs are dynamic and are often condition-specific, these methods classify PPIs as interact or non-interact pairs in a static manner.

Table 1. The list of public databases of protein-protein interactions (the numbers were obtained on Sep, 2012).

Database name	Number of organisms	Number of human PPI	URL
BioGRID	39	75096 (physical interaction)	http://thebiogrid.org/
DIP	504	4794	http://dip.doe-mbi.ucla.edu/dip/
HPRD	1	39194	http://www.hprd.org/
IntAct	186	4578 (direct interaction)	http://www.ebi.ac.uk/intact/
MINT	30 (main)	26700	http://mint.bio.uniroma2.it/

3.2. *Human PPIs databases*

Currently known human PPIs are collected in several databases (Table 1), which are curated from the experimental data and primary biomedical literature. The Biological General Repository for Interaction Datasets (BioGRID) [18] is a public database that collects genetic and protein interaction data from model organisms and humans. The Database of Interacting Proteins (DIP) [19] contains experimentally determined protein-protein interactions of a large number of organisms. The Human Protein Reference Database (HPRD) is a database which integrates the information of protein functions and interaction of human proteins [20]. IntAct [21] is an open-source protein interaction database, where the source code and data are freely accessible. The Molecular INTeraction database (MINT) [22] is a public repository focusing on PPI data reported in peer-reviewed literature. The interaction data contained in these databases are mainly between soluble proteins. Thus, they should be used with caution if interactions between membrane proteins are to be predicted.

3.3. *Datasets for computational study of PPIs*

Known PPIs in these databases are used for training classification methods. To build a method that can reliably classify protein pairs into interacting or non-interacting, we need two datasets, a positive dataset that contains known interacting protein pairs and a negative dataset that consists of non-interacting protein pairs.

To construct a positive dataset, known interacting protein pairs are extracted from some or all of the aforementioned databases. Duplicate interactions from different databases will be deleted. Constructing a negative dataset is not as trivial as a positive dataset, because it is often difficult to distinguish protein pairs that are actually interacting with each other but not yet detected by current experiments from pairs that are truly not interacting. The first strategy to construct a negative set is to randomly select protein pairs that are from different sub-cellular locations so that they are unlikely to physically encounter in a cell [23]. Because the probability that two randomly selected proteins physically interact is low, another approach taken is to randomly pair any two proteins from the positive data set excluding pairs that are actually known to interact [24].

There is a database named Negatome [25], which contains lists of experimentally supported non-interacting protein pairs by manual curation of literature and from analysis of protein complexes with known 3D structure. The database contains 1291 and 809 non-interacting pairs, respectively.

3.4. *Protein features used for predicting PPIs*

A prediction method for PPI considers features of known interacting proteins (from a positive dataset) and known non-interacting protein pairs (from a negative dataset) to build (train) a model. It is a binary classification problem of protein pairs, either to interacting or non-interacting. Features used in existing studies include orthologous relationship to known interacting proteins in another organism (called interolog [26]). This is an effective strategy because protein interactions are often conserved among highly diverged organisms ranging from the model plant *Arabidopsis thaliana* to humans. Gene co-expression data can be also used for predicting interacting proteins because expression profiles of interacting proteins simultaneously rise and fall in different conditions and cell types. Detecting known interacting domains in protein pairs is another strategy to predict their interaction. Functional similarity measured by semantic similarity (which indicate how similar two terms are based on their semantic properties) of Gene Ontology (GO) terms is based on the fact that interacting proteins may function in the same biological process or at the same subcellular location. Some PPI prediction methods exploit comparative genomics features, such as conserved gene orders, gene fusion and phylogenetic profile similarity (i.e. co-occurrence or co-absence of genes in multiple genomes) [27]. Below we discuss three methods using different combination of features and classifiers for integrative analysis and prediction of human PPIs.

3.5. *Case studies on human PPI prediction*

We review three examples of application of classification methods for predicting PPIs in human. The first work [28] used a naïve Bayesian model to integrate four types of features to predict PPIs in human. The second report [17] improved over the first one by using more relevant features and a semi-naïve Bayesian model, which combined dependent features together to construct a naïve Bayesian model. The third paper [16] employed an active learning strategy to guide the selection of training data, which was shown to improve prediction accuracy. The detail comparison of these three case studies is summarized in Table 2.

Table 2. Methodological differences among three case studies.

Difference	**Rhodes *et al.* [28]**	**Scott *et al.* [17]**	**Mohamed *et al.* [16]**
Positive Dataset	11,678 PPIs in HPRD	26, 896 PPIs in HPRD	14, 600 PPIs in HPRD
Negative Dataset	A localization-derived negative dataset	A randomly-generated negative dataset, and a localization-derived negative dataset	A randomly-generated negative dataset
Model	Naïve Bayesian	Semi-naïve Bayesian	Active Learning and Random Forest
Features	interolog, correlation of gene expression, the number of shared biological process GO terms, and the domain enrichment ratio	correlation of gene expression, interolog, sub-cellular localization, co-occurrence of specific InterPro and Pfam domains, co-occurrence of post-translational modifications, presence of disorder regions, and local network topology of PPI network	GO terms in cellular component, molecular function, and biological process, co-occurrence in tissue, gene expression, sequence similarity, interolog, and domain interaction
Evaluation metrics	False positive rate	AUC of ROC	Recall, precision, and F-score

3.5.1. *Application of naïve Bayesian model*

In a study by Chinnaiyan and his colleagues [28], four features were considered to predict PPI using a naïve Bayesian model. A naïve Bayesian model computes the probability that a pair of two proteins are interacting using each feature and multiplicatively combines the probabilities computed for different features. This multiplicative nature requires that the predictive data sets are conditionally independent or nonredundant. The four features they used were interolog (existence of homologous proteins that are interacting), correlation of gene expression, the number of shared biological process GO terms, and the domain enrichment ratio. The domain enrichment ratio is calculated as the probability of observing a pair of domains in a set of known interacting proteins divided by the product of the probabilities of observing each domain of the pair independently. The protein domains were taken from the InterPro database [29]. Biologically, these features have independent information, except for the last two features that are related to protein functions. To avoid bias from the two dependent features, the last two features were analyzed together.

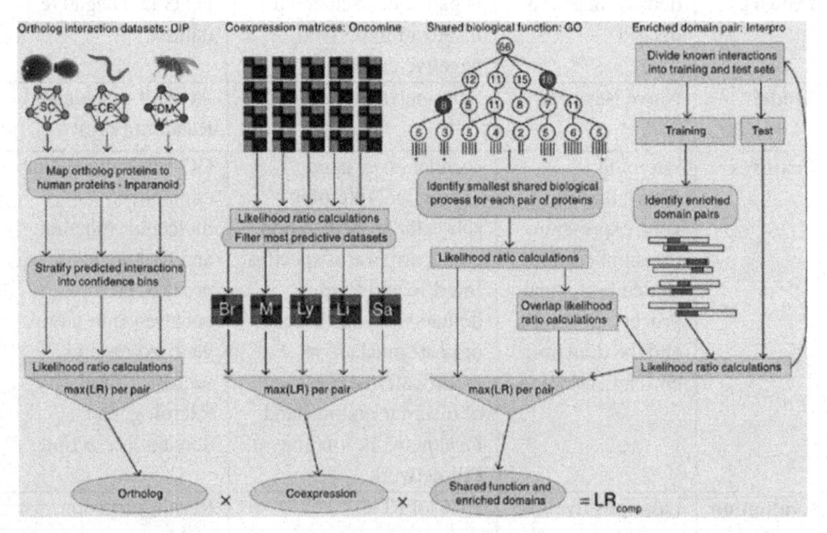

Fig. 6. The flowchart of data integration in a Naïve Bayes model to predict human protein-protein interactions (The figure is taken from [25] with permission from the journal).

Figure 6 shows the naïve Bayes model which combines the features mentioned above. First the authors calculated likelihood ratios (i.e. ratio of the probability that a protein pair is interacting over the probability that they are not) (Eq. 9) using each feature. For the interolog feature, human PPIs were predicted from the interaction data sets of three model organisms, *Sacchromyces cerevisiae, Caenorhabditis elegans,* and *Drosophila melanogaster* (Fig. 6, left branch). A confidence level of interolog assignments were classified by considering several parameters associated with predicted interactions, including the number of independent lines of evidence for a yeast PPI, confidence value of ortholog assignment, etc. Using these parameters, human protein pairs were classified into several confidence level bins using a decision tree (the step of stratifying predicted human PPIs in the figure). Then, Naïve Bayes model was constructed to predict whether a human protein pair interacts or not given the confidence level of the interolog assignment. If a human protein pair has interologs in two or more organisms, it will have multiple likelihood ratio from each of the PPIs. In such case, the maximal ratio was chosen from them (the last step of the PPI branch).

For the gene expression features, human protein pairs are classified into bins by their correlation value of their expression level observed in each of the five expression data sets from different tissues (Fig. 6, second branch from the left). Then, similar to the interolog-based feature, the likelihood ratio of the gene expression-based feature was computed from each of the five expression data sets that contain the given protein pairs. If a protein pair appeared in multiple expression data, the maximum ratio was chosen.

Since they found that the number of shared biological process GO terms and the domain enrichment ratio (two right branches in Fig. 6) were redundant, two features were binned together to compute the likelihood ratio. At last, the likelihood ratios were combined in a Naïve Bayesian model to generate composite likelihood ratios (LR_{comp}).

They used a training dataset of 11,678 known human PPIs from HPRD and 3,106,928 non-interacting protein pairs in human. The trained model predicted 38,986 new PPIs that were not reported in HPRD at a false positive rate of 50% and 9,651 new PPIs at a false positive rate of 20%. The authors claimed that the false positive rate of this classification method is comparable to the results of high throughput experimental approaches in model organisms.

3.5.2. *Application of semi-naïve Bayesian model*

A drawback of applying naïve Bayes for PPI prediction is that it assumes independence of each feature; however, often some features are closely related. If features considered in a model are independent, the likelihood ratio can be calculated as the product of the individual likelihood ratios (naïve Bayesian model, Eq. 9). On the other hand, if features are not independent, all possible combinations of all states of these features must be considered, which can be very computationally intensive (Eq. 8).

The semi-naïve Bayesian model is addressing the drawback of naïve Bayesian model by explicit combination of related features (Eq. 8) while handling independent features (Eq. 9) in the same way as the naïve Bayesian model. In a study by Barton and his colleagues [17], the semi-naïve Bayesian model is used for human PPI prediction using seven features: correlation of gene expression, interolog, sub-cellular localization, co-occurrence of specific InterPro [29] and Pfam [30] domains, co-occurrence of post-translational modifications, presence of disorder regions, and local network topology of PPI network. The local PPI network topology measure reflects the fraction of commonly interacting proteins for a pair of proteins. The sub-cellular localization, protein domain co-occurrence, and post-translational modification co-occurrence are integrated into one combined module because considering all their combinations (dependencies) between them achieved a higher accuracy. That is, the joint probability of all possible combinations of the four localization bins, five chi-square domain-co-occurrence bins, and four post-translational modification co-occurrence score bins, was computed. The rest of four features and the combined module were considered to be independent and integrated in the naïve Bayesian classifier (Fig. 7).

For training and testing, 62,322 human protein sequences were downloaded from the International Protein Index (IPI) database [31]. Among these proteins, 26,896 distinct human protein interactions were identified as the positive dataset from HPRD. Of the 62,322 human proteins, 22,889 human proteins were referred to as the Informative Protein Set (IPS) since they were characterized by at least one of the features mentioned above. Two types of negative datasets were constructed. The first type of negative dataset was generated by selecting protein pairs at random from IPS. The second type of negative dataset was created by selecting protein pairs from IPS for which the HPRD annotates them in two different subcellular locations. The localization-derived negative trained classifier tested on sets containing localization-derived negatives achieves a lower accuracy than that of the random negative trained classifier tested on a test set containing randomly-generated negatives. This is most likely due to the fact that the localization-derived negative trained predictor cannot sample the whole protein pair space well. Their model predicted 37,606 human PPIs, of which 32,892 were not reported in other publicly available large human interaction datasets. The newly discovered interactions thus considerably increased the coverage of the human interaction map.

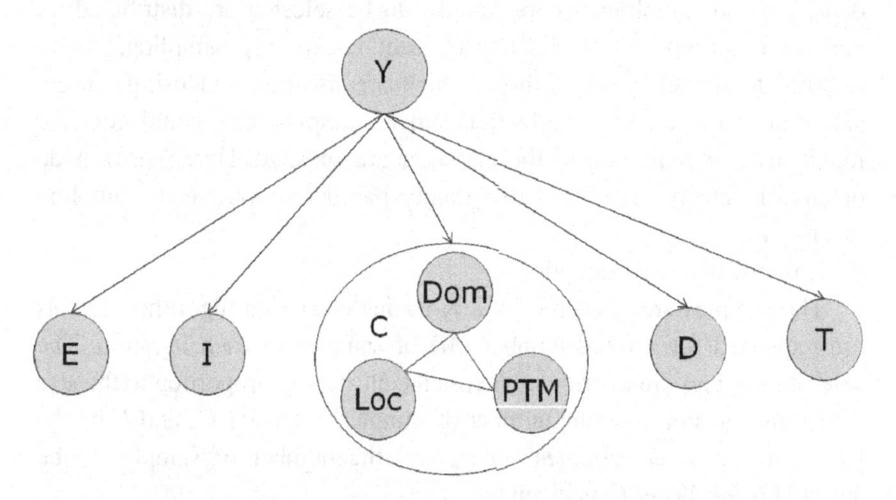

Fig. 7. Overview of semi-naive Bayesian. The variable *Y* (class label) is dependent on the observed variables, i.e. E(Expression), I(Interolog), C(Combined), D(Disorder) and T(Topology), which are treated as independent variables. In the combined module C, localization, domain co-occurrence, and post-translational modification co-occurrence are considered to be dependent on each other.

3.5.3. *Application of Active Learning and Random Forest*

Experimentally verified protein-protein interactions are expensive to obtain; therefore, it would be useful to develop strategies to minimize the number of labeled samples required in the supervised learning task. In comparison to passive learning models (labels of training sample are known prior to training) introduced in the sections above, active learning is allowed to request the label of any particular input sample in the training data. Active learning is a type of iterative supervised learning in which the system selects most informative samples each time to obtain their labels from a large pool of samples. Sampling will be repeated until the obtained samples in the training set are both plentiful and representative to construct a classification model with satisfied performance. The benefit of active learning is to substantially reduce the number of labeled samples required, making the training of a classification model more practical.

In active learning, ideas of selecting informative training data include density based sampling, where samples to be selected are distributed on clusters in proportion to the cluster size; uncertainty sampling, where samples to be selected are those which are mispredicted using current classifier; estimated-error reduction, where samples that would generate maximal error reduction to the classifier are selected. Here, we provide details of density based and uncertainty based (random seed) sampling techniques.

(1) Density based sampling

The samples are clustered by a K-means clustering algorithm. Labels are requested for a fixed number (*M*) of samples in each iteration. The selected samples are distributed across the clusters in proportion to the size of the cluster. Let n_i be the number of samples in cluster C_i, and *N* be the total number of all samples. Then, m_i , the number of samples to be selected from cluster C_i is given by

$$m_i = M * n_i / N \tag{18}$$

and,

$$M = \sum_{i=1}^{K} m_i \tag{19}$$

In each cluster C_i, the algorithm selects m_i unlabelled samples closest to the centroid, and assigns labels on them [16].

(2) Uncertainty based sampling

In this strategy, the samples, whose labels are asked, are randomly selected in the first iteration. For example, a random forest is employed to construct the model with the selected samples. In the following iterations, the samples selected for labeling are those which have maximum disagreement among the decision trees in the random forest. The entropy (confusion) in labeling the sample is measured as

$$e_i = -\sum_{i \in (0,1)} p_i \log(p_i) \tag{20}$$

where, p_0 is the fraction of decision trees in the random forest that label the samples as negative ones, and p_1 is the fraction of decision trees that label the samples as positive.

In each iteration, a fixed number of samples with the maximum confusion are selected and their labels are obtained. A new random forest is trained from the expanding number of samples for selecting the samples with maximal confusion in the next iteration.

In a work by Ganapathiraju and colleagues [16], 14, 600 interacting protein pairs were downloaded from HPRD and a set of 400, 000 non-interacting protein pairs were randomly generated. Features used to characterize protein pairs were GO terms in cellular component category (1), GO terms in molecular function category (1), GO terms in biological process (1), co-occurrence in tissue (1), gene expression (16), sequence similarity (1), interolog (5), and domain interaction (1). The numbers in parentheses show the number of different features in the same category. Not all types of features were available for each protein pair. A homogenous subset of data was built such that every pair had more than 80% feature coverage, which resulted in a total of 55,950 protein pairs for use in the training and testing.

In order to test the active learning model, training samples were selected using different active learning sample selection strategies as described above. Random forest was used for classification. Since some of the 27 features are obviously redundant with each other, a randomly reduced set of features were used to build each decision tree in the random forest. In their work, 20 decision trees were constructed. To split the nodes of the decision trees, a subset of seven feature elements were randomly selected from the total of 27 elements. Of the seven selected features, the feature that offered the maximal information gain (Eq. 2) was used to split each node.

It was shown that using active learning to select training data achieved higher accuracy than the model trained on randomly selected training samples without active learning. The best model achieves an F-score (harmonic mean of recall and precision) of 60% at 3000 labeled samples, with a recall of 51% and precision of 73%. The F-score dropped to 50% when active learning was not used (instead, training data were selected randomly). It was demonstrated that active learning enables better learning with less labeled training data.

4. Conclusions

Data classification is the form of data analysis for extracting models describing important data classes, and predicting a predefined class to which a given sample belongs. Five well known algorithms were introduced in this chapter. There are also a number of other methods, which are gaining increasing popularity in the data mining and bioinformatics fields. These methods include genetic programming-based algorithms [32] and fuzzy set algorithms [33]. Classification and prediction based on classification of known data is an indispensable step for understanding and using a large scale data. With applications to human PPIs discussed here, classification models have increased the coverage of the human interactome. Moreover, classification techniques have been applied for other various types of biological and medical data to discover novel knowledge from them.

Acknowledgments

This work has been supported by grants from the National Institutes of Health (R01GM075004 and R01GM097528 to D.K.), National Science Foundation (EF0850009, IIS0915801, DMS0800568 to D.K., IOS1127027 to D.S. and D.K.), and National Research Foundation of

Korea Grant funded by the Korean Government (NRF-2011-220-C00004) to D.K.

References

1. Han, J. and M. KAMBER, *Data Mining: concepts and techniques*, 2001, Morgan Kaufmann.
2. Golub, T.R., D.K. Slonim, P. Tamayo, C. Huard, M. Gaasenbeek, J.P. Mesirov, *et al., Molecular classification of cancer: class discovery and class prediction by gene expression monitoring.* Science, 1999. **286**(5439): p. 531-7.
3. Nayal, M. and B. Honig, On the nature of cavities on protein surfaces: application to the identification of drug-binding sites. Proteins-Structure Function and Bioinformatics, 2006. **63**(4): p. 892-906.
4. Venkatesan, K., J.F. Rual, A. Vazquez, U. Stelzl, I. Lemmens, T. Hirozane-Kishikawa, *et al., An empirical framework for binary interactome mapping.* Nat Methods, 2009. **6**(1): p. 83-90.
5. Kihara, D., T. Shimizu and M. Kanehisa, Prediction of membrane proteins based on classification of transmembrane segments. Protein Eng, 1998. **11**(11): p. 961-70.
6. Messih, M.A., M. Chitale, V.B. Bajic, D. Kihara and X. Gao, *Protein domain recurrence and order can enhance prediction of protein functions.* Bioinformatics, 2012. **28**(18): p. i444-i450.
7. Xiong, Y., J. Liu and D.Q. Wei, *An accurate feature-based method for identifying DNA-binding residues on protein surfaces.* Proteins-Structure Function and Bioinformatics, 2011. **79**(2): p. 509-17.
8. Zhang, W., Y. Xiong, M. Zhao, H. Zou, X. Ye and J. Liu, Prediction of conformational B-cell epitopes from 3D structures by random forests with a distance-based feature. BMC Bioinformatics, 2011. **12**: p. 341.
9. Chen, X., M.H. Wang and H.P. Zhang, *The use of classification trees for bioinformatics.* Wiley Interdisciplinary Reviews-Data Mining and Knowledge Discovery, 2011. **1**(1): p. 55-63.
10. Quinlan, J.R., *Induction of decision trees.* Machine learning, 1986. **1**(1): p. 81-106.
11. Quinlan, J.R., *C4. 5: programs for machine learning*1993: Morgan Kaufmann.
12. Hecht-Nielsen, R. Theory of the backpropagation neural network. 1988. IEEE.
13. Cortes, C. and V. Vapnik, *Support-vector networks.* Machine learning, 1995. **20**(3): p. 273-297.
14. Breiman, L., *Random forests.* Machine learning, 2001. **45**(1): p. 5-32.
15. Rual, J.F., K. Venkatesan, T. Hao, T. Hirozane-Kishikawa, A. Dricot, N. Li, *et al., Towards a proteome-scale map of the human protein-protein interaction network.* Nature, 2005. **437**(7062): p. 1173-8.

16. Mohamed, T.P., J.G. Carbonell and M.K. Ganapathiraju, *Active learning for human protein-protein interaction prediction.* BMC Bioinformatics, 2010. **11 Suppl 1**: p. S57.
17. Scott, M.S. and G.J. Barton, Probabilistic prediction and ranking of human protein-protein interactions. BMC Bioinformatics, 2007. **8**: p. 239.
18. Stark, C., B.J. Breitkreutz, T. Reguly, L. Boucher, A. Breitkreutz and M. Tyers, *BioGRID: a general repository for interaction datasets.* Nucleic Acids Res, 2006. **34**(Database issue): p. D535-9.
19. Salwinski, L., C.S. Miller, A.J. Smith, F.K. Pettit, J.U. Bowie and D. Eisenberg, *The Database of Interacting Proteins: 2004 update.* Nucleic Acids Res, 2004. **32**(Database issue): p. D449-51.
20. Peri, S., J.D. Navarro, R. Amanchy, T.Z. Kristiansen, C.K. Jonnalagadda, V. Surendranath, *et al.*, *Development of human protein reference database as an initial platform for approaching systems biology in humans.* Genome Res, 2003. **13**(10): p. 2363-71.
21. Kerrien, S., B. Aranda, L. Breuza, A. Bridge, F. Broackes-Carter, C. Chen, *et al.*, *The IntAct molecular interaction database in 2012.* Nucleic Acids Res, 2012. **40**(Database issue): p. D841-6.
22. Licata, L., L. Briganti, D. Peluso, L. Perfetto, M. Iannuccelli, E. Galeota, *et al.*, *MINT, the molecular interaction database: 2012 update.* Nucleic Acids Res, 2012. **40**(Database issue): p. D857-61.
23. Guo, Y., L. Yu, Z. Wen and M. Li, Using support vector machine combined with auto covariance to predict protein-protein interactions from protein sequences. Nucleic Acids Res, 2008. **36**(9): p. 3025-30.
24. Shen, J., J. Zhang, X. Luo, W. Zhu, K. Yu, K. Chen, *et al.*, *Predicting protein-protein interactions based only on sequences information.* Proc Natl Acad Sci U S A, 2007. **104**(11): p. 4337-41.
25. Smialowski, P., P. Pagel, P. Wong, B. Brauner, I. Dunger, G. Fobo, *et al.*, *The Negatome database: a reference set of non-interacting protein pairs.* Nucleic Acids Res, 2010. **38**(Database issue): p. D540-4.
26. Walhout, A.J., R. Sordella, X. Lu, J.L. Hartley, G.F. Temple, M.A. Brasch, *et al.*, *Protein interaction mapping in C. elegans using proteins involved in vulval development.* Science, 2000. **287**(5450): p. 116-22.
27. Hawkins, T. and D. Kihara, *Function prediction of uncharacterized proteins.* J Bioinform Comput Biol, 2007. **5**(1): p. 1-30.
28. Rhodes, D.R., S.A. Tomlins, S. Varambally, V. Mahavisno, T. Barrette, S. Kalyana-Sundaram, *et al.*, *Probabilistic model of the human protein-protein interaction network.* Nat Biotechnol, 2005. **23**(8): p. 951-9.
29. Hunter, S., R. Apweiler, T.K. Attwood, A. Bairoch, A. Bateman, D. Binns, *et al.*, *InterPro: the integrative protein signature database.* Nucleic Acids Res, 2009. **37**(Database issue): p. D211-5.

30. Punta, M., P.C. Coggill, R.Y. Eberhardt, J. Mistry, J. Tate, C. Boursnell, *et al.*, *The Pfam protein families database.* Nucleic Acids Res, 2012. **40**(Database issue): p. D290-301.
31. Kersey, P.J., J. Duarte, A. Williams, Y. Karavidopoulou, E. Birney and R. Apweiler, *The International Protein Index: an integrated database for proteomics experiments.* Proteomics, 2004. **4**(7): p. 1985-8.
32. Liu, K.H. and C.G. Xu, A genetic programming-based approach to the classification of multiclass microarray datasets. Bioinformatics, 2009. **25**(3): p. 331-7.
33. Jin, Y. and L. Wang, *Fuzzy systems in bioinformatics and computational biology.* Vol. 242. 2009: Springer Verlag.

Chapter 10

Trend Analysis

Wen-Chuan Xie[1], Miao He[1,*] and Jake Yue Chen[2]

[1] *School of Life Sciences, Sun Yat-Sen University, Guangzhou 510275, China*
wenchuan.xie@gmail.com
**lsshem@mail.sysu.edu.cn*
[2] *Indiana Center for Systems Biology & Personalized Medicine, Indiana University - Purdue University Indianapolis, Indianapolis, IN 46202,USA*
Jake Yue Chen, Ph.D., Associate professor; Research field: bioinformatics.
jakechen@iupui.edu

Trend analysis often refers to the techniques for extracting an mode of behavior in a time series which would be partly or nearly completely hidden by noise. Now, the techniques of trend analysis have been widely used in detecting outbreaks and unexpected increases or decreases in disease occurrence, monitoring the trends of diseases, evaluating the effectiveness of disease control programs and policies, and assessing the success of health care programs and policies et al. common methods of trend analysis have been introduced in this which include Age-period-cohort (APC) model, Joinpoint regression, Time series analysis, Cox-Stuart test, Runs test and Functional data analysis (FDA) et al. The theoretical foundation of each method has simply described for convenience of readers. Specific details of each method include five parts, such as introduction, mathematical model, empirical study, problems and solutions, and a conclusion. Last, the of methods-related software and the procedure codes are described as supplements.

* Corresponding author.

1. Introduction

Trends in factors such as rates of disease and death, as well as behaviors such as smoking and drinking are often used by public health professionals to assist in healthcare that needs assessments, service planning, and policy development. Data are collected from a group over time to look for trends and changes.

The first step in assessing a trend is to plot the observations of interest by calendar year (or some other time period deemed appropriate). The observations can also be examined in a tabular form. The steps form the basis of subsequent analysis and provide an overview of the general shape of the trend, help identify any outliers in the data, and allow the researcher to become familiar with both the rates being studied. The process may involve comparing past and current rates as they related to various terms in order to project how long the current trend will continue. This type of information is extremely helpful to public health professionals and agencies who wish to make the most from their planning or monitoring.

Generally, the agencies of public health have the responsibilities to monitor trends in rates of disease and death and trends in medical, social, and behavioral risk factors that may contribute to these adverse events. Trends in observed rates can provide invaluable information for further consequence assessment, program planning, program evaluation, and policy development activities. Examining data over time also permits making predictions about future frequencies and rates of occurrence.

Typically in public health, trend data are presented for rates arising from large populations over relatively long periods of time (e.g. ten or more years). For example, the national vital records system is a source for trend analysis of infant mortality and other death rates. The national rates described in these analyses are very reliable and are available over many years insuring a precise characterization of changes over time. These rates are considered as the true underlying population parameters and therefore statistical assessment, which implies that the data are subject to sampling error, is rarely undertaken. If rates are assumed to be error-free, they can be presented "as is" in tables or graphs, and comparisons across populations or predictions of future occurrence can be made intuitively.

The public health community is increasingly interested in examining trends for smaller populations and in smaller geographic areas. In maternal and child health, for example, describing trends in perinatal outcomes in relation to trends in prenatal care utilization within and across local service delivery areas is critical for program planning. There is also interest in examining trends in indicators of emerging health problems which, by definition, are only available for short periods of time. Once the focus of trend analysis is on data from small areas, small populations, or for a narrow range of time, it is necessary to draw from both the classic descriptive methods and the statistical approaches used in research studies. [1]

So, what is trend analysis? As a formal definition, trend analysis can be defined as the practice of collecting data and attempting to spot a certain pattern, or trend in the data. In some fields of study, the term "trend analysis" has more formally definitions. In another way, trend analysis often refers to the techniques for extracting an underlying mode of behavior in a time series which would be partly or nearly completely hidden by noise. A more simple description of these techniques is trend estimation, which can be undertaken within a formal regression analysis.

2. *Methods of trend analysis*

2.1. *Age-period-cohort model*

2.1.1. *Introduction*

The Age-Period-Cohort (APC) models, which could be proposed by Norman Ryder(1965), [2] is a class of models for demographic rates (mortality/morbidity/fertility/ etc) observed for a broad age range over a reasonably long time period, and classified by age, period and cohort. It is used widely in number of disciplines such as sociology, economics and epidemiology. In medical research, epidemiologists always use this model to identify and interpret temporal changes in health characteristics or behaviors, especially in the field of chronic disease. One common aim of fitting an APC model is to assess the effects of the three factors on disease rates. In this section, we will discuss it in the following aspects.

2.1.2. *Theoretical foundation*

Mathematics Model

Before the APC models were constructed, the three variables must meet the following hypothesis:

(1) The number of cases in age group i at time period j is denoted by D_{ij} and is a realization of poisson random variable.

(2) Random variables D_{ij} are jointly independent.

Most of the APC models can be expressed as a Generalized Linear Model (GLM)

$$M_{ij} = \frac{D_{ij}}{P_{ij}} = \mu + \alpha_i + \beta_j + \gamma_k + \varepsilon_{ij} \tag{1}$$

where $\gamma = \beta - \alpha$;

M_{ij} denotes the observed occurrence/exposure rate of incidence/deaths for the i-th age group for i = 1,…, A age groups at the j-th time period for j = 1,…, P time periods of observed data.

D_{ij} denotes the number of new cases/deaths in the ij-th group, P_{ij} denotes the size of the estimated population in the ij-th group.

μ denotes the intercept or adjusted mean effect.

α_i denotes the effect of the i-th age group, differing risks associated with different age groups.

β_j denotes the effect of the j-th time period.

γ_k denotes the k-th cohort effect for k = 1,…,(A+P−1) cohorts, with k=A−i+j.

ε_{ij} denotes the random errors with expectation $E(\varepsilon_{ij}) = 0$.

Conventional APC models represented in formula (1) fall into the class of generalized linear models (GLIM) which can be changed into various forms. It can be transformed to a *log*-linear regression model form via *log* link as

$$log(D_{ij}) = log(P_{ij}) + \mu + \alpha_i + \beta_j + \gamma_k \tag{2}$$

where D_{ij} denotes the expected number of events in cell (i, j) which is assumed to be distributed as a Poisson variant, and $\log(P_{ij})$ is the log of the exposure P_{ij}. This model is most popular in epidemiology where the disease counts follow Poisson distributions. The second alternative formulation is changed from *log* link to a *logit* link

$$\theta_{ij} = \log\left(\frac{m_{ij}}{1-m_{ij}}\right) = \mu + \alpha_i + \beta_j + \gamma_k \tag{3}$$

where θ_{ij} is the log odds of event and m_{ij} is the probability of event in cell (i, j). This model has been implemented more widely in demographic study.

Generally, we can estimate the parameters of APC models via the following methods, maximal likelihood estimates and Bayesian statistics. Additionally, the parameters of APC models can be estimated via centering to satisfy the following:

$$\sum_i \alpha_i = \sum_j \beta_j = \sum_k \gamma_k = 0 \tag{4}$$

Data Description

Table 1 shows such rates data structure which the entries in the age-period array, the rates data can be represented by two separate arrays: one for the number of events (counts) (table 2), the other for the population exposure (table 3), and the ratio of these two is the rate.

Table 1. M matrix

Age	**Period**		
	2000-2004	2005-2009	...
0-14	0.1	0.3	...
15-19	0.5	0.6	...
...	...	...	...

Table 2. D matrix

Age	**Period**		
	2000-2004	2005-2009	...
0-14	1	6	...
15-19	15	30	...
...	...	...	...

Table 3. P matrix

Age	Period		
	2000-2004	2005-2009	....
0-14	10	20	...
15-19	30	50	...
...	...	...	...

2.1.3. *Empirical study*

APC models are widely used by epidemiologists to analyze trends in disease incidence and mortality, such as diagnosis of cancer, stroke, diabetes etc.

Take the Japan's Chronic Obstructive Pulmonary Disease(COPD) for instance, they used APC model to analyze the trend of COPD mortality by age, time period, and birth cohort among adults aged 40 years or older from 1950 to 2004. [3] The age-standardized mortality rates(ASMRs) were tabulated into 10 five-year age groups (from age 40 to 44 years to 85+ years), 11 five-year time periods (from the 1950–1954 year interval to the 2000–2004 year interval), and 20 five-year birth cohorts (from the 1865–1869 year interval to the 1960–1964 year interval), then analyzed the effects of age, time period, and birth cohort on COPD mortality using APC model by means of Poisson regression (the formula 1.2). They applied the Intrinsic Estimator (IE) (see the section 4) Problem/Limits) method to address the identification problem and to provide parameter estimates. Figure 1 depicted the effects of age, time period, and birth cohort, which were expressed in terms of relative risk. The age effects increased with age in both sexes. The period effects rapidly declined in both sexes during the first 10-year period. In men, period effects tended to increase in recent years, while they continued to decrease in women. With regard to birth cohort, an increased effects in both sexes, starting from the 1865–1869 cohort. The highest values were observed in the 1880–1889 cohorts, and they decreased continuously thereafter up to the last birth cohort of 1960–1964.

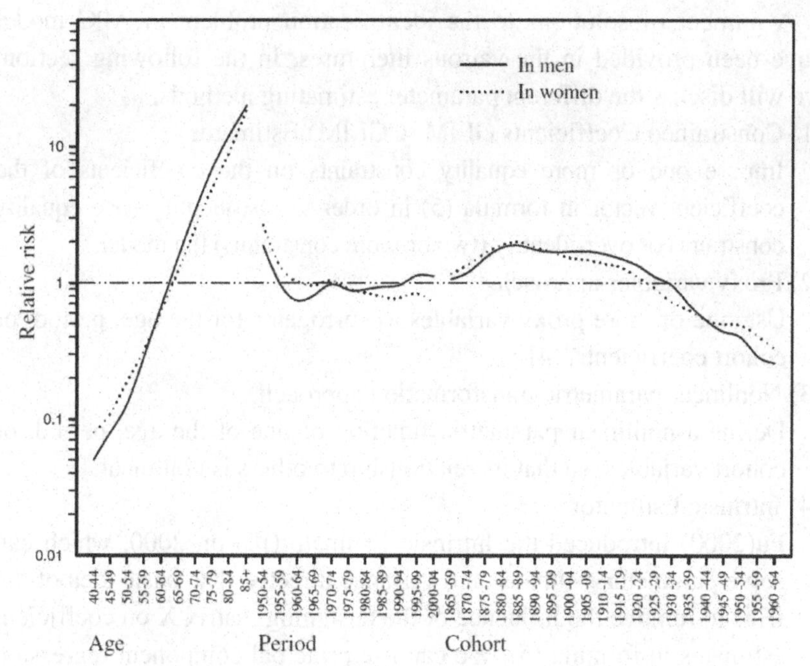

Fig. 1. Effects of age, time period, and birth cohort on chronic obstructive pulmonary disease in men and women [3].

2.1.4. *Problems and solutions*

However, due to the linear dependency of age, period and cohort (Period=Age+Cohort), there is the identification problem, that is, the parameter estimates are not unique. It became a major methodological challenge to APC analysis. Rewriting model (1) in matrix form, we have formula (5):

$$Y = Xb \tag{5}$$

$$b = (\mu, \alpha_1, \dots \alpha_{A-1}, \beta_1, \dots \beta_{P-1}, \gamma_1, \dots \gamma_{A+P-1}) \tag{6}$$

Identification Problem:

$$\hat{b} = (X^T X)^{-1} X^T Y \tag{7}$$

The solution to these normal equations does not exist, because $(X^T X)^{-1}$ does not exist.

A number of solutions to the identification problem in APC model have been provided in the vairous literatures. In the following section, we will discuss the different parameter estimating methods.

(1) Constrained Coefficients GLIM (CGLIM) Estimator

Impose one or more equality constraints on the coefficients of the coefficient vector in formula (5) in order to just-identify (one equality constraint) or over-identify (two or more constraints) the model.

(2) Proxy variables approach

Use one or more proxy variables as surrogates for the age, period, or cohort coefficients. [4]

(3) Nonlinear parametric transformation approach

Define a nonlinear parametric function of one of the age, period, or cohort variables, so that its relationship to others is nonlinear.

(4) Intrinsic Estimator

Fu(2000) introduced the Intrinsic Estimator(IE) on 2000, which can solve parameter estimates and variance directly. The basic idea of the IE is to remove the influence of the designing matrix X on coefficient estimates in formula (5). We can use principal component regression to estimate the IE

2.1.5. *Computation software*

Currently, the software for APC analysis is available only through fairly specialized packages (SAS, R, Stata, Matlab). The *Epi* package mainly focuses on "classical" chronic disease epidemiology analysis in R. It is available from CRAN which can also be found through the R homepage. Moreover, the *epitools* package is with more affinity to infectious disease epidemiology

2.1.6. *Conclusion*

Age-period-cohort models are appropriate to fit observed disease incidence or mortality rates. Because of identifiability problem, the results must be treated with caution.

2.2. *Joinpoint regression*

2.2.1. *Introduction*

Regression analysis is the most common and simple method used to analyze trends over time where there is a linear relationship between rate or frequency and calendar year.

Joinpoint regression is used to find the best-fit line through several years of data; it uses an algorithm that tests whether a multi-segmented line is a significantly better fit then a straight or less-segmented line. This method describes changes in data trends by connecting several different line segments on a log scale at ''joinpoints''. This model assumes a linear trend between joinpoints and continuity at the joinpoints. The method is also known as piecewise regression, segmented regression, [5] broken line regression, or multiphase regression with the continuity constraint. The regression was performed using Joinpoint software from the Surveillance Research Program of the National Cancer Institute in the USA, the latest release: Version 3.5.3(May 14, 2012).

2.2.2. *Mathematical model*

The Joinpoint regression model for the observations: $(x_1, y_1), ...,$ (x_N, y_N), where $x_1 \leq \cdots \leq x_N$ represent the time variable, e.g. calendar year and y_i, $i = 1,2, \ldots, N$ are the response variable, e.g. the annual age standardized rates or frequencies, can be written as: [6]

$$E[y_i|x_i] = \beta_0 + \beta_1 x_i + \gamma_1 (x_i - \tau_1)^+ + \cdots + \gamma_n (x_i - \tau_n)^+ \qquad (1)$$

where $\beta_0, \beta_1, \gamma_1, \ldots, \gamma_n$ are regression coefficients and the $\tau_k, k = 1,2, \ldots, n, (n < N)$, is the k-th unknown joinpoint in which:

$$(x_i - \tau_k)^+ = \begin{cases} (x_i - \tau_k), & if\ (x_i - \tau_k) > 0 \\ 0, & otherwise \end{cases} \qquad (2)$$

Noting that the homogeneous variances and independence assumptions are usually not valid for time series data, heteroscedastic variances are assumed and the weighted least square method is employed for the inference and computation of the Joinpoint linear regression model. The underlying distribution for response variable y or its transformation can be assumed as:

(1) $y_i \sim N(\mu_i, \sigma^2)$ for homoscedastic rates or log-rates;
(2) $y_i \sim N(\mu_i, \sigma_i^2)$ for heteroscedastic rates or log-rates;
(3) $y_i \sim Poi(\mu_i)$ for counts without adjustment (offset) for population size;
(4) $y_i \sim Poi(\mu_i/p_i)$ for counts with adjustment (offset) for population size.

Tests of significance use a Monte Carlo permutation method.

2.2.3. *Empirical study*

Joinpoint analysis has been commonly used to describe the changing trends over distinct periods of time and significant increases or decreases in chronic disease morbidity or mortality.

Now, we used Joinponit regression to analyze the mortality trend of male liver cancer in Hong Kong population during 1983~2010. We estimated the age-standardized mortality rates (ASMR) first. Joinpoint regression was applied to detect significant changes in cancer mortality.

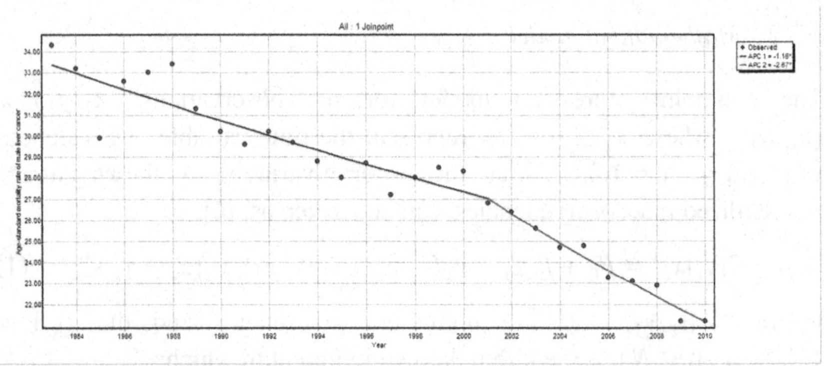

Figure 2. 'Best' Joinpoint Model Estimates for male liver cancer in Hong Kong, 1983-2010

Figure 2 listed the results of the Joinpoint analysis from 1983 to 2010. ASMR of male liver cancer showed a decreasing trend with approximately 1.57% per annum from 1983 to 2010. In the latest 10 years, the annual percent change was down to -2.67% (95%CI: -3.5,-1.9).

Llorca J et al. demonstrated a decline trend in Tuberculosis mortality in Spain from 1971 to 2007 in female and in elderly patients. [7] They used Joinpoint regression estimate the annual percentage change (APC) and average annual percentage change (AAPC) for every age and sex group. They found that Tuberculosis mortality in Spain decreases faster in men than in women, and in young adults than in the elderly.

Forte T et al. used Joinpoint regression to analyze the trend of age-standardized incidence rate (ASIR) in head and neck cancers and human papillomavirus (HPV) associated oropharyngeal cancer in Canada from 1992 to 2009. [8] A significantly decline trend were observed in ASIR of head and neck cancers from 1992 to 1998(APC = -3.0, $p<0.01$), then remained stable through to 2009. But in contrast, the ASIR of HPV-associated oropharyngeal cancer increased significantly during the period (APC = 2.7, $p<0.001$). Their finding highlights the need to survey HPV-associated oropharyngeal cancer separately from other cancers of the head and neck region in order to monitor these emerging trends.

To examine whether the recent flattening of mortality rates for coronary heart disease (CHD) observed among young adults in the UK and the US was also occurring in the Australian population, O'Flaherty M et al. used Joinpint regression to estimate the annual change and detect points in time where significant changes in the trends occur. [9] They found that the CHD mortality decline had slowed since the early 1990s in Australian men and women aged 25-54 years. The most likely explanations for reduction of the CHD mortality decline were attenuations or reversal of the earlier declines in major traditional risk factors (tobacco smoking, serum cholesterol, and blood pressure) and diabetes mellitus.

We analyzed the incidence trends of fifteen common cancers in Hong Kong by Joinpoint regression. Our study suggested that the incidence trends for colorectal, thyroid and sex-related cancers continue to rise from 1983 to 2008. [10]

2.2.4. *Problems and solutions*

It is important for the users to understand that goal of the Joinpoint regression is not to provide models that best fit the data, but models that best summarize the behaviour or the data trend across years. The challenge in Joinpoint regression is to determine the locations of the joinpoints if they exist; and to determine the optimal number of joinpoints for the most appropriate model. Joinpoint is available from the US National Cancer Institute's Statistical Research and Applications Branch.

2.2.5. *Conclusion*

The Joinpoint analysis of the trends in the age-standard disease (mainly cancer) incidence and mortality rates allows the user to more accurately interpret changes over time and, more importantly, to determine if those changes are statistically significant.

2.3. *Time series analysis*

2.3.1. *Introduction*

Time series analysis is a powerful statistical method to analyze data from repeated observations on a single unit or individual at regular intervals over a large number of observations.[11] One goal of the analysis is to identify patterns in the sequence of numbers over time, which are correlated with themselves, but offset in time. Another goal in many research applications is to test the impact of one or more interventions. Time series analysis is also used to forecast future patterns of events or to compare series of different kinds of events.

There are several types of data analysis available for time series which are appropriate for different purposes: general exploration; description; prediction and forecasting. This approach has been widely applied in many fields: financial forecasting, business forecasting, stock market forecasting, chronic/infectious diseases prevention etc.

2.3.2. *Mathematical model*

Mathematical models for time series analysis can have many forms and represent different stochastic processes. Before we discuss these models, we should introduce some definitions first.

Definition 1 The sequence of values $\{x_t\}_{t=1,\cdots,n}$ over time is called a *time series.*

We typically want to forecast the next value in the series (1-step ahead prediction) or the value at k time periods in the future (k-step ahead prediction). An observed time series can be decomposed into three components:

- **Trend (T_t)**:a long term increase or decrease occurs.
- **Seasonal (S_t)**: series influenced by seasonal factors.
- **Cyclical (C_t)**:series rises and falls regularly but these are *not of fixed period.*
- **Residual (e_t)**: this corresponds to random fluctuations that cannot be explained by a deterministic pattern).

Definition 2 The *Sum of Square Errors* (SSE) is defined by:

$$SSE = \sum_{t=1}^{n}(x_t - F_t)^2 \qquad (1)$$

where F_t is the fitted value using in the analysis model.

Definition 3 The *Root Mean Square Error* (RMSE) of the model is:

$$RMSE = \sqrt{\frac{1}{n}\sum_{t=1}^{n}(x_t - F_t)^2} = \sqrt{\frac{SSE}{n}} \qquad (2)$$

Definition 4 The *Mean Absolute Percent Error* (MAPE) is:

$$MAPE = 100\frac{1}{n}\sum_{t=1}^{n}\left|\frac{x_t - F_t}{x_t}\right| \qquad (3)$$

Definition 5 The *AutoCorrelation Function (ACF).* For a time series $\{x_t\}_{t=0,1,\cdots,n}$ the formula of autocorrelation at *lag k* are:

$$ACF(k) = r_k = \frac{\sum_{t=k+1}^{n}(x_t - \bar{x})(x_{t-k} - \bar{x})}{\sum_{t=1}^{n}(x_t - \bar{x})^2} \tag{4}$$

where $\bar{x} = \frac{1}{n}\sum_{t=1}^{n} x_t$ is the mean of the series. The ACF can be plotted reporting the values r_k on the *y-axis* with the *lag k* on the abscissa.

Definition 6 The *Partial AutoCorrelation Function (PACF)* is another useful method to examine serial dependencies. This is an extension of the autocorrelation, where the dependence on the intermediate elements (those within the lag) is removed. The set of partial autocorrelations at different lags is called the *partial autocorrelation function* (PACF) and is plotted like the ACF.

Definition 7 (AIC and BIC criteria) We define two types of information criterion: the *Bayesian Information Criterion* (BIC) and the *Akaike Information Criterion* (AIC). In AIC and BIC, we choose the model that has the minimum value of:

$$\begin{cases} \text{AIC} = -2\log(\text{L}) + 2\text{m} \\ \text{BIC} = -2\log(\text{L}) + \text{mlogn} \end{cases} \tag{5}$$

where L is the likelihood of the data with a certain model; n is the number of observations and m is the number of parameters in the model, the number of parameters is $m = p + q$ for an $ARMA(p, q)$ model.

There are two methods, which were mentioned at different time by different men. One is Exponential Smoothing Algorithm and the other is ARIMA model.

Exponential Smoothing Algorithm

Exponential Smoothing Algorithm was first suggested by Charles C. Holt in 1957. It was originally classified by Pegels'(1969) taxonomy, later extended by Gardner,[12] modified by Hyndman et al. [13] Exponential smoothing forecasts are a weighted sum of all the previous observations.

(1) Single Exponential Smoothing (SES) Algorithm

SES algorithm is designed for time series with no trend nor seasonal patterns. Because we cannot forecast the first term in the series, by convention, we fix:

$$F_1 = x_1;\ large\ sample(n \geq 42) \tag{6}$$

$$F_1 = \frac{(x_1 + x_2 + x_3)}{3};\ small\ sample(n < 42) \tag{7}$$

and the formula is:

$$\begin{aligned} F_{t+1} &= \alpha x_t + (1-\alpha)F_t \\ &= \alpha[x_t + (1-\alpha)x_{t-1} + (1-\alpha)^2 x_{t-2} + \cdots + \\ &(1-\alpha)^{t-1}x_1] + (1-\alpha)^t F_1 \end{aligned} \tag{8}$$

where α is the smoothing factor, and $0<\alpha<1$; we can use SSE or RMSE or MAPE for finding the best parameter α, select the smallest SSE or RMSE or MAPE

(2) Double Exponential Smoothing (DES) Algorithm

SES algorithm does not do well when there is a trend in the data. In such situations, Double Exponential Smoothing (DES) algorithm (also known as Holt's Linear Method) was devised.[14] It allows the estimates of level (L_t) and solpe (b_t) to be adjusted with each new observation.
In it:

$$F_1 = x_1 \tag{9}$$

$$L_1 = x_1 \tag{10}$$

$$b_1 = x_2 - x_1 \tag{11}$$

and choose $0 \leq \alpha \leq 1$ and the trend smoothing factor β $(0 \leq \beta \leq 1)$, compute and forecast:

$$L_t = \alpha x_t + (1-\alpha)(L_{t-1} + b_{t-1}) \tag{12}$$

$$b_t = \beta(L_t - L_{t-1}) + (1-\beta)\, b_{t-1} \tag{13}$$

$$F_{t+1} = L_t + b_t \tag{14}$$

Until no more observation are available then

$$F_{n+k} = L_n + kb_n, \forall\, k \geq 1 \tag{15}$$

F_{n+k} is the output, it is an estimate of the value of x at time $n + k$.

(3) Holt-Winters' Exponential Smoothing with Seasonality

Generally, time series data display behavior that is seasonal, it is defined to be the tendency of the time series data to exhibit behavior that repeats itself every s periods (s is also known as length of the seaaonal cycle). SES and DES algorithms cannot process the time series data with a trend and seasonal behavior. Winters generalized Holt's linear method to come up with such a technique, now called Holt Winters (also known as Triple Exponential Smoothing Algorithm). A seasonal equation is added to Holt's linear method equations. It is done in two ways, additive and multiplicative.

How do we discriminate the two types of seasonal components? In plots of the time series, the distinguishing characteristic between these is that in the additive case, the time series shows steady seasonal fluctuations, regardless of the overall level of the series; in the multiplicative case, the size of the seasonal fluctuations varies, depending on the overall level of the series.

(3.1) Seasonal Holt-Winter's Additive Model Algorithm

Seasonal Holt-Winter's Additive Model Algorithm is noted SHW +. The algorithm can be written as follow,
In it:

$$L_s = \frac{1}{s}\sum_{i=1}^{s} x_i \tag{16}$$

$$b_s = \frac{1}{s}\left[\frac{x_{s+1} - x_1}{s} + \frac{x_{s+2} - x_2}{s} + \cdots + \frac{x_{2s} - x_s}{s}\right] \tag{17}$$

$$S_i = x_i - L_s, i = 1, \ldots s \tag{18}$$

And choose $0 \leq \alpha \leq 1$ and $0 \leq \beta \leq 1$ and $0 \leq \gamma \leq 1$. Compute for $t > s$:

$$\textit{Level} \quad L_t = \alpha(x_t - S_{t-s}) + (1-\alpha)(L_{t-1} + b_{t-1}) \tag{19}$$

$$\textit{Trend} \quad b_t = \beta(L_t - L_{t-1}) + (1-\beta)b_{t-1} \tag{20}$$

$$\textit{Seasonal} \quad S_t = \gamma(x_t - L_t) + (1-\gamma)S_{t-s} \tag{21}$$

$$\textit{Forecast} \quad F_{t+1} = L_t + b_t + S_{t+1-s} \tag{22}$$

Until no more observation available, and subsequent forecasts:

$$\mathrm{F_{n+k} = L_n + kb_n + S_{n+k-s}} \tag{23}$$

(3.2) Seasonal Holt-Winter's Multiplicative Model Algorithm

Seasonal Holt-Winter's Multiplicative Model Algorithm is noted SHW ×. The algorithm can be written as follow,
In it:

$$L_s = \frac{1}{s}\sum_{i=1}^{s} x_i \tag{24}$$

$$b_s = \frac{1}{s}\left[\frac{x_{s+1} - x_1}{s} + \frac{x_{s+2} - x_2}{s} + \cdots + \frac{x_{2s} - x_s}{s}\right] \tag{25}$$

$$S_i = \frac{x_i}{L_s}, i = 1, \dots s \tag{26}$$

and choose $0 \le \alpha \le 1$ and $0 \le \beta \le 1$ and $0 \le \gamma \le 1$. Compute for $t > s$:

$$\textit{Level} \quad L_t = \alpha\frac{x_t}{S_{t-s}} + (1-\alpha)(L_{t-1} + b_{t-1}) \tag{27}$$

$$\textit{Trend} \quad b_t = \beta(L_t - L_{t-1}) + (1-\beta)b_{t-1} \tag{28}$$

$$\textit{Seasonal} \quad S_t = \gamma\frac{x_t}{L_t} + (1-\gamma)S_{t-s} \tag{29}$$

$$\textit{Forecast} \quad S_t = \gamma\frac{x_t}{L_t} + (1-\gamma)S_{t-s} \tag{30}$$

Until no more observation available, and subsequent forecasts:

$$\mathrm{F_{n+k} = (L_n + kb_n)S_{n+k-s}} \tag{31}$$

For a time series data with trend and seasonal patterns, as with the SES and DES methods, we select the best Holt-Winters algorithms with the smallest SSE or RMSE or MAPE.

ARIMA model

As with the Exponential Smoothing algorithm, we introduce some definition before we discuss the ARIMA method.

Definition 8 *Autoregressive model.* Consider a time series $\{x_t\}_{t=1,\cdots,n}$. An *autoregressive model* of order p (denoted $AR(p)$) states that x_i is the linear function of the previous p values of the series plus an error term:

$$x_i = \phi_0 + \phi_1 x_{i-1} + \phi_2 x_{i-2} + \cdots + \phi_p x_{i-p} + \varepsilon_i, i > p \quad (32)$$

where $\phi_1, .. \phi_p$ are weights that we have to define, and $\varepsilon_i \sim N(0, \sigma^2)$.

Definition 9 *Moving Average model.* A *moving average model of order q*, noted $MA(q)$, is a time series model defined as follows:

$$x_t = \psi_0 + \psi_1 \varepsilon_{t-1} + \psi_2 \varepsilon_{t-2} + \cdots + \psi_p \varepsilon_{t-p} + \varepsilon_t \quad (33)$$

where ε_t are independent errors, normally distributed with mean 0 and variance σ^2: $N(0, \sigma^2)$.

Definition 10 *AutoRegressive Moving Average models(ARMA).* combining AR and MA models, we can define $ARMA(p, q)$ models as:

$$\begin{aligned} x_t &= \phi_0 + \phi_1 x_{t-1} + \phi_2 x_{t-2} + \cdots + \phi_p x_{t-p} + \psi_0 \\ &\quad + \psi_1 \varepsilon_{t-1} + \psi_2 \varepsilon_{t-2} + \cdots + \psi_p \varepsilon_{t-p} + \varepsilon_t \\ &= c + \phi_1 x_{t-1} + \phi_2 x_{t-2} + \cdots + \phi_p x_{t-p} + \psi_1 \varepsilon_{t-1} \\ &\quad + \psi_2 \varepsilon_{t-2} + \cdots + \psi_p \varepsilon_{t-p} + \varepsilon_t \end{aligned} \quad (34)$$

Definition 11 *Stationary in mean.* A time series is called stationary in mean if it randomly fluctuates about a constant mean level.

Definition 12 *Stationary in variance.* A time series is said to be stationary in variance if the variance in the time series does not change with time.

Definition 13 *Backshift operator.* In what follows, it will be very useful to denote a lagged series by using the backshift operator B:

$$By_t = y_{t-1} \quad (35)$$

For lags of length k, we apply B k times:

$$B^k y_t = y_{t-k} \quad (36)$$

An ARMA model should only be fitted to time series which are stationary in mean (e.g. no trend or no seasonal pattern) and stationary in variance. Differencing is an operation that can be applied to a time series to remove a trend. If the time series looks stationary in mean and variance after differencing, then an ARMA(p,q) model can be used. Next section presents differencing (of order d) and extending the ARMA(p,q) models to the ARIMA(p,d,q) models.

Definition 14 *Differencing*. Consider the time series $\{x_t\}_{t=0,1,\cdots,n}$, the first order differencing is defined as

$$x'_t = x_t - x_{t-1} \quad (37)$$

We can use the *backshift operator B* to express differencing:

$$x'_t = (1 - B)x_t \quad (38)$$

So, the order d differencing can be defined as

$$x^d_t = (1 - B)^d x_t \quad (39)$$

The ARIMA model is defined with an AutoRegressive part of order p, a moving average part of order q and having applied d order differencing:

$$\begin{aligned}&(1 - \phi_1 B - \phi_2 B^2 - \cdots - \phi_p B^p)(1 - B)^d x_t = c\\&+(1 - \psi_1 B - \psi_1 B^2 - \cdots - \psi_q B^q)\varepsilon_t\end{aligned} \quad (40)$$

Often, there are some criterions to bear in mind. First, the values of p, q *or* d of more than three are very rarely needed. Second, it is often the case that many different ARIMA models give more or less the same predictions, so there is some flexibility in the choice of p, d *and* q.

Seasonal ARIM • $(\mathbf{p}, \mathbf{d}, \mathbf{q})(\mathbf{P}, \mathbf{D}, \mathbf{Q})_s$ model

As we can see that, ARIMA model cannot cope with seasonal patterns, it only models time series data with trends. Now we will incorporate seasonal patterns and present a definition of Seasonal ARIMA models.

Definition 15 *Seasonal ARIMA models.* $ARIMA(p,d,q)(P,D,Q)_s$ is defined as follow,

$$\begin{aligned}(1-\phi_1 B-\phi_2 B^2-\cdots\\ -\phi_p B^p)(1-\beta_1 B^s-\beta_2 B^{2s}-\cdots\\ -\beta_P B^{Ps})(1-B)^d(1-B^s)^D x_t\\ =c+(1-\psi_1 B-\psi_1 B^2-\cdots\\ -\psi_q B^q)(1-\theta_1 B^s-\theta_2 B^{2s}-\cdots\\ -\theta_Q B^{Qs})\varepsilon_t\end{aligned} \tag{41}$$

where

$AR_s(P)=(1-\beta_1 B^s-\beta_2 B^{2s}-\cdots-\beta_P B^{Ps})$ is seasonal autoregressive part of order P.

$\mathrm{MA_s(Q)}=(1-\theta_1 B^s-\theta_2 B^{2s}-\cdots-\theta_Q B^{Qs})$ is seasonal moving average part of order Q.

$\mathrm{I_s(D)}=(1-B^s)^D$ is seasonal differencing of order D.

s is the period of the seasonal pattern.

We can use ACF and PACF to identify P or Q, we just look at the coefficients computed for multiples of s. Sometimes, it is not possible to identify the parameters p,d,q and P,D,Q using visualisation tools such as ACF and PACF. Using the BIC or AIC as the selection criterion, we select the ARIMA model with the lowest value of the BIC or AIC.

Making time series stationary in variance

$ARIMA(p,d,q)(P,D,Q)_s$ can deal with time series data which are stationary in variance. But often, the time series data shows both a trend and seasonal component, suggesting that, it is not stationary in mean. And the amplitude of the variation's increasing overtime from year to year means it is not stationary in variance. Here, we present four transformations to reduce variance by differing amount which depends on how much the variance is increasing with time.

Table 4. Four different transformation forms

Name	Mathematical expression
Square root	$\sqrt{x_i}$
Cube root	$\sqrt[3]{x_i}$
Logarithm	$log(x_i)$
Negative Reciprocal	$-1/x_i$

2.3.3. *Empirical study*

The Box-Jenkins methodology used in analysis and forecasting is widely regarded to be the most efficient forecasting technique, and is used extensively-specially for univariate time series. It has become of increasing importance in such fields as economics and industry, similar applications are also relevant in epidemiology. For example, in some viral infectious diseases,[15] the model-building process is demonstrated in some detail using monthly case reports of the two seasonal endemic diseases chicken-pox and mumps, and the relation between these two time series is investigated. Evaluation of the impact of extreme temperatures on human health (the association between winter temperatures and the mortality of circulatory and respiratory diseases), ARIMA and GAM models were used.[16]

Brazil's epidemiologists use the Box-Jenkins approach to fit an autoregressive integrated moving average (ARIMA) model by software R to dengue incidence in Rio de Janeiro, from 1997 to 2004, and then used to predict dengue incidence for the year 2005.[17] They used the Box-Jenkins approach to ARIMA modeling of time series, which consists of a four-step process. First, they evaluated the need for variance-stabilizing transformations using the mean-range plot. Second, they determined the order of non-seasonal (p, d, q) and seasonal $(P, D, Q)_{12}$ autoregressive (AR) parameters (p and P) and moving average (MA) parameters (q and Q), and the need for non-seasonal and seasonal differencing (d and D), using the following five tools: 1) The plot of dengue incidence, which assists in the needs for non-seasonal and seasonal differencing; 2) The autocorrelation (ACF) and partial autocorrelation (PACF) functions, which indicate the temporal dependence structure in the stationary time series; 3) The Akaike Information Criterion (AIC), which assists in the goodness-of-fit of the model, whereas penalizing for the number of parameters; 4) The Ljung-Box test, which measures the ACF of the residuals; and, 5) the significance of the parameters, which should be statistically different from zero (that is, the t statistic should exceed 2 in absolute value). Third, they estimated the parameters of the ARIMA model by maximum likelihood. Finally, they graphically compared the model's fitted values with the observed data to check if it indeed models dengue incidence.

2.3.4. *Computation software*

Currently, much software can carry out time series analysis. Generally SAS, R, SPSS, Stata and Matlab are commonly used.

2.3.5. *Conclusion*

Exponential Smoothing algorithm and ARIMA model are two types method for analyzing the time series data. Table 5 shows the connection between these two methods.

Table 5. The relation between exponential smoothing and ARIMA methods

Exponential Smoothing algorithms	relations	ARIMA models
Single Exponential Smoothing	$\equiv$	$ARIMA(0,1,1)$
Double Exponential Smoothing	$\equiv$	$ARIMA(0,2,2)$
Holt-Winters' additive method	$\subset$	$ARIMA(0,1,s+1)(0,1,0)_s$
Holt-Winters' multiplicative method	\	*No ARIMA equivalent*

One advantage of the exponential smoothing models is that they can be non-linear. So time series that exhibit non-linear characteristics including heteroscedasticity may be better modeled using exponential smoothing state space models, beside, the results in Hyndman et al.[13] show that the exponential smoothing models performed better than the ARIMA models for the seasonal M3 competition data. (For the annual M3 data, the ARIMA models performed better). The philosophy of exponential smoothing is that the world is non-stationary. So if a stationary model is required, ARIMA models are better.

2.4. *Cox-Stuart trend test*

2.4.1. *Introduction*

In 1955, Cox and Stuart introduced a non-parametric test for increasing or decreasing trend that was based on the sign test, which based on the binomial distribution.[18] The basic principle of this method is that a series of observations is said to exhibit an upward trend if the magnitudes of the later observations tend to be greater than those of the earlier observations, the data exhibit a downward trend if the earlier observations tend to be larger than the later observations.

2.4.2. *Mathematical model*

Cox-Stuart test is designed to test a time series $\{x_t\}_{t=1,2,\cdots,n}$ (n must be even, if not, we delete the middle one, middle one $= x_{\frac{n+1}{2}}$) on whether there is an upward or downward trend. We generate a set of data pairs in the form of $\left(x_1, x_{\frac{n}{2}+1}\right), \left(x_2, x_{\frac{n}{2}+2}\right), \ldots, \left(x_{\frac{n}{2}}, x_n\right)$. Calculation was carried out by identifying the difference between each pair of data $(x_i - x_{\frac{n}{2}-i})$. S_+ is used to mark positive differences and S_- for negative differences, the total $n = S_+ + S_-$. Pairs of data with zero difference are abandoned. Corresponding to different null hypothesis, the selection of statistics is different. If there is no change in the sequence trend, then S_+ and S_- will both obey the binomial distribution $B(n, 0.5)$, where $p = 0.5$. Three different hypotheses have been listed in Table 6.

Table 6. Three different null and alternative hypotheses and statistic

H_0	H_1	statistic
Sequence without a rising trend	Sequence with a rising trend	S_+, one − tailed test
Sequence without a declining trend	Sequence with a declining trend	S_-, one − tailed test
Sequence without a trend	Sequence with a rising or declining trend	$\min(S_+, S_-)$, two − tailed test

In terms of the sample size, the statistical method for different sample size is difference. Generally speaking, when the number of sample is no more than 25, we think of it as a small sample and contrarily as a large sample.

Test for small sample ($n \leq 25$)

The probability for the event X equals to or is greater than k which given by

$$p\_value = f(X \geq k|n) = \sum_{i=k}^{n} \binom{n}{i} p^i q^{n-i} \quad (1)$$

where k is the number of S_+ or S_-. For a 5% level of significance, if p_value is less than 5% significance level, the alternative hypothesis would be accepted. If the p_value is greater than the 5%, the alternative hypothesis is discarded. Note, we test only one-sided here, when we carry out the two-sided test, the p_value should be multiply by 2.

Test for large sample ($n > 25$)

$$Z = \frac{|k - n*p| - 0.5}{\sqrt{n*p*q}} \quad (2)$$

where k is the number of S_+ or S_-, and ± 0.5 is a corrected value, when $k \geq \frac{n}{2}$, $k - 0.5$ is appropriate; when $k < \frac{n}{2}$, $k + 0.5$ is best.

For a 5 percent significance level, if $|Z|$ is less than 1.96, the null hypothesis would be accepted, otherwise, we could discard H_0.

2.4.3. *Empirical study*

Cox–Stuart Test of Trends in mortality of female breast cancer in Hong Kong

Table 7. Cox–Stuart test of trends in mortality of female breast cancer in Hong Kong

Year1	ASMR1	Year2	ASMR2	Difference =ASMR1-ASMR2	Symbol
1983	8.9	1997	10.1	-1.2	-
1984	10	1998	9.7	0.3	+
1985	10.2	1999	9.7	0.5	+
1986	9.2	2000	9.4	-0.2	-
1987	9.6	2001	9	0.6	+
1988	9.5	2002	9.6	-0.1	-
1989	10	2003	9.3	0.7	+
1990	9.5	2004	9.6	-0.1	-
1991	11.2	2005	9.1	2.1	+
1992	10.7	2006	9	1.7	+
1993	9.9	2007	10	-0.1	-
1994	10.1	2008	9.2	0.9	+
1995	9	2009	9.8	-0.8	-
1996	10.3	2010	9.5	0.8	+
					8+ 6-

ASMR: Age-standard mortality rate, per 100,000 people

Consider the need to detect a trend in mortality of female breast cancer in Hong Kong population from 1983 to 2010. Table 7 shows age-standard rate which is calculated based on the 2000 year world standard population used. In the table, the AMSR for year 1 is compared with those for year 2. A + indicates that ASMR in the year 2 is smaller than year 1. For the 14 comparisons, total of 8 + symbols occur. Since the objective was to search for a declining trend, a larger number of + signs indicates that.

H_0: Sequence without a declining trend;

H_1: Sequence with a declining trend.

$$p\_value = f(X \geq k|n) = \sum_{i=k}^{n} \binom{n}{i} p^i q^{n-i}$$

$$= \sum_{i=8}^{14} \binom{14}{i} * 0.5^i * 0.5^{14-i}$$

$$= \binom{14}{8} * 0.5^8 * 0.5^6 + \binom{14}{9} * 0.5^9 * 0.5^5 + \binom{14}{10} * 0.5^{10} * 0.5^4 + \binom{14}{11} * 0.5^{11} * 0.5^3 + \binom{14}{12} * 0.5^{12} * 0.5^2 + \binom{14}{13} * 0.5^{13} * 0.5^1 + \binom{14}{14} * 0.5^{14} * 0.5^0$$

$$= 0.3953$$

Do a two-tailed test, the $p_{value} = 0.3953 * 2 = 0.7906$.

The probability of 8 or more positive sign is 0.7906. Our significance level is 5 percent. The probability is greater than the 5 percent significance level. We do not reject the notion that the ASMR of female breast cancer does not show a downtrend.

Cox-Stuart analysis is an old method, but it is still used in the various fields. In the field of health care, it is used in the incidence/mortality trend analysis of various diseases, for example, typhoid fever and paratyphoid fever.[19] Rajab KE et al. used Cox-Stuart method to test the incidence trend of gestational diabetes mellitus (GDM) of Bahrain's women from 2002 to 2010.[20] Data on the glucose tolerance test (GTT) of pregnant women were retrieved from the database of the Central Biochemistry Laboratory of Salmaniya Medical Complex (SMC). The Cox-Stuart test for trend analysis was used to investigate the changing incidence of GDM, it was performed with SPSS v19.0, and statistical significance was set at a P-value of less than 0.05. Trend analysis suggested that a significant upward trend in the incidence of patients with GDM in government maternity units during the period 2002–2010 ($p < 0.01$).

2.4.4. *Computation software*

In fact, the real substance of Cox-Stuart method is sign test. So basically, all statistical software can carry out Cox-Stuart analysis.

2.4.5. *Conclusion*

Cox and Stuart proposed several versions of a nonparametric trend test that were considered simpler to compute than rank correlations, though admittedly less powerful at detection of trends. These tests were considered useful at the time of publication (1955) because of the computational complexity of rank correlation methods. However, with modern computers and software, rank correlation methods are computationally trivial, and there is no practical bar to their use, so there is no good justification for using less powerful simpler approaches, and these methods are not offered in SPSS (and do not appear to be offered in other major statistical packages such as SAS, Stata or R).

2.5. *RUNS test*

2.5.1. *Introduction*

Runs test is a non-parametric statistical test. It figures among the oldest nonparametric procedures, as evidenced by, e.g. [21]. There have two different Runs tests, one is the Single-Sample Runs Test, and it is used to test the randomness hypothesis for a sequence base upon order of occurrence. Data studied pertain to two category variable (male/female, pass/fail, etc.). The number of runs determines randomness, too many or too few runs causes rejection of the null hypothesis. Another one is Wald-Wolfowitz Runs Test, which tests the null hypothesis that the distribution functions of two continuous populations are the same.

2.5.2. *Mathematical model*

A "run" of a sequence is a maximal non-empty segment of the sequence consisting of adjacent equal elements. A run is also defined as a series of increasing values or a series of decreasing values. For example, the sequence "++++−−−+++−−++++++−−−−" consists of six runs, three of

which consist of +'s and the others of –'s. Let U be the total number of runs, let m and n be the numbers of symbols corresponding to each type of symbol. Note: $m + n = N$.
The hypothesis of two-tailed test:

H_0: The sequence is random.

H_1: The sequence is not random.

For one-sided test, there has two differ alternative hypothesis H_I:

H_1: The sequence exhibits a tendency to mix.

H_1: The sequence exhibits a tendency to cluster.

The Single-Sample Runs Test

For a given α, when $m > 12$, $n > 12$, and $N > 20$, the test statistic:

$$Z = \frac{U-\mu}{\sigma} \sim N(0,1) \tag{1}$$

where $\mu = \frac{2mn}{N} + 1$, $\sigma = \sqrt{\frac{2mn(2mn-N)}{N^2(N-1)}}$. If Z from the test statistics is beyond the critical value of Z_α, the null hypothesis is rejected.

The Single-Sample Runs Test also can be done using median or mode. Runs consist of consecutive outcomes larger or smaller than the median. Outcomes equals to the median are ignored.

When $m \le n \le 12$, or $N \le 20$, $U \sim \gamma$ *distribution.* The table, which containing the critical value of U should be used.

Wald-Wolfowitz Runs Test

Take $\{X_i\}$, $i = 1,2,..,m$ and $\{Y_j\}, j = 1,2,...,n$ and combine them into a single sequence $\{Z_k\} = (\{X_i\},\{Y_j\}) = (X_1, X_2, ..., X_m, Y_1, Y_2, ..., Y_n)$, $k = 1,2,...,m+n$. Now order the data Z_i from smallest to largest (assume no ties) and assign a label "X" or "Y" depending on the population of origin of that data point. For example,

$$Z_{(1)} -> X,\ Z_{(2)} -> Y,\ Z_{(3)} -> Y,\ Z_{(4)} -> X,\ Z_{(5)} -> X,\ ...,\ Z_{(m+n)} -> Y$$

Under H_0, we expect all *Xs* and *Ys* to be well mixed. Define the number of *runs* as the number of sequences of identical symbols preceded and followed by either a different symbol or no symbol at all. In the sequence above, we begin with a run of length 1 "*X*" followed by a run of length 2 "*YY*" and so on. A large number of runs indicates thorough mixing, and supports the null hypothesis.

If the two populations differ among themselves, elements of one type (*Xs* or *Ys*) would be expected to cluster together, tending to make *U* small, whereas if the populations are identical, the arrangement of *Xs* or *Ys* should be random, tending to make *U* large. Thus, small values of *U* do not support the null hypothesis and the appropriate p-value is a left-tailed probability.

2.5.3. *Empirical study*

Take an example, for the given $\alpha = 0.05$, we want to test to determine whether the gender of people walking into a supermarket is a random event. This gender data was collected from supermarket's Sunday morning customers, FFF MM FFFF M FFFFF MMMM F MMMM FFFFF MMMMM FF MM FFF, we can see that there has 13 runs, and M=18 males, F=23 females.

Compute the test statistics

$$Z = \frac{U - \mu}{\sigma} = \frac{13 - 21.195}{\sqrt{651636/_{67240}}} = -2.63$$

for the $\alpha = 0.05$ level of significance, $Z_\alpha = \pm 1.96$ for this two-tail test. $Z = -2.63 < Z_\alpha = -1.96$, we reject the null hypothesis that gender of customers walking into supermarket is not random.

In the field of health care, there are many applications. For example, we test the mortality trend of male lung cancer in Hong Kong from 1983 to 2010. The ASMR is age-standard mortality rate, which is calculated from 2000 word standard population. Take the mean of this sequence as a criterion, the symbols of every year's ASMR are showed in table 8.

$$mean = \frac{sum(ASMRs)}{28} = 58.30357$$

$$Difference = ASMR - mean$$

If difference > 0, the symbol is "+", if difference < 0 and give it "-".

Table 8. Symbols of different year's ASMR

Year	ASMR	Difference	Symbol
1983	63.1	4.7964	+
1984	63.8	5.4964	+
1985	63.4	5.0964	+
1986	64.8	6.4964	+
1987	65.8	7.4964	+
1988	59.9	1.5964	+
1989	66.2	7.8964	+
1990	67.1	8.7964	+
1991	62.4	4.0964	+
1992	64	5.6964	+
1993	64	5.6964	+
1994	61.9	3.5964	+
1995	60.1	1.7964	+
1996	57.9	3.7688	-
1997	56.8	-1.5036	-
1998	57.3	3.86423	-
1999	58.3	-0.0036	-
2000	60.2	7.36667	+
2001	55.5	-2.8036	-
2002	54.9	-3.4036	-
2003	54.9	-3.4036	-
2004	54.4	-3.9036	-
2005	55.2	-3.1036	-
2006	50.9	-7.4036	-
2007	51.5	-6.8036	-
2008	45.7	-12.6036	-
2009	47	-11.3036	-
2010	45.5	-12.8036	-

The ASMR sequence contains six runs including 3 positives and 3 negatives. The shortest and longest runs have length one and fourteen, respectively.

H_0: The AMSR sequence is random.

H_1: The AMSR sequence is not random.

We know that $U = 4; m = 14; n = 14; N = 28; \alpha = 0.05$.

$$\mu = \frac{2mn}{N} + 1 = \frac{2 * 14 * 14}{28} + 1 = 15$$

$$\sigma = \sqrt{\frac{2mn(2mn - N)}{N^2(N-1)}}$$

$$= \sqrt{\frac{2 * 14 * 14 * (2 * 14 * 14 - 28)}{28^2(28-1)}}$$

$$\approx \sqrt{6.741}$$

Statistic:

$$Z = \frac{U - \mu}{\sigma} = \frac{4 - 15}{\sqrt{6.741}} = -4.236$$

for the $\alpha = 0.05$ level of significance, $Z_\alpha = \pm 1.96$ for this two-tail test. $Z = -4.236 < Z_\alpha = -1.96$, we reject the null hypothesis that the ASMR of male lung cancer among Hong Kong population during1983-2010 is not random. It can be seen from the table 8; the ASMR of male lung cancer shows a declined trend.

2.5.4. *Conclusion*

Runs test has a very low power; its asymptotic relative efficiency could compare to the traditional t test for equal variances is zero. Furthermore, it has the least power compared to other nonparametric tests applied to the same data.

2.6. *Functional data analysis*

2.6.1. *Introduction*

Functional Data Analysis (FDA) has been developed into a Multivariate Statistical Analysis (MSA) method based on thoughts of converting discrete data into functional ones since 1980s. It was first brought up by James O. Ramsay and B. W. Silverman. [22] The basic philosophy of functional data analysis is to think of observed data functions as single entities, rather than merely as a sequence of individual observations. This method has been widely used in economics, biology, meteorology, psychology, industry etc. As an example, FDA can be useful for analyzing growth curves of children that are constructed based on body height measurements made over some specific growing period. In this example, FDA would treat one growth curve as one functional data entity. Without any assumption on the parametric forms for growth curves, the discrete measurements for body heights can be transferred into a continuous curve by nonparametric smoothing. With FDA we study many important features of curves such as growth rates, which are derivatives of growth curves. In fact, it is many uses of derivatives that are the central theme of FDA.

The goals of functional data analysis are as following:

(1) to represent the data in ways that aid further analysis

(2) to display the data so as to highlight various characteristics

(3) to study important sources of pattern and variation among the data

(4) to explain variation in an outcome or dependent variable by using input or independent variable information

(5) to compare two or more sets of data with respect to certain types of variation, where two sets of data can contain different sets of replicates of the same functions, or different functions for a common set of replicates.

2.6.2. *Mathematical model*

With functional data analysis methods, data can be smooth curves or functions. Take age-specific mortality data for instance, mortality rates are treated as smooth functions of age.

Let $m_t(x)$ denote the mortality rate for age x and year $t, t = 1, ..., n$. We model the log mortality, $y_t(x) = log[m_t(x)]$, and assume that there is an underlying smooth function $f_t(x)$ that we are observing with error. Thus,

$$y_t(x_i) = f_t(x_i) + \sigma_t(x_i)\varepsilon_{t,i} \quad (1)$$

where x_i is the centre of age-group i $(i = 1, ..., p)$, $\varepsilon_{t,i}$ is an independent and identically distributed standard normal random variable and $\sigma_t(x_i)$ allows the amount of noise to vary with age x.

The error variance $\sigma_t(x_i)$ is computed as follows. Let $N_t(x)$ be the total population of age x at mid-year population in year t (for example at 30 June in year t). Then, $m_t(x)$ is approximately binomially distributed with estimated variance $N_t^{-1}(x)m_t(x)[1 - m_t(x)]$. So the variance of $y_t(x)$ is (via a Taylor approximation):

$$\sigma_t^2(x) \approx [1 - m_t(x)]N_t^{-1}(x)m_t^{-1}(x) \quad (2)$$

The first step is to estimate these smooth functions from the discrete noisy data. Various smoothing techniques are available to estimate the function from the discrete observations, such as penalized regression splines[23] or loess curves. [24] For each year t, we smooth the data $\{y_t(x_i)\}$ over x using weighted local quadratic smoothing with the smoothing parameter (the 'bandwidth') selected using cross-validation and weights set to the inverse variances $\sigma_t^{-2}(x)$.

After constructing the functional observations, we fit the model:

$$f_t(x) = \mu(x) + \sum_{k=1}^{K} \beta_{t,k}\phi_k(x) + e_t(x) \quad (3)$$

where $\mu(x)$ is the mean log mortality rate across years, $\phi_k(x)$ is a set of orthogonal basis functions, and $e_t(x)$ is the model error which is assumed to be serially uncorrelated. We wish to estimate the optimal set of K orthogonal basis functions. Specifically, for a given K, we want to find the basic functions $\{\phi_k(x)\}$ which minimize the mean integrated squared error(MISE):

$$MISE = \frac{1}{n}\sum_{t=1}^{n}\int e_t^2(x)dx \tag{4}$$

This is achieved using functional principal components (PC) decomposition applied to the smooth curves $\{f_t(x)\}$ which gives the least number of basis functions, enables informative interpretations and gives coefficients which are uncorrelated with each other. [25] Hyndman and Ullah proposed a robust method to estimate $\mu(x)$ and a robust approach to obtain PC when computing the basic functions. [26]

We estimate future values of mortality $y_t(x_i)$ by forecasting the entire function $f_t(x)$ for $t = n+1, \ldots, n+h$ and $x_1 < x < x_p$. The coefficients of the fitted function, $\beta_{t,1}, \ldots, \beta_{t,K}$, are forecast using time series models. The forecast coefficients are then multiplied by the basic functions, resulting in forecasts of mortality curves.

Let $\hat{\beta}_{n,k,h}$ denotes the h-step ahead forecast of $\beta_{n+h,k}$, and the h-step ahead forecast of $f_{n+h}(x)$ can be obtained as:

$$\hat{f}_{n+h}(x) = \hat{\mu}(x) + \sum_{k=1}^{K} \hat{\beta}_{n,k,h}\,\hat{\phi}_k(x) \tag{5}$$

where $\hat{\mu}(x)$ and $\hat{\phi}_k(x)$ are the estimates of the mean function and basic functions, respectively. To forecast the coefficients in equation (5), a variety of time series forecasting methods are available.

Take state-space models for exponential smoothing for instance. [13] Forecasts from exponential smoothing methods are estimated recursively where recent observations are given more weight than historical data. The methods accommodate additive and multiplicative trend in the time series. Hyndman and Ullah showed that the forecast variance could be obtained by adding the variances from each of the terms in equations (1) and (3). [26] Therefore,

$$Var[y_{n+h}(x)|\ell,\phi]$$

$$= \hat{\sigma}_\mu^2(x) + \sum_{k=1}^{K} u_{n+h,k}\hat{\phi}_k^2(x) + v(x) + \sigma_t^2(x) \tag{6}$$

where $\ell = \{y_t(x_i); t = 1, \ldots, n; i = 1, \ldots, p\}$ denotes all observed data, $u_{n+h,k} = \mathrm{Var}(\beta_{n+h,k} | \beta_{1,k}, \ldots, \beta_{n,k})$ can be obtained from the time series model, $\hat{\sigma}_{\mu}^2(x)$ is the variance of the smooth estimate, $\hat{\mu}(x)$ can be obtained from the smoothing method used, $\sigma_t^2(x)$ is given by (2) and $v(x)$ is estimated by averaging $\hat{e}_t^2(x)$ for each x. A prediction interval is then easily constructed assuming the forecast errors are normally distributed.

To evaluate the accuracy of the mortality forecasts by computing the mean integrated squared forecasting error (MISFE) that defined as,

$$MISFE(h) = \frac{1}{n-m+1} \sum_{t=m}^{n} \int [y_{t+h}(x) - \hat{f}_{t,h}(x)]^2 \, dx \tag{7}$$

where m is the minimum number of observations used in fitting a model.

2.6.3. *Empirical study*

The main steps in functional data analysis include the following aspects:

(1) Collect, clean, and organize the raw data.

(2) Convert the data into functional form, using the linear combination of basis functions.

The most common basis functions are: Fourier basis, B-spline basis, Monomial basis, Exponential basis, Wavelet-basis, Constant basis.

(3) Explore the data through plots and summary statistics.

(4) Register the data, if necessary, so that important features occur at the same argument values.

(5) Carry out exploratory analysis, such as functional descriptive statistics, functional principal components analysis, functional canonical correlation, etc.

(6) Construct models, if appropriate.

(7) Evaluate model performance.

Now, we give an example of the application of functional data analysis to disease incidence/mortality. Yasmeen F et al. utilized functional data model analyze the breast cancer mortality rate among white and black US women. [27] Mortality data from 1969 to 2004 were obtained from the National Centre for Health Statistics (NCHS) available on the SEER*Stat database, and then computed the age-specific mortality rates in 5-year age groups (45–49, 50–54, 55–59, 60–64, 65–69, 70–74, 75–79, 80–84). Age-specific mortality curves were obtained using nonparametric smoothing methods, then decomposed the curves using functional principal components and fitted functional time series models with four basic functions for each population separately. They adopted a state-space approach and an automatic modeling framework for selecting among exponential smoothing methods. All statistical analyses were performed in R version 2.10.1 using the demography package. An overall decline in future breast cancer mortality rates for both groups of women were observed, it appeared to be steeper among white women aged 55–73 and black women aged 60–84. It suggested, for some age groups, black American women might not benefit equally from the overall decline in breast cancer mortality in the United States.

2.6.4. *Computation software*

Software for functional data analysis was developed by Ramsay and Silverman and freely available on the net from http://www.psych.mcgill.ca/faculty/ramsay/fda.html. They provided functions for R and MATLAB. They also provided written programs for various examples in their book, which could easily be modified to accommodate own needs. In addition, SAS can also perform this analysis, mainly in two aspects: Functional Principal Component Analysis (FPCA) and Functional Linear Regression, which are core techniques for functional data analysis.

There are many R packages that can provide implementation of functional data analysis in different areas(Table 9).

Table 9. Summary of R packages for functional data analysis

R package	Functions	Authors & year
fda	basic analysis for functional data	Ramsay, Wickham, Graves, and Hooker 2012
rainbow	functional data representation	Shang and Hyndman 2012
ftsa	functional time series analysis	Hyndman and Shang 2012
refund	functional penalized regression	Crainiceanu, Reiss, Goldsmith, Huang, Huo, and Scheipl 2012
fpca	restricted maximum likelihood estimation for functional principal components analysis	Peng and Paul 2011
MFDF	modeling functional data in finance via generalized linear models	Dou 2009
fdaMixed	for FDA in a mixed model framework	Markussen 2011
geofd	spatial prediction for function value data	Giraldo, Delicado, and Mateu 2010
PACE	for FDA and empirical dynamics(written in Matlab)	Liu B 2011
fda.usc	provide a broader, flexible tool for the analysis of functional data	Manuel Febrero-Bande, Manuel Oviedo de la Fuente 2011

2.6.5. *Conclusion*

A potential limitation of FDA is the assumption that there are only period effects and no cohort effects. In future work, we will extend this model to allow for cohort effects.

3. *Summary*

Trend analysis mainly refers to techniques for extracting underlying characters of time-trend of population or time series data. Now, the techniques of trend analysis have been widely used in detecting outbreaks and unexpected increases or decreases in disease occurrence, monitoring disease trends, evaluating the effectiveness of disease control programs and policies, and assessing the success of health care programs and policies et al.

Trends in factors such as rates of disease and death, as well as behaviors such as smoking are often used by public health professionals to assist in healthcare needs assessments, service planning, and policy development. Examining data over time also makes it possible to predict future frequencies and rates of occurrence.

Studies of time trends may focus on any of the following:

- Patterns of change in an indicator over time – for example whether usage of a service has increased or decreased over time, and if it has, how quickly or slowly the increase or decrease has occurred;
- Comparing one time period to another time period – for example, evaluating the impact of a smoking cessation programme by comparing smoking rates before and after the event;
- Comparing one geographical area or population to another;
- Making future projections – for example to aid the planning of healthcare services by estimating likely resource requirements.

In this chapter, the common techniques of trend analysis have been introduced, which include Age-period-cohort model, Joinpoint regression, Time series analysis, Cox-Stuart trend test, Runs test, Functional data analysis. And some of the successful applications have also been presented in the chapter. Generally speaking, Runs test and Cox-Stuart trend test were less recommended to use, because their weak effects. Joinpoint regression was recommended by the National Cancer Institute of USA, but it cannot simultaneously analyze more than two diseases or display them in one graph. If you want to get a intuitively result of trend about two diseases' or above , you have to process these graph later, which maybe result in some errors. Although there exist intrinsic estimator problem, Age-period-cohort analysis has been a popular tool to identify temporal trend in age, period and cohort of chronic diseases in epidemiology or of social events in social studies or demography. Experts has developed a great number of improvements versions, such as

Bayesian Age-Period-Cohort Modeling and Prediction (BAMP). [28] Time series analysis is always a popular topic. It is used in various fields. In healthcare, the calendar year accumulation of disease data fitting this model is recommended. Compared with the methods described above, functional data analysis is a relatively new method, which was developed in 1980s and spreaded quickly. Meanwhile, this method belongs to high-dimensional data analysis model.

But, the advantages of trend analysis can be summarized as:

- reveal potentially fruitful areas of investigation;
- detect significant variations over time;
- be easily understood and communicated;
- be readily accepted due to its widespread use.

The disadvantages of trend analysis can also be summarized as:

- provide little insight into the root causes of variations;
- fail to indicate what the entity's normal or benchmark position is;
- be heavily influenced by the choice of the base period.

References

1. Rosenberg D: Trend analysis and interpretation. *Key concepts and methods for maternal and child health professionals Rockville: Division of Science, Education and Analysis Maternal and Child Health Information Resource Center* 1997.
2. Ryder NB: The cohort as a concept in the study of social change. *American sociological review* 1965, 30(6):843-861.
3. Pham T, Ozasa K, Kubo T, Fujino Y, Sakata R, Grant E, Matsuda S, Yoshimura T: Age-period-cohort analysis of chronic obstructive pulmonary disease mortality in Japan, 1950-2004. *Journal of epidemiology* 2012, 22(4):302.
4. O'Brien RM: Age period cohort characteristic models. *Social science research* 2000, 29(1):123-139.
5. Lerman PM: Fitting segmented regression models by grid search. *Applied Statistics* 1980, 29(1):77-84.

6. Kim HJ, Fay MP, Feuer EJ, Midthune DN: Permutation tests for joinpoint regression with applications to cancer rates. *Statistics in medicine* 2000, 19(3):335-351.
7. Llorca J, Dierssen-Sotos T, Arbaizar B, Gómez-Acebo I: Mortality From Tuberculosis in Spain, 1971 to 2007: Slow Decrease in Female and in Elderly Patients. *Annals of Epidemiology* 2012, 22(7): 474-479.
8. Forte T, Niu J, Lockwood GA, Bryant HE: Incidence trends in head and neck cancers and human papillomavirus (HPV)-associated oropharyngeal cancer in Canada, 1992–2009. *Cancer Causes & Control* 2012, 23(8): 1343-1348.
9. O'Flaherty M, Allender S, Taylor R, Stevenson C, Peeters A, Capewell S: The decline in coronary heart disease mortality is slowing in young adults (Australia 1976-2006): A time trend analysis. *International journal of cardiology* 2012, 158(2):193.
10. Xie WC, Chan MH, Mak KC, Chan WT, He M: Trends in the Incidence of 15 Common Cancers in Hong Kong. *Asian Pacific Journal of Cancer Prevention* 2012, 13:3911-3916.
11. Velicer WF, Fava JL: Time series analysis. *Handbook of psychology* 2003, John Wiley & Sons, Inc.
12. Gardner Jr ES: Exponential smoothing: The state of the art. *Journal of Forecasting* 1985, 4(1):1-28.
13. Hyndman RJ, Koehler AB, Snyder RD, Grose S: A state space framework for automatic forecasting using exponential smoothing methods. *International Journal of Forecasting* 2002, 18(3):439-454.
14. Kalekar PS: Time series forecasting using Holt-Winters exponential smoothing. *Kanwal Rekhi School of Information Technology* 2004, Tech. Rep.
15. Helfenstein U: Box-jenkins modelling of some viral infectious diseases. *Statistics in medicine* 2006, 5(1):37-47.
16. Díaz J, García R, López C, Linares C, Tobías A, Prieto L: Mortality impact of extreme winter temperatures. *International Journal of Biometeorology* 2005, 49(3):179-183.
17. Luz PM, Mendes BVM, Codeço CT, Struchiner CJ, Galvani AP: Time series analysis of dengue incidence in Rio de Janeiro, Brazil. *The American journal of tropical medicine and hygiene* 2008, 79(6):933-939.
18. Cox DR, Stuart A: Some quick sign tests for trend in location and dispersion. *Biometrika* 1955, 42(1/2): 80-956.
19. Wu WS, Xie XH, Shan AL, Liu H, He HY, Xia WD: Analysis on the epidemic trend of typhoid fever and paratyphoid fever in Tianjin during 1990-2005. *Modern Preventive Medicine* 2007, 34(17): 3304-3305.
20. Rajab KE, Issa AA, Hasan ZA, Rajab E, Jaradat AA: Incidence of gestational diabetes mellitus in Bahrain from 2002 to 2010. *International Journal of Gynecology & Obstetrics* 2012, 117(1): 74-77.
21. Fisher RA: On the random sequence. *Quarterly Journal of the Royal Meteorological Society* 1926, 52(250).

22. Ramsay JO, Wickham H, Graves S, Hooker G: Functional data analysis: Wiley Online Library; 2005.
23. Ruppert D, Wand MP, Carroll RJ: Semiparametric regression, 2003, New York: Cambridge University Press.
24. Cleveland WS, Devlin SJ: Locally weighted regression: an approach to regression analysis by local fitting. *Journal of the American Statistical Association* 1988, 83(403):596-610.
25. Ramsay JO, Dalzell CJ: Some tools for functional data analysis. *Journal of the Royal Statistical Society Series B (Methodological)* 1991, 53(3): 539-572.
26. Hyndman RJ, Ullah S: Robust forecasting of mortality and fertility rates: a functional data approach. *Computational Statistics & Data Analysis* 2007, 51(10):4942-4956.
27. Yasmeen F, Hyndman RJ, Erbas B: Forecasting age-related changes in breast cancer mortality among white and black US women: A functional data approach. *Cancer epidemiology* 2010, 34(5):542-549.
28. Schmid V, Knorr-Held L: BAMP: Bayesian age-period-cohort modelling and prediction. 2001, Software Documentation. Department of Statistics, Ludwig-Maximilian-University, Munich.

Chapter 11

Data Acquisition and Preprocessing on Three Dimensional Medical Images

Yuhua Jiao[1], Liang Chen[2] and Jin Chen[3*]

[1]*MSU-DOE Plant Research Laboratory, Michigan State University, East Lansing, MI 48824*
yuhjiao@msu.edu

[2]*Department of Neurosurgery, Huashan Hospital, Fudan University, Shanghai, P.R.China 200433*
clclcl95@shmu.edu.cn

[3]*MSU-DOE Plant Research Laboratory and Department of Computer Science and Engineering,*
Michigan State University, East Lansing, MI 48824
jinchen@msu.edu

Three dimensional (3D) medical imaging technique is a new field that is rapidly evolved over the last years, leading to a major improvement in patient care. Starting from 3D models of the patient, anatomical structures can be identified and extracted, and diagnosis and surgical simulation shall be supported. Given that the computational reconstruction of 3D models is critical for medical diagnosis and treatment, and the complex imaging processing requires considerable resources and advanced training, we introduce 3D image acquisition, segmentation and registration in this chapter, for all of which, the main purpose is to effectively transform the unstructured image data to structured numerical data for further data mining tasks.

1. Introduction

Data acquisition and preprocessing are the essential steps in data mining process, especially when applying data mining to 3D images. The quality of 3D imaging in healthcare fields is inferior to that in other computer vision fields in following aspects: (1) Automatic data preprocessing is highly required to obtain high quality data for biomedical data mining,

*Corresponding author.

since biomedical 3D images collection systems are primarily designed for doctor's manual diagnosis; (2) Much of healthcare data are often inconsistent or non-standardized, such as pieces of medical 3D images in different formats from different data sources. These data cannot be directly used for data mining without significant efforts on data preparation, including the utilization of two main techniques: 3D image segmentation and 3D image registration.

Image segmentation is the task of partitioning the data into contiguous regions representing individual anatomical objects.[1] It is a prerequisite for further investigations in many computer-assisted medical applications in that it can classify different tissues to different logical classes in medical images. A central problem in 3D medical image segmentation is how to effectively distinguish interested objects from noisy background. To integrate different forms of 3D image-derived information to serve further data mining tasks, 3D datasets need to be automatically registered. 3D image registration is the process of determining the point-by-point correspondence between two 3D images of a scene. Two medical images may differ by any amount of rotation or translation in any direction, and they may also differ in scale. A central problem in 3D medical image registration is to define the search space of the registration problem, ranging from nonlinear transformation with a virtually infinite degree of freedom, to rigid registration with six degrees of freedom.[2] By registering two medical images, the fusion of multi-modality information becomes possible, and regions of abnormal function can be recognized.

2. Three Dimensional Image Acquisition Techniques

Data processing programs based from radiologic examinations have been widely adopted in clinical application to better visualize specific structures or region of interests in the human body.[3] One of the best examples of such applications is to demonstrate the intracranial arteries and veins in 3D or even four dimensional patterns.[4] That is extremely valuable for doctors to find out the vascular abnormalities and give treatment.

Catheterized cerebral angiography[5] is the most often used imaging technique under plane film x-rays to demonstrate the intracracial vessels inside patients. A catheter is placed in the bilateral external carotid artery, internal carotid artery and the vertebral artery one by one via femoral artery or radial artery puncture. Then a jet of contrast medium from the catheter will make its vascular pathway opaque to x-ray, thus we can see the vessel shadow in the x-rays machine. Vascular abnormalities in the neck and brain will be realized after all the 6 vessels have been catheterized and angiographed. But the irregular skull base bony structures and the bone fissures will also be demonstrated as linear or massive shadows, making the angiograph obscure.[6] The resulted ambiguous diagnosis would put the treatment in dilemma. To properly settled this problem, digital subtraction angiography (DSA) has been developed.[6,7] X-ray features both before and after the jet of contrast medium are acquired and stored in the digital manner. Then the shadows before contrast admission are taken as the noise background and subtracted from the angiography. Theoretically, only the shadow of vessels is left. So DSA is the gold-standard for the diagnosis of vascular abnormalities. But pitfalls for this technique are still prominent: both the patient and the doctor are exposed under x-ray during the process, which may continue for a few minutes or over an hour. The dose of radiation is cumulatively limited and it is harmful especially for a pregnant woman or a child. The technique to safely put the catheter in the right place needs years of training, but unexpected complications might still occur even under those experienced hands.[8] Second of all, blood clot from the catheter might result in cerebral infarction and bleeding from puncture site is still not rare to see. Putting together, less invasive techniques are needed to visualize brain vessels.

Computerized tomography (CT)[9] developed by Hounsfield and Cormack, Nobel Prize winners in 1979, divides a human head into a series of slides and each slide into properly arranged small cubes, namely matrix. The x-ray absorption value of each cube is calculable and each slide is shown as a grey-level image. The obvious enhancement of vessels makes them distinguishable from surrounding structures. But those are only short segments of vessels. With the availability of multi-detector-row CT scanner and development of data processing programs, 3D vascular reconstructions can be acquired, which is called computerized tomographic angiography (CTA).[4] Its non-invasive, less radiation hazard, fast and convenient, and is now replacing traditional DSA.[10] Similar slides of vascular images can be obtained from magnetic resonance (MR) scans and 3D reconstructed at the image post-processing workstation. This technique is called MR angiography (MRA).[11] The examples of the CT image and MR image are show in Figure 1.

In comparison of CTA and MRA, CTA is usually clearer because of higher resolution of raw scan images. Another priority of the CTA is that the vessels and surrounding other structures can be simultaneously demon-

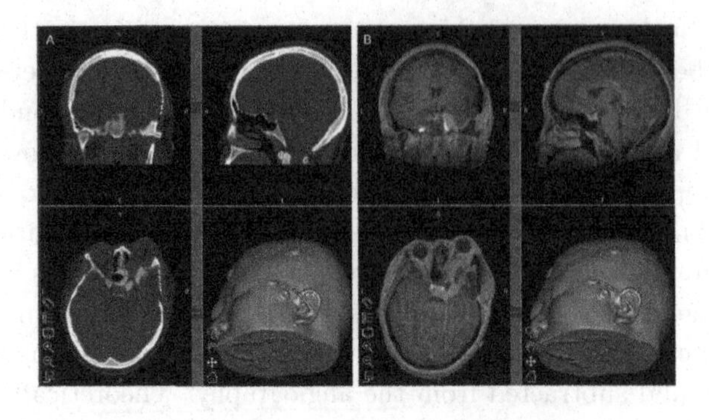

Fig. 1. Examples of (A) the CT image and (B) the MR image.

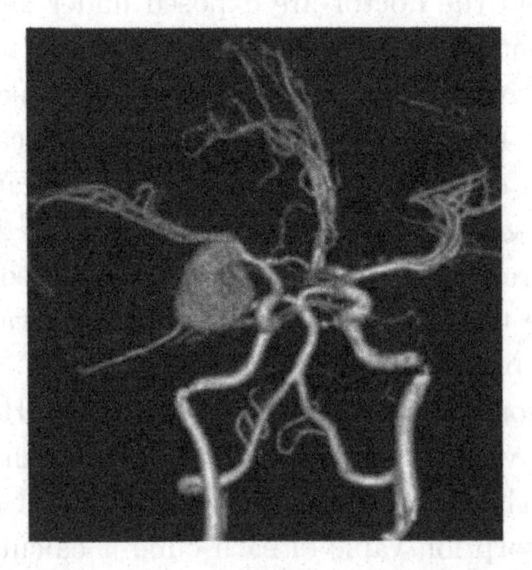

Fig. 2. Three dimensional image of a localized, blood-filled balloon-like sac (aneurysm).

strated after proper threshold adjustment, which is valuable for doctors to choose a proper way to access the vascular disease. But when the vessel is close to bony structure, such as the internal carotid artery near the anterior clinoid process, or the vertebral artery close to the posterior fossa, partial volume effect would make the artery indistinguishable from the bone as they are both of high density. To adjust the window center or window width cant settle this problem. Shorter scan time is also an important priority of the CTA. MRA has the priority of non-radiation hazard, being unaffected by bony structures, but the image is influenced by the direction of blood flow .[12] Turbulent flow in the stenosed vessel often makes the vessel seem to be more severely stenosed or even occluded.

In Figure 2, there is a localized, blood-filled balloon-like sac, which is named as aneurysm in medicine. An aneurysm is an abnormal bulge in the wall of an artery due to its weakness and hemodynamic force. Rupture of the aneurysm can result in high rate of morbidity and mortality. As the 2D CT or MRI slice images are acquired in a regular pattern (*e.g.*, one slice every millimeter) and all image pixels are regularly numbered, they actually compose a complete 3D data set. Actually, we always need better image processing programs to maximally utilize the raw data acquired from CT or MR scans. Maximum intensity projection (MIP) and volume rendering (VR) are the most often used techniques to render a 2D projection of the 3D data set. The capability of four-dimensional (4D) time-resolved CTA and MRA will be helpful in hemodynamic analysis.

Besides the 3D techniques of inside objects, 3D surface imaging techniques, for example, 3D digital stereo-photogrammetry, have been popularly applied to reconstruct the 3D coordinates of points on a body from a pair of stereo images,[13] especially in craniofacial clinics,[13,14] such as orthodontics, oral and maxillofacial surgery.[13,14] Structured light[15] is another technique to detect the depth and surface information of an object. Different to stereophotogrammetry, only one camera is needed in structured light 3D scanner. Note that calibration is required in both techniques, and the reconstructed data should be evaluated to guarantee the optimal data collection.[16]

The formats of the 3D image data can be generally categorized into two categories: 1) the measured surface is usually represented as a point cloud[17] or polygonal mesh[18] and is displayed as a smooth picture with surface rendering algorithms; and 2) volumetric data is a common representation of 3d-CT[19] and 3d-MRI[20] images. Medical images are usually coded in the DICOM (digital imaging and communications in medicine) format.

Graph-based mining algorithms could be applied directly on the point cloud or polygonal mesh,[21] but the acquired 3D images are often noisy and unstructured for most data mining algorithms. To make the 3D data ready for knowledge discovery, segmentation and registration algorithms, the two fundamental 3D image preprocessing steps, are commonly applied.

3. Three Dimensional Image Segmentation

To extract the "ready-to-data-mine" regions from 3D images, computer-assisted automatic segmentation need to be accurate, repeatable and quantitative. In particular, medical images are often corrupted by noise, which can cause considerable difficulties when applying traditional low-level segmentation techniques such as edge detection and thresholding. Consequently, classical image segmentation tends to generate infeasible object boundaries.

3.1. *Deformable Model*

To overcome these difficulties, deformable models have been extensively studied and widely used in medical image segmentation, with promising results. 3D deformable models are object outlines that move under the influence of forces and constraints minimizing internal energy, which is defined within the outline itself, and external energy, which is computed from the image data. The internal energy is designed to keep the model close to the shape during deformation. The external energy is defined to move the model toward an object boundary. By constraining boundaries to be smooth within the range of an explicit domain learned from a training set, deformable models are robust to both image noise and boundary gaps, and boundary elements in models are in coherent and consistent mathematical description which can then be readily used by subsequent applications, such as object tracking. For example, Xu *et al* reconstructed human cerebal cortex from magnetic resonance images using a deformable model.[22]

Deformable models have been highly recognized with the seminal paper "Snakes: Active Contours" by Kass, Witkin, and Terzopoulos[23] and have grown to be one of the most active and successful research areas in image segmentation. Different names, such as snakes, active contours or surfaces, balloons, and deformable contours or surfaces, have been used in literatures to refer to deformable models. The models are generally grouped into two types: parametric deformable models[23–26] and geometric deformable models.[27–30] Parametric deformable models represent object outlines explicitly in their parametric forms during deformation, so that direct interaction with the model is feasible and a compact representation for fast real-time implementation can be developed. However, model topology operation, such as splitting or merging parts during the deformation, can be difficult when using parametric models. The segmentation result shown in Figure 3 is a parametric surface. Geometric deformable models, on the other hand, can handle topological changes naturally. These models, based on the theory of curve evolution[31–34] and the level set method,[35,36] represent curves and surfaces implicitly as a level set of a higher-dimensional scalar function. Their parameterizations are computed only after complete deformation, thereby allowing topological adaptivity to be easily accommodated. Despite this fundamental difference, the underlying principles of both methods are very similar.

3.2. *Three Dimensional Image Segmentation with Deformable Model*

Given the successful 2D deformable models in 2D image segmentation, segmenting 3D image volumes can be done by applying 2D deformable model slice by slice. However, it is labor intensive and requires a accurate post-processing step to connect the sequence of 2D contours into a continuous surface, because the resulting surface reconstruction would contain inconsistencies if the post-precessing is not well done. True 3D deformable surface model is faster and more robust resulting in a globally smooth and coherent surface between image slices.

Deformable surface models in 3D were first proposed by Terzopoulos[37] for computer vision. For the demanding needs in medical image analysis, the use of deformable surface models have explored for segmenting structures in medical image volumes. Miller[38] constructs a polygonal approximation to a sphere and geometrically deforms this balloon model until the balloon surface conforms to the object surface in 3D CT data. The segmentation process is the minimization of cost functions each of which measures costs with local deformation. The cost function that associates with every vertex of the polygonal is a weighted sum of three terms: a deformation potential that expands the model vertices towards the object boundary, an image term that identifies features such as edges and opposes the balloon expansion, and a term that maintains the topology of the model by constraining each vertex to remain close to the centroid of its neighbors.

Cohen and Cohen[39,40] and McInerney and Terzopoulos[41] use finite element and physics-based techniques to implement an elastically deformable cylinder and sphere, respectively. The models are used to segment the inner wall of the left ventricular (LV) of the heart from MR or CT image volumes[41] (as shown in Figure 3).

These deformable surfaces are based on a thin-plate under tension surface spline, which controls and constrains the deformation of the surface. Lagrangian equations of motion through time is used to fit the models to data dynamically and to adjust the deformational degrees of freedom. In addition, the models are represented with the finite element method as a continuous surface in the form of weighted sums of local polynomial basis functions. Unlike Miller's[38] polygonal model, the finite element method provides an analytic surface representation which uses high-order polynomials so that fewer elements are required to accurately represent an object. Pentland and Sclaroff[42] and Nastar and Ayache[43] also develop physics-based models but use a reduced modal basis for the finite elements. Staib and Duncan[44] describe a 3D surface model used for geometric surface matching to 3D medical image data. The model uses a Fourier parameterization that decomposes the surface into a weighted sum of sinusoidal basis functions. Several different surface types are developed such as tori, open surfaces, closed surfaces and tubes. Surface finding is an optimization problem in gradient ascent that attracts the surface to strong image gradients in the vicinity of the model. A wide variety of smooth surfaces can be described with a small number of parameters by using the Fourier parameterization. Because the basis functions in a Fourier representation are orthonormal and the higher indexed basis functions represent higher spatial variation. Therefore, the series can be truncated and still represent relatively smooth objects accurately.

Besides aforemetioned approaches, Szeliski et al.[45] use a dynamic, self-organizing oriented particle system to model surfaces of objects. Small flat disks defined as the oriented particles evolve according to Newtonian mechanics and interact through external and internal forces. The external forces keep the particles to the data, while internal forces attempt to group the particles into a coherent surface (model). Objects reconstructed with the particles are with complex shapes and topologies by flowing over the data, extracting and conforming to meaningful surfaces. A triangulation is then performed to connect the particles into a continuous global model that

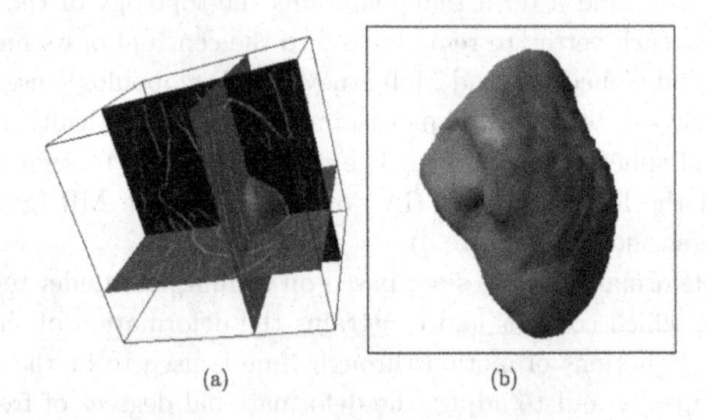

(a) (b)

Fig. 3. (a) Deformable balloon model embedded in volume image deforming towards LV edges. (b) Reconstruction of LV.[41]

is consistent with the inferred object surface. Other notable work involving 3D deformable surface models and medical image applications can be found in.[30,46–48]

3.3. *A Case Study in Segmentation with Active Appearance Models*

Active Appearance Models (AAM) is a group of highly flexible deformable models firstly introduced in 1998.[49] AAMs model an 3D object as an combination of a model of shape and a model of texture;[50] where "shape" is the vertex locations of the 3D mesh, and "texture" also being termed as "appearance" can be pattern of intensities or colors across an image patch. Both shape and texture are represented in AAMs as means with their eigen variations, by applying Principal Component Analysis (PCA) on training data that have been hand annotated by experts.

Mathematically, a labelled training example i is defined by its shape $s_i = (x_0, y_0, z_0, ..., x_{S-1}, y_{S-1}, z_{S-1})$ containing S corresponding 3D surface points (x, y, z) and its texture $g_i = (g_0, ..., g_{T-1})$ containing T corresponding appearance patterns, such as colors. PCA decomposes shape and texture into their means and eigenvectors, given in Eq.1 and Eq.2, where, $\bar{s}$ is the mean shape, $\bar{g}$ is the mean texture in a mean shaped patch, and Q_s, Q_g are eigen matrices that describe the modes of variation derived from the training data. c_s and c_g are parameter vectors for parameterizing shape and texture.

$$s = \bar{s} + Q_s c_s \tag{1}$$

$$g = \bar{g} + Q_g c_g \tag{2}$$

AAM based segmentation is to fit an AAM to an 3D object by iteratively updating parameter vectors, minimizing the error between the input 3D object and the closest model instance; i.e. solving a nonlinear optimization problem. Texture residual (defined in Eq.3) is a commonly used error measure, where p are the parameters of the model, g_m is the modeled texture as a function of the current parameters p (Eq.2) and g_s is the texture, sampled from the current object.

$$r(p) = g_s - g_m \tag{3}$$

AAM based segmentation is fully automated and successful in practice. The detected contours follow the ground truth quite well. In Leung *et al*,[51] the authors developed a 3D echocardiography segmentation approach, which has been evaluated on 99 patients.

4. Three Dimensional Image Registration

3D images are often acquired with different image modalities (as introduced in Section 2). To integrate different forms of 3D image-derived information to serve the further data mining tasks, the point-by-point correspondence between two or multiple 3D medical images of a scene need to be automatically determined. Such process is called 3D image registration.

4.1. *Image Registration Methods*

For the two 3D images to register, the image that is kept unchanged is called "reference image", whereas the image that is resampled to register the reference image is called "sensed image". Mathematically, let T denote the spatial transformation that maps coordinates (spatial locations) from a sensed image A to a reference image B, and P_A and P_B denote coordinate points (pixel locations) in images A and B, respectively, the image registration problem is defined as to determine T such that the mapping $T : p_A \rightarrow p_B \Leftrightarrow T(p_A) = p_B$ results in the best alignment of A and B.[2] Two medical images may differ by any amount of rotation or translation in any direction, and they may also differ in scale. The nature of the T transformation characterizes the search space of the registration problem, ranging from nonlinear transformation with a virtually infinite degree of freedom, to rigid registration with six degrees of freedom.[2] By registering two medical images, the fusion of multi-modality information becomes possible, and regions of abnormal function can be recognized. A registration algorithm usually has three components: a matching criterion of how well two images match; the transformation model, which specifies the way in which the sensed image can be modified to match the reference image; the optimization process that varies the parameters of the transformation model to maximize the matching criterion.

For DICOM images, a preliminary registration can be performed using the geometrical features such as the position and orientation of the image with respect to the acquisition device and the patient, as well as to the voxel size.[52] To align the corresponding features of two or more images, the preliminary feature based registration approach, consisting of two components which are a number of features selected from the images and the correspondence established between them, can be applied. Knowing the correspondences, a transformation function is then found to resample the sensed image to the geometry of the reference image. The live example of image registration of the CT and MR sample images in Figure 1 is shown in Figure 4.

When domain knowledge or known corresponding geometric function is not applicable, the voxel intensity-based registration approach, developed mainly based on the concept of mutual information (MI),[53] can be applied, implying the comparison of image gray levels to be registered. If one image provides some information about the second one, then $MI(A, B) > 0$, otherwise, we have $MI(A, B) = 0$, meaning that the two images are independent. The MI is related to the image entropy by $MI(A, B) = H(A) + H(B) - H(A, B)$, where $H(x)$ is the entropy of x. The voxel intensity-based registration aligns the images by optimizing the MI values of the corresponding voxel pairs. Because no assumption is made about the nature of the relation between the image intensities, MI is robust for both multimodal and unimodal registration, and it does not depend on the specific dynamic range or intensity scaling of the images.[52]

Inspired from the optical flow equations, Thirion proposed the demons registration[54] that considers a diffusion process to best match the boundary of a reference image to that of a sensed image. Demons (entities) are defined in the reference image, typically on its contour points, to attract the deformation of a subject image for best alignment. The demon registration can be thought of as an approximation to fluid registration.[55]

For large-scale applications such as brain image registrations, images may have different orientations, brightness, sizes, evenness of staining, morphological damage and other types of image noise, which requires a robust algorithm. However, such image registration approaches usually require a lot of computational power and computational time. This means that these algorithms are difficult to use where real time constraints are introduced, as in virtual reality applications. In addition, in one comparison[56] of several widely used methods for registration, all the tested methods yielded unsatisfactory alignments at a rate that make them unsuitable for use in a pipeline that involves thousands of high-resolution 3D laser scanning microscope images. In the end, a major problem of the existing registration approaches is the lack of an effective quality index to be used when assessing the quality of the registration process. To tackle these problems, a non-linear non-rigid mapping schemes called BrainAligner was proposed in Peng et al.[56]

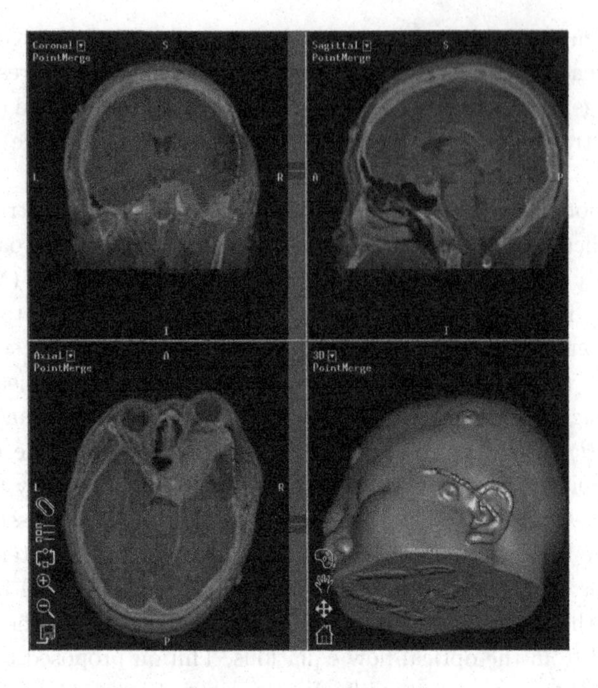

Fig. 4. Example of the registered CT and MR image.

4.2. *BrainAligner: A Case Study in 3D Image Registration*

Much of the current work on 3D medical image registration involves non-rigid registrations,[55] which is needed to take into account the differences of two patients or one patient and an atlas, especially in acquisition protocols, resolution, *etc.* BrainAligner is a 3D image non-rigid registration program that automatically finds the corresponding landmarks in a sensed brain image and maps it to the coordinate system of the reference brain via a deformable warp.[56] With BrainAligner, the reference channel for each sensed brain image is mapped to a reference image using a nonlinear geometrical warp, so that the patterns in multiple brain images can be compared in the same coordinate space for the identification of intersecting patterns in various anatomical structures. Note that, in BrainAligner, the percentage of the reference landmarks that are automatically reliably matched can be used to score how many image features are preserved in the automatic registration.

BrainAligner has two steps: global 3D affine transformation and nonlinear local 3D alignment. The global alignment process sequentially optimizes

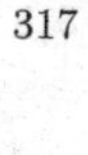

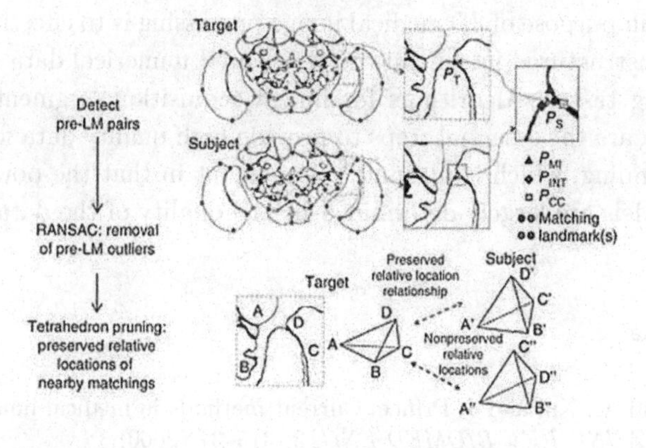

Fig. 5. Outline of the RLM algorithm for detecting corresponding feature points in sensed and reference images.[56]

the displacement, scaling and rotation parameters of an affine transform from sensed to reference to maximize the correlation of voxel intensities between two images. First, the center of mass of a sensed image is aligned to that of the reference image. Then a sensed image is rescaled proportionally, so that its principal axis had the same length with that of the reference image. Finally, a sensed image is rotated around its center of mass and thus detected the angle for which the reference image and the rotated subject image had the greatest overlap.

In the local alignment step, a reliable landmark matching (RLM) algorithm was designed to detect corresponding 3D feature points (*i.e.* a landmark) in every reference-sensed image pair, as shown in Figure 5. For each reference landmark, at least two independent matching criteria are used in RLM to locally search for matching landmarks in the sensed image. These criteria include mutual information, inverse intensity difference, correlation and similarity of invariant image moments. A match confirmed by a consensus of these criteria is defined as a preliminary landmark match (pre-LM). Next RLM uses a random sample consensus algorithm to remove the outliers that violate the smoothness constraint or the relative location relationship.

In summary, BrainAligner is one of the state-of-the-art automated 3D image registration methods on large-scale datasets.[56] In BrainAligner, the RLM method, which compares the results produced using different criteria and only uses results that agree with each other, can be viewed as an optimized combination of several existing methods.[56]

The main purpose of 3D medical image processing is to effectively transform the unstructured image data to structured numerical data for further data mining tasks. Algorithms for image acquisition, segmentation and registration are the essential steps to provide high quality data for biomedical data mining, which is ultimately important in that the power of data mining models are largely determined by the quality of the data.

References

1. D. Pham, C. Xu, and J. Prince, Current methods in medical image segmentation, *ANNU REV BIOMED ENG.* **2**, 315–37 (2000).
2. L. Landini, V. Positano, and M. Santarelli, 3d medical image processing, *Image Processing in Radiology: Current Applications.* pp. 67–85 (2007).
3. C. Metz, Roc methodology in radiologic imaging, *INVEST RADIOL.* **21** (1986).
4. M. Matsumoto, M. Sato, M. Nakano, Y. Endo, Y. Watanabe, T. Sasaki, K. Suzuki, and N. Kodama, Three-dimensional computerized tomography angiographyguided surgery of acutely ruptured cerebral anuerysms, *J NEUROSURG.* **94**(5), 718–727 (2001).
5. G. Hankey, C. Warlow, and R. Sellar, Cerebral angiographic risk in mild cerebrovascular disease, *Stroke.* **21**, 209–222 (1990).
6. E. Chappell, F. Moure, and M. Good, Comparison of computed tomographic angiography with digital subtraction angiography in the diagnosis of cerebral aneurysms: a meta-analysis, *Neurosurgery.* **52**, 624–631 (2003).
7. W. Brody, Digital subtraction angiography, *IEEE T NUCL SCI.* **29**, 1176–1180 (1982).
8. J. Fifi, P. Meyers, S. Lavine, V. Cox, L. Silverberg, S. Mangla, and J. Pile-Spellman, Complications of modern diagnostic cerebral angiography in an academic medical center, *J Vasc Interv Radiol.* **20**(4), 442–447 (2009).
9. G. Herman, *Fundamentals of computerized tomography: Image reconstruction from projection.* Springer (2009).
10. H. Westerlaan, J. van Dijk, M. Jansen-van der Weide, J. de Groot, R. Groen, J. Mooij, and M. Oudkerk, Intracranial aneurysms in patients with subarachnoid hemorrhage: Ct angiography as a primary examination tool for diagnosis - systematic review and meta-analysis, *Radiology.* **258**(1), 134–145 (2011).
11. C. Dumoulin and H. H. Jr, Magnetic resonance angiography, *Radiology.* **161**, 717–720 (1986).
12. G. Fung, B. Krishnapuram, J. Bi, M. Dundar, V. Raykar, S. Yu, R. Rosales, S. Krishnan, and R. Rao. Mining medical images. In *SIGKDD*, Paris, France (June, 2009).
13. C. Heike, K. Upson, E. Stuhaug, and S. Weinberg, 3d digital stereophotogrammetry a practical guide to facial image acquisition, *Head and Face Medicine.* **6**(18) (July, 2010).

14. B. Shetty, A. Raju, and V. Reddy, Three dimensional imaging an impact in orthodontics - a review, *Annals and Essences of Dentistry.* **3**(3), 88–91 (September, 2011).
15. S. Treuillet, B. Albouy, and Y. Lucas, Three-dimensional assessment of skin wounds using a standard digital camera, *IEEE T MED IMAGING.* **28**(5), 752 –762 (May, 2009). ISSN 0278-0062.
16. U. Ozsoy, B. Demirel, F. Yildirim, O. Tosun, and L. Sarikcioglu, Method selection in craniofacial measurements advantages and disadvantages of 3d digitization method, *J CRANIO MAXILL SURG.* **37**(3), 285–290 (July, 2009).
17. In ed. W. Osten, *Fringe 2005* (2006). ISBN 978-3-540-26037-0.
18. C. Bourne, W. Kerr, and A. Ayoub, Development of a three-dimensional imaging system for analysis of facial change, *Clinical Orthodontics and Research.* **4**(2), 105–111 (2001). ISSN 1600-0544.
19. C. Chen, J. Luo, K. Parker, and T. Huang, Ct volumetric data-based left ventricle motion estimation: an integrated approach, *COMPUT MED IMAG GRAP.* **19**(1), 85–100 (1995). ISSN 0895-6111.
20. J. Brewer, Fully-automated volumetric mri with normative ranges: Translation to clinical practice, *BEHAV NEUROL.* **21**(1-2), 21–28 (2009).
21. A. Lund, C. Bilgin, M. Hasan, L. McKeen, J. Stegemann, B. Yener, M. Zaki, and G. Plopper, Quantification of spatial parameters in 3d cellular constructs using graph theory, *J BIOMED BIOTECHNOL* (2009).
22. C. Xu, D. L. Pham, M. E. Rettmann, D. N. Yu, and J. L. Prince, Reconstruction of the human cerebral cortex from magnetic resonance images, *IEEE Trans. Med. Imag.* **18**, 467–480 (1999).
23. M. Kass, A. Witkin, and D. Terzopoulos, Snakes: active contour models, *INTERNATIONAL JOURNAL OF COMPUTER VISION.* **1**(4), 321–331 (1987).
24. A. A. Amini, T. E. Weymouth, and R. C. Jain, Using dynamic programming for solving variational problems in vision, *IEEE Trans. Patt. Anal. Mach. Intell.* **12**(9), 855–867 (1990).
25. L.D.Cohen, On active contour models and balloons, *CVGIP:Imag.Under.* **53** (2), 211–218 (1991).
26. T. McInerney and D. Terzopoulos, A dynamic finite element surface model for segmentation and tracking in multidimensional medical images with application to cardiac 4d image analysis, *Comp. Med. Imag. Graph.* **19**(1), 69–83 (1995).
27. V. Caselles, F. Catte, T. Coll, and F. Dibos, A geometric model for active contours, *Numerische Mathematik.* **66**, 1–31 (1993).
28. V. Caselles, F. Catte, T. Coll, and F. Dibos, Shape modeling with front propagation: a level set approach, *IEEE Trans. Patt. Anal. Mach. Intell.* **17** (2), 158–175 (1995).
29. V.Caselles, R.Kimmel, and G.Sapiro. Geodesic active contours. In *ICCV*, pp. 694–699, Boston, MA, USA (Jun, 1995).
30. R.T.Whitaker, Volumetric deformable models: active blobs, *Visualization in Biomedical Computing.* **2359**, 121–134 (1994).

31. G. Sapiro and A. Tannenbaum, Affine invariant scale-space, *Intl. J. Comp. Vis.* **11**(1), 25–44 (1993).
32. B. B. Kimia, A. R. Tannenbaum, and S. W. Zucker, Shapes, shocks, and deformations i: the components of two-dimensional shape and the reaction-diffusion space, *Intl. J. Comp. Vis.* **15**, 189–224 (1995).
33. R.Kimmel, A.Amir, and A.M.Bruckstein, Finding shortest paths on surfaces using level sets propagation, *IEEE Trans. Patt. Anal. Mach. Intell.* **17**(6), 635–640 (1995).
34. L.Alvarez, F.Guichard, P.L.Lions, and J.M.Morel, Axioms and fundamental equations of image processing, *ARCH RATION MECH AN.* **123**(3), 199–257 (1993).
35. S. Osher and J. A. Sethian, Fronts propagating with curvature-dependent speed: algorithms based on hamilton-jacobi formulations, *J COMPUT PHYS.* **79**, 12–49 (1988).
36. J. A. Sethian, *Level Set Methods and Fast Marching Methods: Evolving Interfaces in Computational Geometry, Fluid Mechanics, Computer Vision, and Material Science*, 2nd edn. Cambridge University Press, Cambridge, UK (1999).
37. D. Terzopoulos and K. Fleischer, Deformable models, *VISUAL COMPUT.* **4** (6), 306–331 (1988).
38. J. Miller, D. Breen, W. Lorensen, R. O'Bara, and M. Wozny. Geometrically deformed models: A method for extracting closed geometric models from volume data. In *SIGGRAPH*, pp. 217–226, Las Vegas, NV (July, 1991).
39. I. Cohen, L. Cohen, and N. Ayache, Using deformable surfaces to segment 3d images and infer differential structures, *CVGIP: Image Understanding.* **56**(2), 242–263 (1992).
40. I. Cohen and L. Cohen, Finite element methods for active contour models and balloons for 2d and 3d images, *IEEE T PATTERN ANAL.* **15**(11), 1131–1147 (1993).
41. T. McInerney and D. Terzopoulos, A dynamic finite element surface model for segmentation and tracking in multidimensional medical images with application to cardiac 4d image analysis, *COMPUT MED IMAG GRAP.* **19** (1), 69–83 (1995).
42. A. Pentland and S. Sclaroff, Closed-form solutions for physically based shape modelling and recognition, *IEEE T PATTERN ANAL.* **13**(7), 715–729 (1991).
43. C. Nastar and N. Ayache. Fast segmentation, tracking, and analysis of deformable objects. In *ICCV*, pp. 275–279, Berlin, Germany (May, 1993).
44. L. Staib and J. Duncan. Deformable fourier models for surface finding in 3d images. In *VBC*, pp. 90–104, Chapel Hill, NC (October, 1992).
45. R. Szeliski, D. Tonnesen, and D. Terzopoulos. Modeling surfaces of arbitrary topology with dynamic particles. In *CVPR*, pp. 82–87, New York, NY (June, 1993).
46. H. Delingette, M. Hebert, and K. Ikeuchi, Shape representation and image segmentation using deformable surfaces, *IMAGE VISION COMPUT.* **10**(3), 132–144 (1992).

47. H. Tek and B. B. Kimia. Shock-based reaction-diffusion bubbles for image segmentation. In *CVRMed*, pp. 434–438 (1995).
48. C. Davatzikos and R. Bryan. Using a deformable surface model to obtain a mathematical representation of the cortex. In *Proc. International Symp. on Computer Vision*, p. 212?17, Coral Gables, FL (November, 1995).
49. G. J. Edwards, C. J. Taylor, and T. F. Cootes. Interpreting face images using active appearance models. In *Proceedings of the 3rd. International Conference on Face & Gesture Recognition*, FG '98, pp. 300–305, IEEE Computer Society, Washington, DC, USA (1998). ISBN 0-8186-8344-9.
50. T. Cootes, G. Edwards, and C. Taylor, Active appearance models, *IEEE T PATTERN ANAL.* **23**(6), 681 –685 (jun, 2001). ISSN 0162-8828.
51. K. Leung, M. van Stralen, G. van Burken, N. de Jong, and J. Bosch. Automatic active appearance model segmentation of 3d echocardiograms. In *I S BIOMED IMAGING*, pp. 320 –323 (april, 2010).
52. A. Goshtasby, *2-D and 3-D Image Registration: for Medical, Remote Sensing, and Industrial Applications*. Wiley-Interscience (2005). ISBN 978-0471649540.
53. J. Pluim, J. Maintz, and M. Viergever, Mutual-information-based registration of medical images: a survey, *IEEE T MED IMAGING.* **22**, 986–1004 (2003).
54. J. Thirion, Image matching as a diffusion process: an analogy with maxwell's demons, *Med Image Anal.* **2**, 243–260 (1998).
55. W. Crum, T. Hartkens, and D. Hill, Non-rigid image registration: theory and practice, *Br J Radiol.* **77**, S140–S153 (2004).
56. H. Peng, P. Chung, F. Long, L. Qu, A. Jenett, A. Seeds, E. Myers, and J. Simpson, Brainaligner: 3d registration atlases of drosophila brains, *NAT METHODS.* **8**, 493–498 (2011).

4. Text Mining and its Biomedical Applications

Chapter 12

Text Mining in Biomedicine and Healthcare

Hong-Jie Dai[1], Chi-Yang Wu[2], Richard Tzong-Han Tsai[3] and Wen-Lian Hsu[2]

[1]Graduate Institute of Biomedical Informatics, College of Medical Science and Technology, Taipei Medical University
250 Wu-Xin Street, Taipei, Taiwan, ROC
[2]Information Science, Academia Sinica
128 Academia Road, Sec.2, Nankang, Taipei, Taiwan, ROC
[3]Department of Computer Science and Information Engineering, National Central University, 300 Jhong-Da Road, Jhong-Li, Tao-Yuan, Taiwan, ROC
E-mail: hjdai@tmu.edu.tw, celestial114@gmail.com, thtsai@csie.ncu.edu.tw, hsu@iis.sinica.edu.tw

The massive flow of scholarly publications from traditional paper journals to online outlets has benefited scientists by providing ease to access to published work. Many publishers are now making the full texts of articles publicly available online. However, due to the sheer volume of available literature, researchers are finding it increasingly difficult to locate information of interest. Recent advances in text-mining technology have aimed to accelerate literature curation, maintain the integrity of information, and ensure proper linkage of data to other resources. This chapter will give an overview of essential text-mining technologies that can be applied in biomedicine and healthcare including: entity recognition, entity linking, relation extraction, and co-reference resolution. Different perspectives on the application of these techniques for researchers, clinicians and patients will be presented, and we will see how these technologies can be integrated in healthcare administration. Finally, a case study of a biomedical text mining database system that automatically recognizes and collects cardiovascular disease related genes will be demonstrated to illustrate the application of some of the principles discussed in this chapter.

1. Introduction to Text Mining

Looking for ways to automatically gather and utilize the enormous amount of biomedical knowledge buried within literature has always been an intriguing topic. Effective text mining systems should be able to extract and exploit not only explicitly stated information but also implied and inferred data. During the past few years, text mining has shown promising results in enabling scientists to efficiently and systematically recognize, collect and interpret knowledge required for research or education, making literature more accessible and useful [1,2].

A prevalent definition of text mining is provided by Hearst's essay [3]:

> *Text Mining is the discovery by computer of new, previously unknown information, by automatically extracting information from different written resources. A key element is the linking together of the extracted information together to form new facts or new hypotheses to be explored further by more conventional means of experimentation.*

This statement emphasizes a major difference between regular data mining and text mining. In text mining, the source of information is "natural language text" instead of "structural data", such as that in database. Structural data are designed for computer programs to process automatically. In contrast, text is written in a comprehensive manner for humans to read and understand and is thus much more difficult for machines to deal with. A classic example in data mining is using the association rule mining method to learn patterns from transaction data, then using the acquired patterns to predict which products to group together on shelves, to offer coupons for, and so on [4].

In order to obtain high-quality information from unstructured text, text mining employs a broad range of techniques from natural language processing (NLP, also known as computational linguistics). NLP focuses on several tasks in text analysis and requires a massive amount of knowledge on a number of levels. For example, deep and advanced knowledge of how phrases are combined (syntactic) and interpreted (semantic) is necessary to understand why a sentence like "Consistent with this, we found that an FtsZ mutant unable to interact with both ZipA and FtsA was unable to assemble into Z rings" implies an existing interaction between FtsZ and ZipA/FtsA. Consider another two sentences "P53 is able to inhibit the metastasis of liver cancers with overexpressed Bcl2" and "Myc and Bax-induced apoptosis is down-regulated by the treatment of doxorubicin". Both sentences are equally syntactically ambiguous, with two possible phrasal structures for each sentence, depending on whether P53 is in collaboration with Bc12, and whether Myc worked together with Bax. NLP researchers focus on the analysis of these natural-language ambiguities and try to figure out ways to transform them into machine-readable forms [5].

Text-based mining analogies can also be found in data mining literatures. For instance, the process of text classification is exploited to designate text samples into one or more of predefined sets of categories. Its primary usage was to index scientific literatures according to controlled vocabularies. The length of present text samples varies in the form of sentences, abstracts and full texts. Other feasible applications include spam filtering, hierarchical catalogue-dependent web page categorization, automatic generation of metadata, detection of text genre, and many others. Text classification mainly consists of two approaches [6]. The first is established from the knowledge engineering point of view, in which the expert's domain knowledge is directly encrypted into the system, either in a declarative manner or in the form of procedural classification rules. The other is the machine learning-based approach, in which a classifier is developed by learning the knowledge of a set of pre-classified data. Accurate text classification systems are highly beneficial database curators, who have to browse through an enormous collection of data in search of the information of interest [7]. Due to the increasing growth of biomedical texts, database curation has advanced into a more time and effort consuming process. Text classification plays an important role in distinguishing the information of interest out from the massive amount of literature. Instead of a comprehensive overview of the methodologies employed in text classification, we will introduce some openly available text classification tools in Section 4.4. Readers who are interested in more details may consult references [6,8-10].

Nevertheless, from a theoretical point of view, each of the aforementioned tasks alone is not considered to be text mining. Hearst

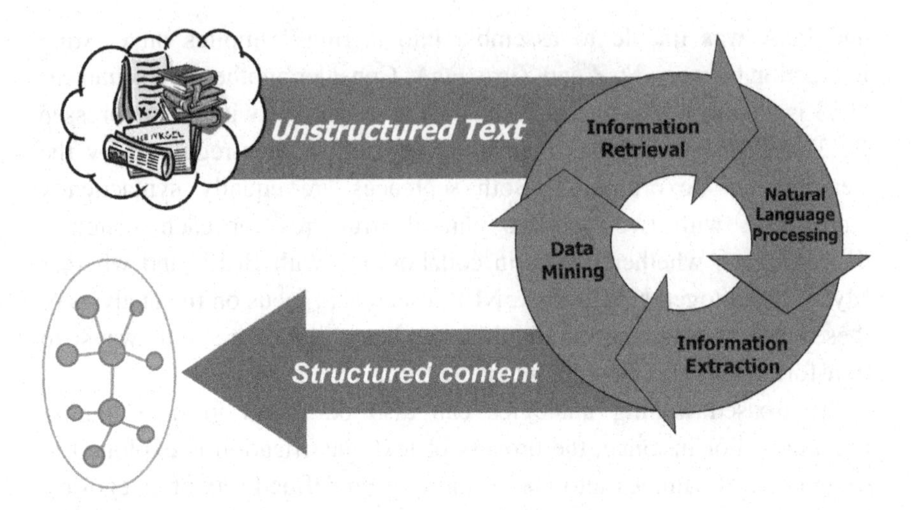

Fig. 1. From text to knowledge through text mining.

pointed out another key attribute which distinguishes text mining from those analogies—the ability to link the extracted information together and come up with new facts or hypotheses that can be explored further by more conventional means of experimentation [11]. Text classification and summarization boil down a document into one (or more) set of predefined labels or texts that conveys important information of the original document. Both results in compact summaries of known knowledge, but cannot lead to the discovery of new information.

Hence, text mining involves revealing novel understanding through analyses of large collections of natural language document. This understanding may include relationships or patterns concealed within documents, which would otherwise be extremely difficult, if possible at all, to be disclosed without comprehensive examination of the literatures. Text mining refers to an effectual approach with an integration of various techniques. Figure 1 delineates the four major courses in text mining in the transformation of implicit knowledge into explicit knowledge. These steps can be incorporated into a single workflow to form a text mining pipeline.

(i) Information retrieval: Collects texts that are relevant to the user's query. *e.g.* Through Google and PubMed.

Hypertension was **induced** by continuous **angiotensin II (Ang II)** infusion via osmotic mini-pumps over 4 weeks.	Structured Context
	Relation: induce
	Disease: hypertension
	Gene: angiotensin II EntrezID: 183

Fig. 2. Information extraction from text to structured content.

(ii) Natural language processing: Provides prerequisite linguistic data for information extraction.
(iii) Information extraction (IE): Identify and extract a range of information nuggets from texts of interest.
(iv) Data mining (often known as knowledge discovery): Recognize patterns in large sets of data extracted from the third step to uncover latent meaningful knowledge.

Swanson *et al.*'s pioneering work on hypothesis generation [12] shows a potential of employing text mining in synthesizing separated literatures for the assembly of hypotheses from vast amounts of text. Assuming that we are interested in mining genes relevant to the onset and development of diseases, a text mining system will first retrieve related articles from data sources such as PubMed, then utilize NLP to perform text analysis to a certain extent. IE follows by performing tasks like named entity recognition to identify named entities in a document, and generate structured outputs as shown in Figure 2. In this incident, the IE process must be capable of identifying that *induce* is a unique type of disease-gene relation, and that hypertension and angiotensin II (Ang II) are appropriate terms referring to a disease and a gene, respectively. This kind of knowledge can be stored in a database or an ontology, which defines the terms in a particular field and their relationship with one another. Based on the large number of disease-gene relations extracted from excessive collections of documents, applying data mining to this database can dig out patterns which facilitate new discoveries about the types of interactions that may occur, and the relationship between these interactions with individual diseases, and so on.

Text mining has been applied in many different areas of biomedicine and healthcare, such as finding functional relationships among genes, determining protein-protein interactions, interpreting array experiments, associating genes and phenotypes, gene clustering, protein structure prediction, spike signal detection, clinical diagnosis, biomedical hypothesis generation, measurement of patient care quality, and evidence-based medicine. The growing attention of applying text mining technologies in biomedicine and healthcare is also evident through numerous dedicated workshops, tutorials, and special tracks at major conferences in bioinformatics and natural language processing. Examples include the Pacific Symposium on Bio-computing, the international conference on Intelligent Systems for Molecular Biology, the annual meeting of the Association for Computational Linguistics, the BioCreative workshop and the i2b2 shared task for clinical data. In the following subsections, we will illustrate how text mining can help users with different demands in biomedicine and healthcare, along with several real case examples.

2. Natural Language Processing Techniques Used in Text Mining

Referring again to Figure 1, text mining itself is an interdisciplinary research that involves information retrieval, NLP, IE, and data mining techniques. These techniques are usually integrated into a single workflow and form a text mining pipeline. Among these techniques, NLP and IE play an important role in dealing with unstructured texts. In this subsection, we provide a general introduction to NLP, since a basic understanding of these techniques is necessary to fully appreciate follow-up discussions.

2.1. *Introduction of Natural Language Processing*

Interactions between computers and natural languages are the main concerns of NLP. Due to the inherent complexity of natural language, the analysis of unstructured texts usually requires a multi-step action. Comparable with the divide and conquer algorithm design paradigm in computer science, which works by recursively breaking down a very hard problem into two or more sub-problems, until these become simple enough to be solved directly, NLP deals with natural languages through several layers of sub-processing. A common decomposition can

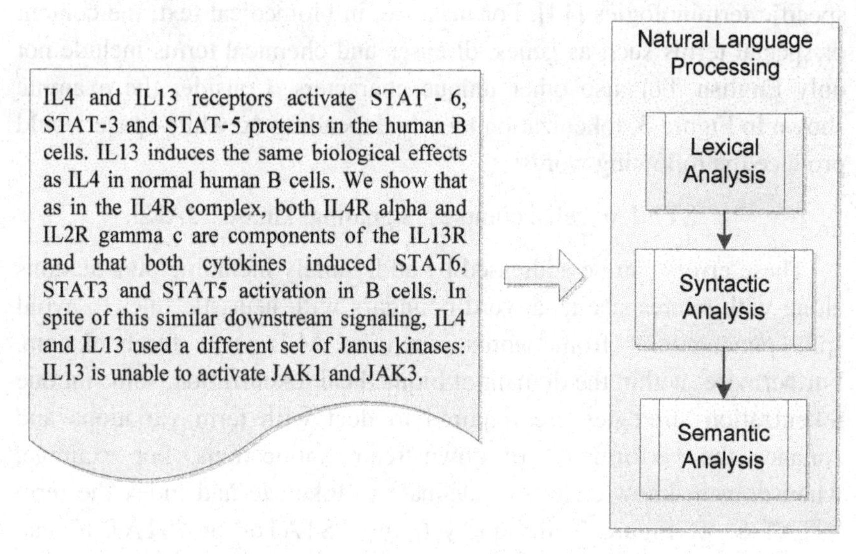

IL4 and IL13 receptors activate STAT - 6, STAT-3 and STAT-5 proteins in the human B cells. IL13 induces the same biological effects as IL4 in normal human B cells. We show that as in the IL4R complex, both IL4R alpha and IL2R gamma c are components of the IL13R and that both cytokines induced STAT6, STAT3 and STAT5 activation in B cells. In spite of this similar downstream signaling, IL4 and IL13 used a different set of Janus kinases: IL13 is unable to activate JAK1 and JAK3.
Natural Language Processing
Lexical Analysis
Syntactic Analysis
Semantic Analysis

Fig. 3. Main processing levels in natural language processing.

discriminate between the consideration of words (the lexical level), the aggregation of words as sentences, phrases or clauses (the syntactic level), and the meaning that can be ascribed to these entities at the content layer (the semantic level) [13]. In this section, we will use the example shown in Figure 3 to elaborate the subtasks of each level.

2.2. *Lexical Analysis*

Before any linguistic analysis can take place, the basic tokens involved in the natural language have to be identified. *Tokenization*, which segments the input text into linguistically plausible units, is an elementary preprocessing step in any NLP system. This step is also crucial for information retrieval, because it generally relies on matching between a token in the query and a token in a document to increase the confidence of the document's relevance to the query.

Word tokenization may seem quite straightforward in a language like English, which separates words via a special "space" character. However, a closer examination clarifies the fact that white-space is not sufficient by itself, especially for languages in special domains due to the domain specific terminologies [14]. For instance, in biomedical text, the content of special terms such as genes, diseases and chemical terms include not only English, but also other unique characters. Consider the example shown in Figure 3, tokenization merely depending on white-space would produce the following words:

STAT-6, cells. complex, signaling, kinases: JAK3.

These errors can be addressed by additionally including punctuations along with whitespace as a word boundary with heuristic rules to avoid split punctuations from words, such as M.D. and facebook.com. Furthermore, within the domain of biomedical tokenization, some unique tokenization strategies are required to deal with term variations and enhance the performance of downstream applications. For example, when domain knowledge is inadequate to tokenize and index the term "STAT-6" of Figure 3, the query terms, "STAT6" or "STAT 6" can possibly lead to a mismatch for the article shown in Figure 3. In order for tokenization to be carried out without problems, approaches such as finite-state automata, lexicon-based approaches, machine learning or a mixture of those mentioned above are employed. He and Kayaalp [15] gave a comprehensive survey for 13 open available tokenizers developed for processing biomedical articles, and pointed out important factors which needs to be taken into account when a user judges on choosing the right tokenizer that can effectively utilize the resulting tokens with the minimum loss of information.

2.3. *Syntactic Analysis*

At this level, syntactic analysis assembles sequences of tokens from a sentence syntactically into larger units, such as phrases or clauses. The knowledge required to designate tokens into their proper syntactic structures can be found in two resources —grammars and treebanks.

2.3.1. *Grammar*

In natural language, groups of words may behave as a single unit or phrase, called a *constituent*. For instance, a *noun phrase* (NP) is a sequence of words surrounding at least one noun, which often acts as a unit. It may include single words like "IL13" and phrases like "the human B cells", and "IL4 and IL13 receptors".

Context-free grammar is the most commonly used mathematical system for modeling constituent structure in natural languages. It comprises a set of production rules, each expressing the circumstances that symbols of the language can be grouped and ordered together, along with a *lexicon* of words and symbols. The symbols used in a production rule are divided into two classes:

(i) Terminal symbols: words in the language ("a", "the", "IL13") are called *terminal* symbols.
(ii) Non-terminals: other types of symbols that only denote groups or vague notions of terminal symbols.

Every context-free rule is of the form $V \rightarrow w$. V stands for a single non-terminal symbol, while w represents an ordered list of one or more terminals/non-terminals. Parentheses symbol "( )" is used to mark optional constituents.

For example, the following production rules express that a NP can be composed of a proper noun, a coordination consisted with two other NPs or a nominal. A *Nominal* can be one or more nouns, which may be conjoined with conjunctions like "and".

Rule 1: $NP \rightarrow ProperNoun$
Rule 2: $NP \rightarrow NP \quad Conj \quad NP$
Rule 3: $NP \rightarrow (Det) \quad Nominal$
Rule 4: $Nominal \rightarrow Noun \mid Nominal \quad Noun \mid ProperNoun$
Rule 5: $Nominal \rightarrow Nominal \quad Conj \quad Nominal$
The following rules express facts about the lexicon:
Rule 6: $ProperNoun \rightarrow$ IL13 | IL4
Rule 7: $Noun \rightarrow$ receptors | proteins
Rule 8: $Conj \rightarrow$ and | or

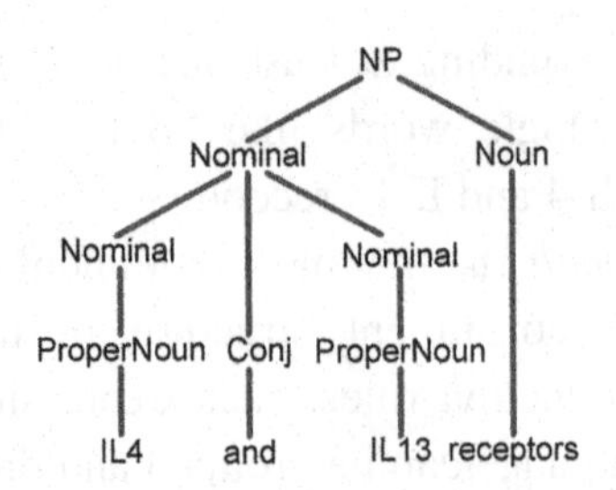

Fig. 4. A parse tree for "IL4 and IL13 receptors".

Based on these rules, we use them for generating sentences, as well as designating the structure of given sentences. So starting from the symbol: *NP*, we can

(i) use Rule 3 to rewrite *NP* as: (*Det*) *Nominal*,

(ii) and then Rule 4 to rewrite the *Nominal* in (i) to get: (*Det*) *Nominal Noun*,

(iii) and then Rule 5 as: (Det) Nominal Conj Nominal Noun,

(iv) and then Rule 4 as: (Det) ProperNoun Conj ProperNoun Noun.

(v) Finally via Rule 6, 7 and 8 as: IL4 and IL13 receptors.

We can then say the string "IL4 and IL13 receptors" can be *derived* from the non-terminal *NP*. Such a derivation is commonly represented by a *parse tree*, as shown in Figure 4.

In short, context-free grammar is merely a simplified model of how natural languages actually work. In linguistics, the action of modeling natural languages on the basis of formal languages is called *generative grammar*, due to the fact that the language is defined by the set of probable sentences "generated" by the grammar.

The advancement of grammar leans on the linguistic perception of grammar writers to discern and come up with a system of inconspicuous linguistic entities. In general, linguistic constructions that exceed the descriptive power of context-free grammar are rare. The method is computationally tractable, and techniques such as Cocke-Kasami-Younger [16] and Earley [17] algorithms for processing context-free grammars are well understood. However, a huge amount of grammar rules or lexical entries require sophisticated exertions, and it is rather difficult for grammar writers to keep up the consistency of detailed constraints due to the demand of massive manpower [18]. Grammars also get very cumbersome once we try to deal with things like transitivity. It has also been argued that certain languages contain constructions that are provably beyond the descriptive capacity of context-free grammars [19].

2.3.2. *Treebank*

The context-free grammar rules we have examined can be applied to assign a parse tree to any sentence, meaning that a corpus can be created if all sentences are syntactically annotated with a parse tree. Such a corpus is called a *treebank*. Treebanks are crucial in parsing and various pragmatic investigations of syntactic phenomena. They provide a "gold standard" for the judgment of a created parser and a starting point for researchers relying on supervised learning approaches, such as probabilistic context-free grammar parsers, so that the parser can be created with less human intervention. Furthermore, they make a paradigm shift from the manually constructed linguistic grammars to "learn" a context-free grammar from a treebank, the *treebank grammar* [20]. In contrast to using a given grammar to learn the probabilities for each rule, treebank grammar approaches deal with an infinite set of rules by learning those actual rules as well as their probabilities from the treebank corpus.

Generally, the induction of grammar from treebank is a search space problem. The annotated treebank provides great assistance by inducing the grammar of a language with high confidence. Nevertheless, such an abundant data source is not attainable for many languages, and its assembly is quite a time and effort consuming process. As a result, a supervised grammar induction will confront a barrier for languages with less developed data sources. Albeit the above limitations, evidences show that treebank grammars perform better than one may initially expect. In fact, they turn out to outperform other non-word-based grammars/parsers [20,21].

Corpus-oriented grammar development is a relative new tactic for establishing a linguistics-based grammar based on an annotated treebank. This idea is articulated within the context of lexicalized grammar formalism, including lexicalized tree adjoining grammar [22], combinatory categorial grammar [23] and head-driven phrase structure grammar [24]. A large lexicon and a small number of grammar rules are engaged in formulating lexicalized grammars. Therefore, while grammar rules can be manually composed, a lexicon can be obtained automatically. Nonetheless, it is beyond the scope of this chapter. The developments of publicly available biomedical syntactic parsers dependent on lexicalized grammar include the C&C Parser [25] and Enju [26].

For the resources of treebank, a wide variety of treebanks have been created through the general strategy of parsing each sentence automatically with parsers, followed by the manual correction of the parses by linguists. The Penn Treebank project [27] has produced treebanks from the Brown, Switchboard, ATIS, and Wall Street Journal corpora of English, as well as treebanks in Arabic and Chinese. In biomedical text mining, the GENIA corpus [28] contains annotations for parts-of-speech and a treebank [29]. The Brown-GENIA Treebank [30] contains the syntactic structures of 21 abstracts (215 sentences) taken from the GENIA corpus, and is not overlapped with the GENIA treebank (beta version, 500 abstracts). The PennBioIE [31] CYP corpus contains 1,100 PubMed abstracts on the inhibition of cytochrome P450 enzymes, and the abstracts are annotated with paragraph, sentence boundary, and part-of-speech information. Besides, 324 of the abstracts are syntactically annotated. Another PennBioIE corpus, the PennBioIE Oncology corpus, contains similar annotations but includes supplementary cancer-related abstracts that focused on molecular genetics.

2.4. *Semantic Analysis*

Semantic analysis is motivated by the assumption that the interpretation of sentences can be captured in proper formats. The core of the semantic structure of natural languages is a form of predicate-argument arrangement. For instance, one way to scrutinize the relation between the verb and the other constituents, such as NP, is to think of the verb as a logical *predicate* and the constituents as *arguments* of the predicate. Thus, the first sentence of Figure 3, which describes a molecular activation process, can be represented by the following *predicate-argument structure*,

Activate ("IL4 and IL13 receptors",
"STAT6, STAT3, and STAT5 proteins"),

in which "Activate" is the predicate, and the two other strings between the quotation marks are arguments that represent the "Acting Agent" and the "Affected Object" of an "Activation Activity," respectively. This structure asserts that specific relationships, or dependencies, are held among the various concepts underlying the constituent words and phrases, which make up the framework of sentences.

Organizing predicate-argument structures is one of the objectives of grammar. Consider the grammar rules for the verb phrase (VP), which consists of the verb and a number of other constituents, including NPs, prepositional phrases (PPs) and combinations of both:

(i) $VP \rightarrow Verb\ NP$
(ii) $VP \rightarrow Verb\ PP$
(iii) $VP \rightarrow Verb\ NP\ VP$

and the following sentences:

- This **leads** to the nuclear association of the cytosolic component of NF-ATc.
- Presenters **lead** the discussion.
- He **led** me to believe that the story was true.

The above examples can be said to have the following syntactic argument frames:

(i) *NP* lead *PP*
(ii) *NP* lead *NP*
(iii) *NP* lead *NP infinitive-VP*

These syntactic frames illustrate the number, position and syntactic category of the arguments that are anticipated to accompany a verb. For example, the frame for the first "lead" specifies the following facts:

- There are two arguments to this predicate.
- The first argument must be a *NP*, which plays the role of the subject.

- The second must be PP, which plays the role of results, actions or attributes of the predicate's object (referring to "the cytosolic component of NF-ATc" in this case).

Information like this is quite valuable in retrieving a diversity of necessary facts about syntax. Through the analysis of appreciable semantic information associated with these frames, we can also gain significant insight into natural language comprehension.

Relations between concepts are commonly linguistically intervened through verbal, nominal or adjectival expressions, such as *X inhibits Y* or *the inhibitory effect of X on Y*. At the semantic level, these relations are lexically represented as predicates with corresponding argument frames as shown in the aforementioned examples. This process is based on structural information from syntactic analysis; by properly relating the subject ("IL4 and IL13 receptors") and direct object ("STAT6, …") of an active-voice clause as the agent and the patient. At times, a semantic inference mechanism can be built on the on the basis of known truth, which can lead to the discovery of new relations between factual assertions that are completely apart. For example, from the statements of Figure 3, we can get the following predicate-argument structures, which represent facts in the paragraph:

- SimilarDownstreamSignaling("IL4", "IL13")
- NotActivate("IL13", "JAK1 and JAK3")

According to the two structures, we know that although IL4 and IL13 possess similar downstream signaling, the transduction of signaling is done through different JAK kinases. Plus the fact that IL13 did not activate JAK1 and JAK3, it would be plausible to speculate that the mechanism of IL4 signaling may constitute JAK1 and JAK3. Inherently, this is what text mining is all about. We either search for relevant and new entities for specific relations, or inspect unknown relations between specific entities.

The semantic roles of predicates in sentences are traditionally determined through the output of a full parser, along with the application of a set of manually designed rules. These rules define precise syntactic relations, which are attainable from syntactic analysis, plus lexical information about the mapping of lexical items to terms. Supposing that a syntactic pattern is matched, the rules will then map the semantic coordinates to a predicate-argument frame. As the importance of uncovering relations among concepts rises, semantic role labeling (SRL) has the potential to augment the performance of any tasks with respect to language understanding. In the following section, we will elucidate the major concepts of SRL.

2.4.1. *Semantic Role Labeling*

In SRL, sentences are represented by one or more predicate-argument structures. Each predicate-argument structure is composed of a predicate (*e.g.*, a verb) and several arguments (*e.g.*, noun phrases) that have different semantic roles, including main arguments such as an agent[a] and a patient[b], as well as adjunct arguments, such as time, manner, and location. Here, the term *argument* refers to a syntactic constituent of the sentence related to the predicate; and the term *semantic role* refers to the semantic relationship between a predicate (e.g., a verb) and an argument (e.g., a noun phrase) of a sentence. For example, in Figure 3, the first sentence describes a molecular activation process. It can be represented by a predicate-argument structure in which "activate" is the predicate, "IL4 and IL13 receptors" comprise the agent, "STAT6, STAT3, and STAT5 proteins" comprise the patient, and "in the human B cells" is the location. Thus, the agent, patient, and location are the arguments of the predicate.

2.4.2. *Proposition Bank*

A collection of predicate-argument structures forms a *proposition bank*, which is essential in building a machine learning based SRL system. In 2006, Chou *et al.* [32] constructed the first BIOmedical PROPosition bank, BioProp, by annotating semantic role information on the full parse

---

[a] agent: deliberately performs the action (*e.g.*, **Bill** drank his soup quietly).
[b] patient: experiences the action (*e.g.*, The falling rocks crushed **the car**)

trees of the GENIA corpus [33]. BioProp includes 82 predicates[c]. With the release of the large noun argument structure newswire corpus, NomBank [34], Ozyurt introduced a biomedical noun argument structure corpus named BioNom [35]. It is constructed by annotating sentences from BioProp that contain noun phrases with verb nominalizations following the NomBank annotation guidelines. *Nominalization* is defined as the use of a verb, an adjective, or an adverb as the head of a noun phrase, with or without morphological transformation. Below is an example from the BioNom corpus:

the [constitutive]$_{\text{ARGM-MNR}}$
expression [of the T24 Ha-ras oncogene]$_{\text{Arg1}}$
[in EBV-immortalized B lymphoblasts]$_{\text{ARGM-LOC}}$

Here, the predicate noun "expression" is formed from the verb "express". BioNom's noun phrase annotations complement BioProp's verb coverage and the combination of both provide researchers with a complete source of event information for biomedical text mining[d].

Several biomedical text mining tools have been developed using proposition banks as corpora. For instance, trained on the BioProp corpus, BIOSMILE web search [36] can label biomedical entities in sentences and summarize recognized predicate-argument structures in table formats. The offline tool BioKIT [37] is trained on the newswire PropBank corpus [38] and small portion of BioProp using the authors' domain adaptation method. It achieves comparable performance to BIOSMILE.

3. Information Extraction

Refining information such as names and dates from natural language text is a non-trivial task. In this section, a series of techniques that can be used to extract various kinds of semantic content from text is presented.

---

[c] BioProp is available at http://www.ldc.upenn.edu/Catalog/catalogEntry.jsp?catalogId=LDC2009T04. Its frameset definition is available at http://bws.iis.sinica.edu.tw/bioprop/.
[d] BioNom is available at http://bionom.s3.amazonaws.com/bionomkit-dist.tar.gz.

This process of transforming unstructured data buried within text into structured data is called IE. As we advance through this section, it will become apparent that the robust solutions for IE problems are in fact clever consolidations of the NLP techniques mentioned earlier in this chapter.

Generally speaking, the first step of most information IE tasks is to distinguish and classify all existing proper names in a text — a task generally known as named entity recognition(also acknowledged as entity identification and entity extraction).

3.1. *Named Entity Recognition*

The goal of named entity recognition is to locate and classify entity mentions in text into predefined categories. Named entities are words or word sequences which usually cannot be found in common dictionaries, and yet encapsulate important information that can be useful for the semantic interpretation of texts. In general-language domains, the set of entities can range from names of individuals to monetary amounts, whereas in the biomedical domain, it is obvious that names of genes, proteins, chemical substances, diseases, drugs etc., are of special importance, which is why we have to begin by focusing on such entities if we want to do text mining in that domain. Regardless of its type, ambiguity has always been the major issue of recognizing entities. This subsection uses the task of biomedical entity recognition as an example to elaborate the challenges as well as the standard approaches toward entity recognition.

3.1.1. *Challenges in Named Entity Recognition*

Although many well-known nomenclatures for biomedical entities, including the one published by the HGNC (http://www.genenames.org/genefamilies/a-z) for human genes [39] have been released, recognition of biomedical named entities is still considered a nontrivial task. Moreover, ambiguities caused by gene names and their relative aliases are not addressed in these resources. Moreover, life scientists often come up with personalized names in articles, despite the standard nomenclatures available [40]. Biomedical named entities can comprise long compound words and short abbreviations [41]. Some entities contain various symbols and other spelling variations [42]. Irregularities are as follows: A biomedical named entity can have unknown acronyms and can contain hyphens, digits, letters, and Greek letters; adjectives preceding a named entity may or may not be part of that named entity, depending on the context and application; named entities with the same orthographical features may fall into different categories; a named entity may also belong to multiple categories intrinsically, and a named entity of one category may contain a named entity of another category inside it. Biomedical entity recognition demands abundant linguistic analysis of these entities. In contrast to text retrieval systems, named entity recognition in IE systems needs to handle diverse linguistic expressions to a greater extent. Furthermore, IE systems merely fill in the slot or templates of the knowledge of interest with determined entities in text, while text mining systems establish index of unique entities, along with their relevant context and provide these information upon queries. For example, a text mining system for discharge summary analysis should identify not only the patient names described in the summary, but also relevant anaphoric expressions such as "he/she" and "the patient", which refer to the same person. Such anaphoric expressions may be subsequently replaced with their antecedent patient name defined in the preceding context when inserted into relevant slots of medical record templates.

Early attempts at biomedical entity recognition used handcrafted patterns [43] to recognize the various entity forms. However this approach suffered from lack of portability and scalability. Later, machine learning models were introduced to tackle the entity recognition problem—first simple classifiers [44,45] and then more complex probabilistic sequence models [46-50]. These approaches either predict a tag for each token in the input sequence or predict a tag sequence for the whole input sequence. In the following subsection, we first introduce the concept of machine learning, and then describe a standard way to approach the problem of named entity recognition.

3.1.2. *Machine Learning*

As a branch of artificial intelligence, machine learning involves the design and development of algorithms that takes empirical data, such as that from annotated corpus, medical images or databases as inputs, and produce patterns or predictions that are acknowledged as features of the intrinsic mechanism responsible for data generation. A learner can take advantage of the knowledge given by the sample data, and acquire better understanding of the attributes and probability distribution of the information of interest. Data can be viewed as instances of the potential relations between observed variables. One of the most important issues of machine learning is the ability to recognize complex patterns of knowledge expression, and be able to make perceptive decisions based on the input data. It is impractical to try and include all possible occurrences of the inputs, since this will result in an excessive amount of samples to be observed (training data). Hence, the learner must organize the training data in a more generalized manner to produce rational outputs when dealing with new cases. For example, in biomedical named entity recognition, the numerical normalization step has been used by several machine learning-based entity recognition systems [50] to deal with some proteins or genes of the same family that usually differ in their numerical parts, such as members of the interleukin family - interleukin-2 and interleukin-3. Theoretically, value of the numerals is not related to their named entity types. Therefore, we can normalize all numerals into one, like interlukin-1. The advantages of numerical normalization include: (1) the number of training examples can be greatly reduced; (2) concealed training examples can be transferred to seen instances. Take gene names IL2, IL3, IL4, and IL5 for example. IL2, IL3, IL4 are in the training set, and IL5 is not. If we apply numerical normalization to these terms, they are all normalized to IL1. Therefore, the numbers of training instances corresponding to the first three terms are reduced to a minimum of 1/3. In addition, since IL5 is treated as IL1 and share the same weights after training, this unseen training instance becomes seen instance.

Machine learning systems rely on many features and functions indicating the properties of tokens or the input sequence to make a prediction. Researchers have introduced various linguistic features generated by NLP to their machine learning models. These linguistic features include neighbour tokens' affixes, part-of-speech tags, chunk tags, and surface words. We will see some examples of these features in the next subsection. The weight of each feature comes from the corpus it is trained on. That is to say, the feature weights of a model are simultaneously trained on the same set of data, and then we apply the models to perform some other task using novel data, and observe their performance. The idea of training the system on certain data, and testing it on other data can be characterized by referring to these two data sets as a training set (training corpus) and a test set (test corpus). When given a corpus of relevant data, it is first dissected into training and test sets. Features of the model and their weights are trained using the training set. Afterwards, the model is used to compute the probabilities on the test set. This training-and-testing scheme is a widely accepted concept for evaluating disparate machine learning systems.

Several machine learning techniques have been employed in the text mining field, including hidden Markov models [47], naïve Bayesian approaches, support vector machines [44], maximum entropy and maximum entropy Markov models [49], conditional random fields [51], Markov logic network [52] and decision trees.

3.1.3. *Named Entity Recognition as a Sequence Labelling Task*

Named entity recognition is customarily approved as a word-by-word sequence labelling task, and both the boundary and the type of any detected named entities can be obtained by the assigned tags. As for the sequence labelling approach, which utilizes the style of the IOB2 encoding, trained classifiers label the tokens in a text with unique tags that indicate the existence of specific kinds of named entities. In IOB2 format, each word in a sentence is regarded as a token. Each token is associated with a tag that indicates the category of the named entity and the location of the token within the named entity, for example, *B_c*, *I_c* where *c* is an entity category. These two tags denote respectively the start token and the following token of an entity in category *c*. In addition, we use the tag *O* to indicate that a token is not part of a named entity. The

Tokens	Label
IL4	B-Protein
and	O
IL13	B-Protein
receptors	I-Protein
activate	O
STAT6	B-Protein
,	O
STAT3	B-Protein
and	O
STAT5	B-Protein
proteins	O
in	O
the	O
human	B-Cell
B	I-Cell
cells	I-Cell
.	O

Fig. 5. IOB2 encoding example.

named entity recognition problem can then be phrased as the problem of assigning one of $2m + 1$ tags to each token, where m is the number of named entity categories. Figure 5 shows an example of standard word-by-word IOB2-style tagging that captures the same information of the following annotated phrase in XML format for the first sentence of Figure 3:

"<PROTEIN>IL4</PROTEIN> and <PROTEIN>IL13 receptors</PROTEIN> activate <PROTEIN>STAT6</PROTEIN>, <PROTEIN>STAT3</PROTEIN> and <PROTEIN>STAT5</PROTEIN> proteins in the <CELL>human B cells</CELL>."

After our training data is encoded with the IOB2 tags, we should then select a set of features to associate with each input token as shown in Figure 5. The chosen features should be credible predictors of the class label, and should be reliably extracted from the source text with ease. Such features can depend on not only the characteristics of the target token, but also its surrounding context as well. Briefly speaking, they can be categorized into two types: singleton features and conjunction features [53]. Singleton features are conditioned on only one linguistic property and joined by at least one entity class tag. For example, a simple binary singleton feature for token-tagging might be "if the current word = X_1 AND current tag = C_1 THEN feature value = 1." Take the sentence "The IL-2 gene localizes to bands BC on mouse Chromosome 3" as another example. If the target word is "IL-2," the following word "gene" will help the machine learning model to distinguish IL-2 gene from the protein of the same name.

Conjunction features, however, are conditioned on multiple linguistic properties, instead of only one linguistic property as in singleton features, and joined by at least one class tag. A conjunction feature similar to the above singleton feature would be "if current word = X_1 AND previous word = X_2 AND current tag = C_1 THEN feature value = 1," in which the multiple linguistic properties are current word = X_1 and previous word = X_2. In most systems singleton features far outnumber conjunction features due to the fact that the latter occupy a lot of memory.

3.2. *Entity Linking*

Entity linking is the task to link textual entity mentions into databases [54], such as linking a person or organization's name to its Wiki infobox [55], or normalizing a gene mention to its database identifier [56]. For many applications of text mining, entity linking is as important as entity recognition. Many biomedical entities, including genes and proteins, have large numbers of synonymous terms. Without normalizing them to a standard representation, different mentions of the same entity are treated as distinct items, which can distort statistical and other analyses. Entity linking can aggregate references to a given named entity and can therefore increase the sample size. With linked to databases or ontologies, identities of entity mentions are resolved, which can facilitate database indexing and provide an accurate search of the biological literature.

Figure 6 presents a specific application of entity linking to a biomedical abstract. The abstract discusses the relationship between genes "urocortin" and "CRF receptors". Using the abstract as a disambiguation reference, an entity linking system must link a given

TITLE: Cloning and characterization of human **urocortin**.
ABSTRACT:
Urocortin, a new member of the CRF peptide family which also includes *urotensin I* and *sauvagine*, was recently cloned from the rat midbrain. The synthetic replicate of **urocortin** was found to bind with high affinity to **type 1 and type 2 CRF receptors** and, based upon its anatomic localization within the brain, was proposed to be a natural ligand for the **type 2 CRF receptors**. Using a genomic library, we have cloned the human counterpart of rat **urocortin** and localized it to human chromosome 2. Human and rat **urocortin** share 95% identity within the mature peptide region. Synthetic human **urocortin** binds with high affinity to **CRF receptor types 1**, stimulates cAMP accumulation from cells stably transfected with these receptors, and acts in vitro to release **ACTH** from dispersed rat anterior pituitary cells.
In addition, the **CRF-binding protein** binds human **urocortin** with high affinity and can prevent **urocortin**-stimulated **ACTH** secretion in vitro. The inhibitory effect of the **CRF-binding protein** on human **urocortin** can be blocked by biologically inactive **CRF** fragments, such as CRF(9-33).

Fig. 6. An example of entity linking (PMID: 8612563).

gene name "urocortin" to the Entrez Gene database ID: 7349 if one only considers the human gene. In addition, an EL system must return "NIL" for "urotensin I" and "sauvgine" in the first sentence because these mentions are peptides, which are not Entrez Gene's curation targets.

3.2.1. *Challenges in Entity Linking*

The traditional entity linking task complements entity recognition by addressing the following issues.

(i) *Name variation*: An entity may be named in multiple forms, including abbreviations, shortened forms, alternative spellings, and aliases. Take the name John A. Smith as an example. The name may appear in free text as John Smith, John Alexander Smith, J. Smith, John, or Smith. EL must link all these instances to the same database entry.

(ii) *Entity ambiguity*: A single mention can match multiple database entries. The name John A. Smith may appear in a database as John Alexander Smith, John Blair Smith, or John D. Smith. Notably, EL applies a disambiguation process to determine which of all possible John Smith entries a given "John Smith" refers to.

(iii) *Absence*: The issue of absence was introduced by the EL task at TAC. When entities have no corresponding entry in a database, a NIL[e] should be returned.

(iv) *Error Propagation*: In contrast to the distinct boundary between person names in the general domain, several research results have shown that gene mentions have subtle or even conflicting boundaries. For example, the gene name "insulin" with or without the term "receptor" as its suffix refer to two different genes (human "insulin": Entrez Gene ID 3630, human "insulin receptor": Entrez Gene ID 3643). However, the receptor of the human gene name "IL3" (Entrez Gene ID 3562), "IL3 receptor", does not have any corresponding Entrez Gene entry. Subtle factors like these can affect the results of entity recognition, thereby causing error propagation to entity linking.

(v) Insufficient Information: When we try to link each individual entity, the contextual information surrounding the entity is usually obscure. For example, consider the second sentence "The synthetic replicate of **urocortin** was found to bind with high affinity to type 1 and type 2 CRF receptors and, based upon its anatomic localization within the brain, was proposed to be a natural ligand for the type 2 CRF receptors." in Figure 6. The sentence does not explicitly indicate the identity of the gene mention "urocortin", which has at least 8 ambiguous Entrez Gene IDs. One approach is to expand the context window used for disambiguation to the paragraph or the section level. However, this abstract mixes together the profile information of the

[e] "NIL" is a word commonly used to mean nothing or zero. Here, it used to indicate an invalid/uninitialized database entry.

two urocortin genes in two different species in its description, which leads traditional entity linking approaches to fail. Few recent works [57-59] have started to deal with this issue by leveraging the relations among entities to improve entity linking performance.

In addition to the above challenges, several previous works [60-62] assumed that the same surface name described in an article always refers to the same instance. This assumption might be true in encyclopedia-style articles, such as Wikipedia, but is not suitable for biomedical articles. Based on the analysis [54], the same surface name annotated with more than one linked database entries only occupies 6% of the articles of Cucerzan's dataset[f]. However, in the gene mention linking corpus compiled by Dai *et al.* [59], 14.9% of the articles contain entity mentions with the same surface name but linked with an average of 2.93 different database identifiers. The "urocortin" and "type 1/2 CRF receptor genes" in Figure 6 are an example, which are linked to two different Entrez Gene IDs.

3.2.2. *Linking of Gene Mention in BioCreative Workshop*

BioCreative is a community-wide effort that promotes the development and evaluation of IE systems applied in the biomedical domain. As one of the largest public biomedical text-mining competitions in biomedical fields, BioCreative has conducted several challenges and has released standard evaluation datasets for different tasks. To spur development in regards to the name variation and the ambiguity issues, BioCreative has held several open competitions for the gene mention linking task [56,63-65], which evaluates the ability of automated systems to generate a list of unique gene identifiers from PubMed abstracts.

[f] In average, those names are linked with 2.09 Wikipedia entries.

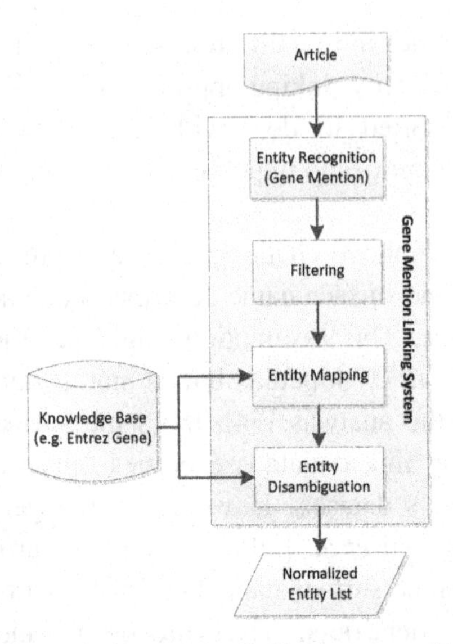

Fig. 7. General gene mention linking system architecture.

In general, after gene mention recognition, the current top-performing gene mention linking systems [66-68] include three main steps as shown in Figure 7: (1) filtering: filter out false positives or NILs, (2) entity mapping: generate candidate database identifiers and (3) entity disambiguation. Some studies only focused on improving one of these steps. For example, Hakenberg *et al.* [68] employed an isolated stage to filter out erroneous non-gene terms recognized by the gene mention recognizer, including protein families, groups or complexes. Tsuruoka *et al.* [69] utilized logistic regression to improve the accuracy of entity mapping. Xu *et al.* [70] proposed a knowledge-based disambiguation approach that combines features from text and knowledge sources via an information retrieval method. Crim *et al.* [71] used the maximum entropy model to classify valid identifiers from candidate identifier lists. Dai *et al.* [66] collected external knowledge for each gene, such as chromosome locations, gene ontology terms, etc., and calculated the likelihoods stating the similarity of the current text with the knowledge to improve the disambiguation performance. Wang *et al.* [72] focused on one source of entity ambiguity, the model organism, and developed a corpus for organism disambiguation.

3.3. Relation Extraction

Relation extraction is to identify relations among entities embedded within sentences, paragraphs, or entire documents. In the biomedical field, researchers are interested in protein-protein interactions, gene-disease interactions and drug-drug interactions. A relation usually contains more than one name entity; however, name entities co-occurring in the text do not mean they have certain relations. Relation extraction classifies the relations between name entities whether they have positive relation (ex: "increase"), negative relation (ex: "decrease"), no relation (ex: "cannot increase") or co-occur (ex: abbreviation-full name pair).

In the past, issues concerning relation extraction have been proposed and widely investigated. Currently, popular relation extraction approaches include rule-based [73,74], kernel-based [75,76], and co-occurrence-based [77,78] methods. Most of the studies focus on identifying the relations between proteins [78-80]. Craven and Kumlien [81] identified the relations between proteins and sub-cellular locations; while Rindflesch *et al.* [82] extracted relations between cancer-related genes, drugs, and cell lines. Fewer researches have worked on the extraction of gene-disease relations [83,84], but this topic has gained increasing attention over the years.

In the rule-based approaches, Feng *et al.* [85] presented an automatic relation extraction system which utilized simple clauses separated from each sentence within PubMed abstracts, and identified interactions between chemicals and CYP3A4 by pattern matching via manual-defined templates and filter rules. Chang *et al.* [86] presented a new database, named as AutoBind, based on information retrieval which automatically extracted information of protein-ligand binding affinity from full-text articles via manual-defined sentence patterns and sentence ranking techniques.

Comparatively, relation extraction using machine learning-based approaches construct semantic trees such as dependency trees, parsing trees, and phrase structure trees. The predicate argument structures will utilize semantic/syntactic features or relation rules to discover the potential relations buried in each sentence. Herein, Bui *et al.* [87] introduced a novel algorithm conducted by hybrid approach to extract protein-protein interactions from biomedical corpora. The proposed algorithm consists of two stages. The first stage extracted candidate protein-protein interaction pairs using parsing trees introduced in Section 2.2 and predefined semantic rules. The second stage made use of the feature sets extracted from parsing trees and follows by training a support vector machine model to classify protein-protein interaction pairs. Jiao *et al.* [88] presented a relation extraction system which automatically extracted CYP protein and chemical interactions from the PennBioIE corpus. With syntactic, semantic and lexical features identified from dependency parsing tree and full context, a maximum entropy model was used for predicting the potential interactions.

Moreover, Jang *et al.* [89] proposed a protein-protein interaction validation system conducted on PubMed abstracts using Penn Treebank syntactic tree and its syntactic rules to extract crucial protein interaction information. Among existing methods, employing parsers to analyze syntactic and semantic structures is considered to be beneficial. Yusuke *et al.* [90] performed a comparative evaluation of the state-of-the-art syntactic parsing methods, including dependency parsing, phrase structure parsing and deep parsing, and their contribution to protein-protein interaction extraction. This study provides researchers with a good reference for adapting the appropriate parsers for their work. Nonetheless, there is no guarantee that the results reported by Yusuke *et al.* can be generalized to other datasets and tasks. The results of the BioCreAtIvE II protein-protein interaction task [91] demonstrate that current text mining systems can detect binary relations in abstracts reasonably well [74,92], but are not as efficient in extracting significant relations from full-text articles. Three main explanations were found to account for this phenomenon [2].

First, biomedical terms, such as gene names, may have different meanings in full texts depending on their context or the section in which they appear. The same gene name in one section may in fact be used to denote a certain gene of different species. Second, the frequent use of synonyms, abbreviations, and acronyms in biomedical texts hinders semantic analysis. For instance, extracting facts from the Results section may require resolving acronyms or synonyms that were only mentioned previously in the Introduction. Third, biomedical texts usually contain several compound nouns as well as noun phrases linked by prepositions. Fourth, text mining systems have problems dealing with cross-sentence interactions that involve more than one protein and its anaphoric expressions, as shown in the following example:

> **Human growth hormone** (**hGH**) binds to *its* receptor (**hGHr**) in a three-body interaction: one molecule of *it* and two identical monomers of the receptor form a trimer.

Many papers have addressed relation extraction, summarization, and evaluation issues, but few have investigated complications caused by co-reference (anaphora) resolution [93], possibly due to the few publicly available datasets for system building and evaluation. Despite the substantial amount of annotation work carried out on co-referencing in molecular biology, few biomedical corpora with co-reference annotations are currently available [94]. Recently, the GENIA corpus was annotated with co-references. Nguyen *et al.* [95] conducted a pioneering study on the differences between newswire and biomedical co-reference annotated corpora. In the next subsection, we will elaborate on the co-reference resolution task.

3.4. Co-reference Resolution

Co-reference resolution is the task of determining whether or not two noun phrases are used to refer to the same thing i.e. finding expressions that co-refer. We call the set of co-referring expressions a co-reference chain. Here, we use the co-reference resolution track in the 2011 i2b2/VA/Cincinnati Challenge as an example to illustrate the idea of co-reference resolution [96]. The goal of this task is to determine the anaphoric relation between existing named entities in patient discharge summaries, such as the patient mentioned in the patient's discharge summary, including the patient's name and all pronouns referring to the patient, and the medications and states describing the patient (e.g. "back pain"). Figure 8 shows a sample discharge summary de-identified by the i2b2 challenge. In this example, the co-reference chain of the patient described in the clinical record includes "70 y/o male" and "who" in the

70 y/o male with h/o CAD ***s/p LAD PTCA*** 33 yrs ago , COPD , T2DM , and AICD pocket infection **who** presented to the Juan on 2017-09-03 with ***worsening SOB*** x 5 days and feeling " cloudy ."
The patient was currently at Hubbard Regional Hospital and began to develop ***SOB*** for the past 5 days .
This AM while talking with **his daughter** on the phone , **he** was noted to desaturate into the high 80s and **he** was brought to the The Hospital for Orthopedics Juan for further evaluation .

Fig. 8. Co-referents in a discharge summary.

first sentence, "the patient" and the two "he" in the second and third sentence, respectively. The two medical problems, "worsening SOB" and "SOB", are also a co-referent. On the other hand, the person concept "his daughter" and the treatment concept "s/p LAD PTCA" are singletons because both are not involved in any co-reference chain. These concepts are common notions of clinical documents, and resolving their co-references is essential to get a full view of the clinical condition. For instance, even if a natural language processing system could recognize the phrase "it reduced his pain immediately", that information would not be very useful if the system could not also distinguish what "it" was that resolved the pain, and which pain "it" reduced.

The 2011 i2b2/VA challenge systems were grouped with respect to their use of external resources, involvement of medical experts, and methods (see online supplements for definitions). Seven systems were described by their authors as rule-based, eight systems as supervised, and three as hybrids. Two systems were declared to have utilized external resources, and two systems were designed under the supervision of medical experts.

Simply put, the systems in i2b2 2011 i2b2/VA challenge created distinct modules to solve co-reference for the person concepts, pronoun concepts, and the non-person concepts (*i.e.*, problem, test, treatment, etc.), respectively. In order to assist the co-reference resolution of the person category, most systems had ways to differentiate between the patient and non-patient entities, and basically all systems explored the mentions' surrounding context. Rule-based, supervised learning or hybrids of both methods are proposed to manipulate the co-reference resolution task in discharge summaries. In most systems, the section-oriented co-reference strategy is used; they assumed that two mentions were more likely to co-refer if located within the same section. For personal pronouns, rule-based approaches assumed to co-refer them to the nearest person mentions, while the non-personal pronouns were classified based on their form and syntactic dependency relations. In the supervised co-reference resolution systems, they tend to formulate the problem as a classification problem for a given candidate pair, determining whether or not the pair is a co-referent.

4. Case Study Using a Real Database

In this section, we use a database constructed by employing text mining technology to review the aforementioned activities in text mining.

4.1. *Background*

Hypertension, obesity, and diabetes (HOD) are three well-known components of metabolic syndromes which are associated with numerous degenerative complex diseases. The study of HOD diseases has become increasingly difficult because of the diverse factors in disease progression, like gene variation, chromosomal defects, genetic variations, environmental factors, family history, etc. In most cases, development of these diseases is modulated by the variations of multiple genes and their interactions with environmental factors. Therefore, it is challenging to elucidate the pathogenic mechanisms of HOD. Recently, researchers have been using various high-throughput experimental platforms such as microarrays in transcriptomics, and co-immunoprecipitation purification and mass spectrometry in proteomics to screen all possible candidate genes, generating large amounts of data. To study HOD genetics systematically, it is necessary to integrate the finding of both small-scale studies and high-throughput research.

In this subsection, we introduce the text-mined hypertension, obesity and diabetes database T-HOD [97], which employed state-of-the-art text-mining technologies, including a NER-NEL system and a disease-gene

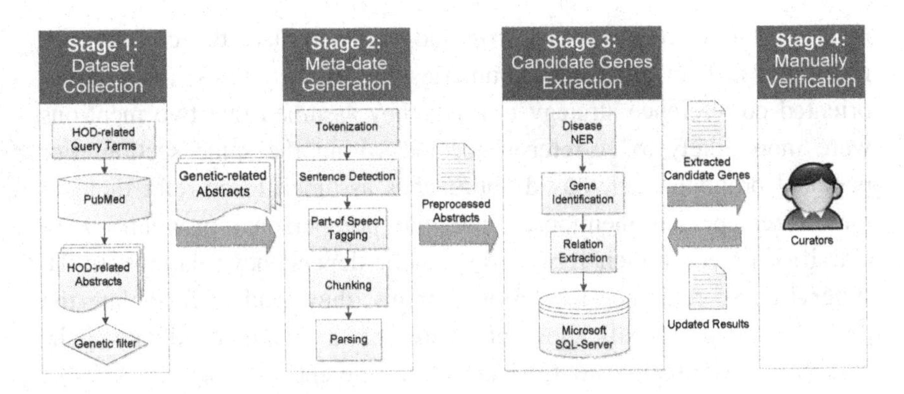

Fig. 9. The flowchart of T-HOD database construction.

(D-G) relation extraction system, to automatically extract HOD-related genes from literature. The system is developed by collaborated with Dr. Wen-Harn Pan, a biomedical researcher who has interests in identifying genes that cause genetic diseases. The user interface of T-HOD includes a network-based viewer via which researchers can observe the interactions among all extracted genes, established with information in the Human Protein Reference Database (HPRD). By using the viewer of T-HOD, biologists can easily cross-compare the extracted genes in terms of number of publications and protein-protein interactions (among all extracted genes). This system allows users to distinguish genes that are studied to a different degree in the genetic level and in the proteomic level. The interface of T-HOD also facilitates easy browsing, and allows users to provide feedback about the extracted information to improve the database. We will also present how T-HOD can be used for experimental decision making by biomedical researchers.

4.2. *Text Mining-based Database Curation*

Figure 9 shows the flowchart for constructing the T-HOD database. It is comprised of four stages: (1) Dataset Collection, (2) Meta-data Generation, (3) Candidate Gene Extraction, and (3) Manual Verification. We collected abstracts on HOD from PubMed, and used IE techniques to extract HOD candidate genes and single nucleotide polymorphisms (SNPs) from them. The T-HOD curators verify the extracted list and curate the knowledge into the T-HOD database. In the following sub-sections, we will describe each stage in detail.

4.2.1. *Stage 1: Dataset Collection*

This stage employs information retrieval techniques to collect HOD-related abstracts from PubMed. HOD-related terms, such as hypertension, blood pressure, obesity and diabetes were used to retrieve related articles from PubMed[g]. Non-genetic articles within the collection were filtered out using a list of keywords that are frequently mentioned in abstracts of genetic research, such as "polymorphism", "alleles" and "variant". Table 1 shows the numbers of collected/filtered articles.

Table 1. The numbers of collected/filtered HOD-related articles from PubMed from 1997 to 2011.

	Collected	Filtered
Hypertension	227,996	20,708
Obesity	98,993	7,337
Diabetes	139,940	11,655

4.2.2. *Stage 2: Metadata Generation*

The filtered dataset is then pre-processed by several NLP components arranged in a pipeline to generate metadata for further IE. For example, a NLP component deals with the problem of identifying words from a given document — the tokenization problem. After tokenization, a sentence splitter detects candidate positions for splitting using selected delimiters, such as periods, commas, etc. Then, it classifies whether the positions really split sentences or not. After the NLP pipeline, the genetic-related abstracts were split into sentences associated with part-of-speech tags and paring information. In addition to the tools employed in T-HOD, we will list some openly available text mining tools in Section 4.4.

---

[g] The full list of the query terms can be found at the "Supplementary Data" tab in T-HOD web site: http://bws.iis.sinica.edu.tw/thod/.

4.2.3. *Stage 3: Candidate Genes Extraction*

Stage 3 extracts HOD-related candidate genes from the pre-processed dataset through the following steps. First, NER systems are employed to recognize disease and protein/gene terms in a sentence. Second, an EL system is used to link mentioned genes to their corresponding Entrez Gene IDs. Based on the results of the previous steps, if a disease term and a gene are present in the same sentence, they are extracted as a D-G candidate pair. The D-G relation extraction system determines whether a relation indeed exists within this D-G pair. Finally, the extracted D-G pairs are stored in the SQL database ready for manual verification and data mining analysis.

4.2.4. *Stage 4: Manual Verification*

While the employed text mining components have shown satisfactory scores, the text mined candidate genes are examined by all T-HOD curators in this stage to further ensure the quality of the curated content. In this stage, newly extracted candidate genes and their corresponding evidence sentences and abstracts are presented to the T-HOD curators. T-HOD curators review each extracted candidate gene and remove the incorrect results. Based on the verified results, applying data mining to this database may be able to find the relationship between types of interactions and particular diseases and so on.

4.3. *T-HOD Content and Analyses*

Figure 10 shows the number of newly discovered HOD candidate genes by year, which was generated by the T-HOD statistics viewer. The steeply climbing curves observed in Figure 10 represent the steadily increasing number of HOD-related genetic studies over the past 10 years.

This version of T-HOD was constructed from abstracts recorded in PubMed from 1970 to 2011. Currently, there are 991, 893, 829 candidate genes and 255, 277, 228 rs numbers recorded in T-HOD for hypertension, obesity and diabetes, respectively. The rs number indicates genes that are officially registered and given a reference SNP identifier by dbSNP. This

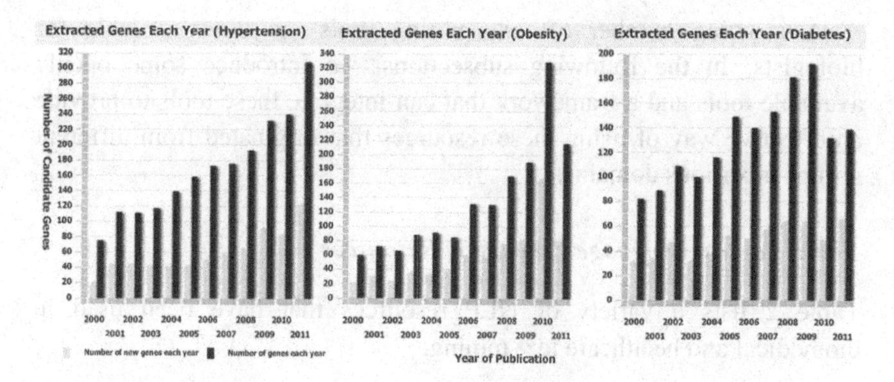

Fig. 10. Statistics of the extracted hypertension, obesity, diabetes candidate genes in T-HOD since 2000 to 2011/07.

result reveals that T-HOD contains more candidate genes and SNP sites than other related databases or review papers. One reason is that T-HOD included the most recently published candidate genes. According to the statistics shown in Figure 10, new HOD-related genes are constantly being discovered. Therefore, a continuously updated database is essential. The other is that the candidate gene extraction method employed by T-HOD does not rely on the frequency of gene-disease co-occurrences, which can improve the chance of finding promising but infrequent candidate genes supported by few papers.

4.4. *Useful Text Mining Resources*

T-HOD employs several openly available text-mining tools, including:

(i) GENIA Tagger: A text-mining tagger that can tokenize English sentences and output tags such as part-of-speech and chunk tags.
(ii) LingPipe: A Java-based programming toolkit for processing natural language texts. It contains several features such as text classification, entity recognition and sentence boundary detection.
(iii) NERBio: A web-based gene mention recognition and protein-protein interaction article classification tool.

A growing number of text mining tools are now available for biologists. In the following subsections, we introduce some openly available tools and a framework that can integrate these tools to provide an effective way of using these resources that originated from different groups in various domains.

4.4.1. *Natural Language Processing Resources*

Table 2 lists a variety of NLP resources that have been used in biomedical and healthcare text mining.

Table 2. Openly available natural language processing tools or libraries.

Name	Type	URL
BWS	Semantic	http:/bws.iis.sinica.edu.tw/bws
Enju	Syntactic	http://www.nactem.ac.uk/enju/
GENIA Sentence Splitter	Lexical	http://www.nactem.ac.uk/y-matsu/geniass/
GENIA Tagger	Lexical/Syntactic	http://www.nactem.ac.uk/GENIA/tagger
LingPipe	Java Library	http://alias-i.com/lingpipe/
OpenNLP	Java Library	http://http://opennlp.apache.org/
Stanford Biomedical Event Parser	Syntactic	http://nlp.stanford.edu/software/eventparser.shtml
Self-trained parser	Syntactic	http://nlp.stanford.edu/~mcclosky/biomedical.html

4.4.2. *Information Extraction Resources*

Several researchers have released tools that provide entity recognition or linking functions as summarized in Table 3. The table also contains Web-based annotation services dedicated to serving the function of entity recognition. Reflect [98], AIIAGMT [99] and NERBio [100] are easy-to-use online tools for detecting gene and gene product names in free text. PubTator [101] and BWS [36] can augment abstracts retrieved from PubMed by identifying gene/protein, disease and their semantic relations. Some other existing tools can directly integrate text mining features into the standard search interface of PubMed, such as PubMed-EX [102] and CDAPubMed [103]. For medical entity recognition, MetaMap [104] is a reference tool that can map medical text to UMLS concepts. Besides recognizing gene mentions, Moara [105] and GeneTUKit [106] possess the ability to link recognized entities to database entries.

Table 3. Openly available named entity recognition tools or services.

Name	Entity Type	URL
ABNER	Protein/Gene/DNA/RNA/Cell	http://pages.cs.wisc.edu/~bsettles/abner/
AIIAGMT	Gene/Protein	http://search.cpan.org/dist/AIIA-GMT/
BANNER	Gene/Protein	http://banner.sourceforge.net/
BWS	Protein/DNA/RNA/Cell/Disease	http://bws.iis.sinica.edu.tw/BWS/
CDAPubMed	MeSH	http://porter.dia.fi.upm.es/cdapubmed/downloads/
ChemSpot	Chemicals	https://www.informatik.hu-berlin.de/forschung/gebiete/wbi/resources/chemspot/chemspot/
GeneTUKit	Linked Gene/Protein	http://www.qanswers.net/GeneTUKit/.
Gimli	Protein/DNA/RNA/Cell	http://bioinformatics.ua.pt/gimli
MetaMap	UMLS concepts	http://metamap.nlm.nih.gov/
Moara	Linked Gene/Protein	
NERBio	Gene/Protein	http://bws.iis.sinica.edu.tw/NERBio/
OSCAR	Chemicals	https://bitbucket.org/wwmm/oscar4/wiki/Home
PubMed-EX	Gene/Protein/Disease	http://bws.iis.sinica.edu.tw/PubMed-EX
PubTator	Gene/Chemical/Disease/Mutation/Species	http://www.ncbi.nlm.nih.gov/CBBresearch/Lu/Demo/PubTator/
Reflect	Protein/Molecule	http://reflect.ws/

For article retrieval, biologists are now able to search through a massive volume of online articles with the aid of IE. For example, BioText [107] serves as a new way to access scientific literatures by enabling biologists to search and browse the figures and captions in biomedical articles. The iHOP service [108] retrieves sentences containing specified genes, labels the biomedical entities associated with these genes, and provides graphs of the co-occurrences among all entities. MEDIE can identify subject-verb-object relations and biomedical entities in sentences, contributing to the ability of semantic query for finding biomedical relations. BWS can annotate entities as well as a wider range of semantic relation types. For researchers interested in protein-protein interaction, BWS classifies articles as either protein-protein interaction-relevant or –irrelevant. ASCOT (Assisting Search and Creation Of clinical Trials) is a text mining-based search application [109]. It adds different types of metadata to clinical trial collections, and these metadata can help the system to further narrow down search results. Textpresso [110] enhances the capability of full text searching of scientific articles by identifying entities in articles based on an ontology. Table 4 lists the URLs for the aforementioned resources. A more comprehensive review of systems for attaining information in biomedical literatures to date can be found in Lu's work [111]. He also constructed a web site[h] dedicated to tracking existing systems and future advances in the field of biomedical literature search.

Table 4. Openly available web searching tools enhanced by text mining.

Name	URL
ASCOT	http://www.nactem.ac.uk/clinical_trials/
BioText	http://biosearch.berkeley.edu/
BWS	http://bws.iis.sinica.edu.tw/BWS/
Chilibot	http://www.chilibot.net/
iHOP	http://www.ihop-net.org/UniPub/iHOP/
MEDIE	http://www.nactem.ac.uk/tsujii/medie/
PubTator	http://www.ncbi.nlm.nih.gov/CBBresearch/Lu/Demo/PubTator/
Textpresso	http://www.textpresso.org/

4.4.3. *Integration of Text Mining Resources*

As shown in Table 3, at least ten tools can be used for the recognition of protein mentions. These tools are developed by different researchers, programming languages and operation systems. The interoperability issues between these resources are becoming significant obstacles which hinder effective usage. Fortunately, U-Compare [112] can cat as a middleware role to facilitate the integration of these resources.

U-compare is a robust system based on a general software framework—Unstructured Information Management Architecture (UIMA) [113]. This framework provides standard computer software interfaces to facilitate the analysis process of unstructured data. If the text mining components described in the previous section implements the interface defined by the framework and offers a self-describing metadata via a XML file, the framework can then manage these components and the data flow between them renders the interoperability feasible.

---

[h] http://www.ncbi.nlm.nih.gov/CBBresearch/Lu/search

U-Compare has now provided the world's largest set of text mining UIMA components. Some of the resources shown in Table 2 and 3 have already been included in U-Compare, including ABNER, Enju, GENIA tools, LingPipe, Moara and OpenNLP. It also contains a number of predefined workflows that can demonstrate the range of possible combinations of these text mining components.

5. Conclusions and Outlook

Researches within the biomedical field have evolved with time. Advancement of experimental techniques, accumulation of past experiences and the ease of access to publications around the world nowadays have all contributed to the acceleration of biomedical studies, resulting in enormous repositories of scientific journals and papers. Data processing has always been a challenging task, especially with such a massive amount of data. Therefore, traditional methods employed in data processing are no longer competent enough to meet the requirements of researchers. The field of biomedical text mining has gained increasing interest, as it has proven to be an efficient method of deriving high-quality information from text.

For text mining researchers, Thamrongrattanarit *et al.* [114] recently reported an analysis of current research trends on NLP applications in the medical domain. Their report unveils the relationship among tasks introduced in this chapter and the correlations between their employed approaches. Through mature topics, such as named entity recognition, the ascending trends in biomedical text mining researches can be perceived, which includes the extraction of biomedical events and their triggers [115], steps toward the needs of pathway extraction [116], and the curation for domain databases [117].

While current technologies themselves remain imperfect, with the aid of text mining, biomedical researchers can retrieve knowledge of interest not only more accurately, but also with less time and effort. Furthermore, text mining has been adapted to quantify relevance between biomedical entities by using information extracted from the literature. Through the great utility of text mining, researchers can rapidly obtain functional information about genes, including protein-protein interactions, gene function annotations, and the measurement of gene-gene similarity [118,119].

Text mining also exhibits great potential when used in data fusion, in which mined information is integrated, thereby gaining more insight into biological literatures. For example, ten out of 13 SNPs identified by Raychaudhuri *et al.* [120]'s method has been associated with Crohn's disease, and were later validated by follow-up genotyping [121]. The results from high-throughput experiments can be integrated with the evidences extracted from literatures to allow the prediction of novel protein-protein interactions by transferring annotations to orthologous protein pairs [122].

However, a common problem faced by biologists or bio-curators upon using text mining systems is the lack of a good interactive user interface that can be easily adopted [123]. To resolve this problem, researchers must design applications with intuitive interfaces that require little or even no knowledge of text-mining and NLP technology. The objective is to provide bioinformatics, biological, biomedical, and pharmacological researchers with a high-level view of biological interactions and help them come up with new hypotheses [2]. Some pioneers have started and showed their devotion in this respect, an example being the U-Compare system introduced in Section 4.4.3. The recent interactive task proposed in BioCreative [124] is another example that serves to address the utility and usability of text mining tools for real-life bio-curation tasks. Their report showed that text mining developers cannot work behind closed doors; they should collaborate with real users throughout the process of system development. Zhang *et al.* [125]'s Web 2.0 model represents a shift in focus from working locally to working in networked settings. Under this new approach, the Web is seen as a social, collaborative, and collective space. The model envisions the future of text mining, where annotation will be performed collaboratively with the support of innovative web tools. We believe that further development of tools like WikiProtein [126] and PubTator [101], is essential in supporting collaborative annotation.

Biomedical text mining is an extremely active research area, and the outlook for continued progress is encouraging. It can be foreseen that the texts of articles will be systematically mined by computer programs, allowing the interrelation of journal texts and the vast repository of knowledge to be stored semi-automatically in databases. We optimistically anticipate that all life scientists will learn the convenience of text mining tools and apply them to assist their research in the future.

References

1. Hahn U, Wermter J, Blasczyk R, Horn PA (2007) Text mining: powering the database revolution. Nature 448: 130-130.
2. Dai H-J, Chang Y-C, Tsai RT-H, Hsu W-L (2010) New challenges for biological text-mining in the next decade. Journal of Computer Science and Technology 25: 169-179.
3. Hearst M (2003) What is text mining (http://people.ischool.berkeley.edu/~hearst/text-mining.html).
4. Agrawal R, Imieliński T, Swami A. Mining association rules between sets of items in large databases; 1993; New York. ACM. pp. 207-216.
5. Hunter KBCL (2004) Natural Language Processing and Systems Biology. Artificial Intelligence Methods and Tools for Systems Biology.
6. Aggarwal C, Zhai C (2012) A Survey of Text Classification Algorithms. In: Aggarwal CC, Zhai C, editors. Mining Text Data: Springer US. pp. 163-222.
7. Cohen AM, Hersh WR (2005) A survey of current work in biomedical text mining. Briefings in Bioinformatics 6: 57-71.
8. Sebastiani F (2002) Machine learning in automated text categorization. ACM Comput Surv 34: 1-47.
9. Sehgal AK, Das S, Noto K, Saier MK, Elkan C (2011) Identifying Relevant Data for a Biological Database: Handcrafted Rules versus Machine Learning. Computational Biology and Bioinformatics, IEEE/ACM Transactions on 8: 851-857.
10. Qi X, Davison BD (2009) Web page classification: Features and algorithms. ACM Comput Surv 41: 1-31.
11. Hearst MA. Untangling text data mining; 1999; Stroudsburg, PA, USA. Association for Computational Linguistics. pp. 3-10.
12. Swanson DR (1990) Medical literature as a potential source of new knowledge. Bulletin of the Medical Library Association 78: 29.
13. Ananiadou S, McNaught. J (2005) Text mining for biology and biomedicine.
14. Jiang J, Zhai C (2007) An empirical study of tokenization strategies for biomedical information retrieval. Information Retrieval 10: 341-363.
15. He Y, Kayaalp M (2006) A Comparison of 13 Tokenizers on MEDLINE. THE LISTER HILL NATIONAL CENTER FOR BIOMEDICAL COMMUNICATIONS.
16. Younger DH (1967) Recognition and parsing of context-free languages in time n3. Information and Control 10: 189-208.

17. Earley J (1970) An efficient context-free parsing algorithm. Commun ACM 13: 94-102.
18. Miyao Y, Ninomiya T, Jun, Tsujii i (2005) Corpus-Oriented grammar development for acquiring a head-driven phrase structure grammar from the penn treebank. Proceedings of the First international joint conference on Natural Language Processing. Hainan Island, China: Springer-Verlag. pp. 684-693.
19. Culy C (1985) The complexity of the vocabulary of Bambara. Linguistics and Philosophy 8: 345-351.
20. Charniak E. Tree-bank grammars; 1996. pp. 1031-1036.
21. Charniak E (1997) Statistical parsing with a context-free grammar and word statistics. Proceedings of the National Conference on Artificial Intelligence: JOHN WILEY & SONS LTD. pp. 598-603.
22. Schabes Y, Abeille A, Joshi AK (1988) Parsing strategies with 'lexicalized' grammars: application to tree adjoining grammars. Proceedings of the 12th conference on Computational linguistics - Volume 2. Budapest, Hungry: Association for Computational Linguistics. pp. 578-583.
23. Steedman M (2000) The syntactic process. Cambridge, MA: The MIT Press.
24. Pollard C, Sag I (1994) Head-Driven Phrase Structure Grammar: University of Chicago Press and CSLI Publications.
25. Rimell L, Clark S (2009) Porting a lexicalized-grammar parser to the biomedical domain. J of Biomedical Informatics 42: 852-865.
26. Miyao Y, Tsujii Ji (2008) Feature forest models for probabilistic hpsg parsing. Comput Linguist 34: 35-80.
27. Marcus M, Kim G, Marcinkiewicz MA, MacIntyre R, Bies A, *et al.* (1994) The Penn Treebank: annotating predicate argument structure. Proceedings of the workshop on Human Language Technology: 114-119.
28. Kim JD, Ohta T, Tateisi Y, Tsujii J (2003) GENIA corpus--a semantically annotated corpus for bio-textmining. Bioinformatics 19: 180-182.
29. Tateisi Y, Yakushiji A, Ohta T, Tsujii J (2005) Syntax Annotation for the GENIA corpus. Proc IJCNLP 2005, Companion volume: 222–227.
30. Lease M, Charniak E (2005) Parsing biomedical literature. Proceedings of Second International Joint Conference on Natural Language Processing. pp. 58–69.
31. Kulick S, Bies A, Liberman M, Mandel M, McDonald R, *et al.* Integrated annotation for biomedical information extraction; 2004. pp. 61-68.
32. Chou W-C, Tsai RT-H, Su Y-S, Ku W, Sung T-Y, *et al.* (2006) A Semi-Automatic Method for Annotating a Biomedical Proposition Bank. Proceedings of ACL Workshop on Frontiers in Linguistically Annotated Corpora. Sydney, Australia. pp. 5-12.
33. Tateisi Y, Ohta T, Tsujii J-i (1997) Annotation of Predicate-argument Structure on Molecular Biology Text. Nature 386: 296-299.

34. Meyers A, Reeves R, Macleod C, Szekely R, Zielinska V, *et al.* (2004) The NomBank Project: An Interim Report. Proceedings of the NAACL/HLT Workshop on Frontiers in Corpus Annotation.
35. Ozyurt IB (2012) Automatic Identification and Classification of Noun Argument Structures in Biomedical Literature. IEEE TRANSACTIONS ON COMPUTATIONAL BIOLOGY AND BIOINFORMATICS.
36. Dai H-J, Huang C-H, Lin RTK, Tsai RT-H, Hsu W-L (2008) BIOSMILE web search: a web application for annotating biomedical entities and relations. Nucl Acids Res 36: W390-W398.
37. Dahlmeier D, Ng HT (2010) Domain adaptation for semantic role labeling in the biomedical domain. Bioinformatics 26: 1098-1104.
38. Palmer M, Gildea D, Kingsbury P (2005) The proposition bank: An annotated corpus of semantic roles. Computational Linguistics 31: 71-106.
39. Wain HM, Bruford EA, Lovering RC, Lush MJ, Wright MW, *et al.* (2002) Guidelines for Human Gene Nomenclature. Genomics 79: 464-470.
40. Shatkay H, Feldman R (2003) Mining the biomedical literature in the genomic era: an overview. Journal of Computational Biology 10: 821-855.
41. Pakhomov S. Semi-supervised maximum entropy based approach to acronym and abbreviation normalization in medical text; 2002.
42. Hanisch D, Fluck J, Mevissen H, Zimmer R. Playing biology's name game: identifying protein names in scientific text; 2003.
43. Fukuda K, Tsunoda T, Tamura A, Takagi T. Toward information extraction: identifying protein names from biological papers; 1998.
44. Kazama J, Makino T, Ohta Y, J. Tsujii. Tuning support vector machines for biomedical named entity recognition; 2002.
45. Lee K-J, Hwang Y-S, Rim H-C. Two phase biomedical NE Recognition based on SVMs; 2003.
46. Settles B. Biomedical Named Entity Recognition Using Conditional Random Fields and Rich Feature Sets; 2004.
47. Zhao S. Named Entity Recognition in Biomedical Texts using an HMM Model; 2004.
48. Zhou G, Su J. Exploring Deep Knowledge Resources in Biomedical Name Recognition; 2004.
49. Finkel J, Dingare S, Nguyen H, Nissim M, Manning C, *et al.* Exploiting Context for Biomedical Entity Recognition: From Syntax to the Web; 2004.
50. Tsai RT-H, Sung C-L, Dai H-J, Hung H-C, Sung T-Y, *et al.* (2006) NERBio: using selected word conjunctions, term normalization, and global patterns to improve biomedical named entity recognition. BMC Bioinformatics 7: S11.
51. Lafferty J, McCallum A, Pereira F (2001) Conditional random fields: Probabilistic models for segmenting and labeling sequence data. Proceedings of the 18th International Conference on Machine Learning (ICML). pp. 282–289.

52. Domingos P, Kok S, Poon H, Richardson M, Singla P (2006) Unifying logical and statistical AI. Proceedings of the 21st National Conference on Artificial Intelligence. pp. 2-7.
53. McCallum A. Efficiently Inducing Features of Conditional Random Fields; 2003.
54. Dai H-J, Wu C-Y, Tsai RT-H, Hsu W-L (2012) From Entity Recognition to Entity Linking: A Survey of Advanced Entity Linking Techniques. Proceedings of the 26th Annual Conference of the Japanese Society for Artificial Intelligence. Yamaguchi, Japan.
55. Ji H, Grishman R, Dang HT, Griffitt K, Ellis J (2010) Overview of the TAC 2010 Knowledge Base Population Track. Proceedings of the Third Text Analysis Conference (TAC 2010). Gaithersburg, Maryland USA: National Institute of Standards and Technology (NIST). pp. 1-10.
56. Morgan AA, Lu Z, Wang X, Cohen AM, Fluck J, *et al.* (2008) Overview of BioCreative II gene normalization. Genome Biology 9: S3.
57. Rastogi V, Dalvi AN, Garofalakis AM (2011) Large-scale collective entity matching. Proceedings of the VLDB Endowment. Seattle, Washington. pp. 208-218.
58. Han X, Sun L, Zhao J (2011) Collective entity linking in web text: a graph-based method. Proceedings of the 34th international ACM SIGIR conference on Research and development in Information Retrieval. Beijing, China: ACM. pp. 765-774.
59. Dai H-J, Chang Y-C, Tsai RT-H, Hsu W-L (2011) Integration of gene normalization stages and co-reference resolution using a Markov logic network. Bioinformatics 27: 2586-2594.
60. Cucerzan S (2007) Large-scale named entity disambiguation based on Wikipedia data. Proceedings of the 2007 Joint Conference on Empirical Methods in Natural Language Processing and Computational Natural Language Learning. Prague, Czech Republic. pp. 708-716.
61. Mihalcea R, Csomai A (2007) Wikify!: linking documents to encyclopedic knowledge. Proceedings of the sixteenth ACM conference on Conference on information and knowledge management. Lisbon, Portugal: ACM. pp. 233-242.
62. Kulkarni S, Singh A, Ramakrishnan G, Chakrabarti S. Collective annotation of wikipedia entities in web text; 2009; Paris, France. ACM. pp. 457-466.
63. Hirschman L, Colosimo M, Morgan A, Yeh A (2005) Overview of BioCreAtIvE task 1B: normalized gene lists. BMC Bioinformatics 6: S11.
64. Leitner F, Mardis SA, Krallinger M, Cesareni G, Hirschman LA, *et al.* (2010) An Overview of BioCreative II.5. IEEE/ACM TRANSACTIONS ON COMPUTATIONAL BIOLOGY AND BIOINFORMATICS 7: 385-399.
65. Lu Z, Kao H-Y, Wei C-H, Huang M, Liu J, *et al.* (2011) The gene normalization task in BioCreative III. BMC Bioinformatics 12: S2.
66. Dai H-J, Lai P-T, Tsai RT-H (2010) Multistage Gene Normalization and SVM-Based Ranking for Protein Interactor Extraction in Full-Text Articles. IEEE/ACM TRANSACTIONS ON COMPUTATIONAL BIOLOGY AND BIOINFORMATICS 7: 412-420.

67. Wermter J, Tomanek K, Hahn U (2009) High-performance gene name normalization with GENO. Bioinformatics 25: 815-821.
68. Hakenberg J, Plake C, Leaman R, Schroeder M, Gonzalez G (2008) Inter-species normalization of gene mentions with GNAT. Bioinformatics 24: 126-132.
69. Tsuruoka Y, McNaught J, Tsujii J, Ananiadou S (2007) Learning string similarity measures for gene/protein name dictionary look-up using logistic regression. Bioinformatics 23: 2768-2774.
70. Xu H, Fan J-W, Hripcsak G, Mendonça EA, Markatou M, *et al.* (2007) Gene symbol disambiguation using knowledge-based profiles. Bioinformatics 23: 1015-1022.
71. Crim J, McDonald R, Pereira F (2005) Automatically Annotating Documents with Normalized Gene Lists. BMC Bioinformatics 6: S13.
72. Wang X, Tsujii Ji, Ananiadou S (2010) Disambiguating the species of biomedical named entities using natural language parsers. Bioinformatics 26: 661-667.
73. Saric J, Jensen LJ, Ouzounova R, Rojas I, Bork P (2006) Extraction of regulatory gene/protein networks from Medline. Bioinformatics 22: 645-650.
74. Ono T, Hishigaki H, Tanigami A, Takagi T (2001) Automated extraction of information on protein-protein interactions from the biological literature. Bioinformatics 17: 155-161.
75. Kim S, Yoon J, Yang J (2008) Kernel approaches for genic interaction extraction. Bioinformatics 24: 118.
76. Bunescu R, Mooney R (2006) Subsequence kernels for relation extraction. ADVANCES IN NEURAL INFORMATION PROCESSING SYSTEMS 18: 171.
77. Barnickel T, Weston J, Collobert R, Mewes H, Stumpflen V (2009) Large Scale Application of Neural Network Based Semantic Role Labeling for Automated Relation Extraction from Biomedical Texts. PLoS ONE 4.
78. Ramani A, Bunescu R, Mooney R, Marcotte E (2005) Consolidating the set of known human protein-protein interactions in preparation for large-scale mapping of the human interactome. Genome Biology 6: R40.
79. Bunescu R, Ge R, Kate R, Marcotte E, Mooney R, *et al.* (2005) Comparative experiments on learning information extractors for proteins and their interactions. Artificial Intelligence in Medicine 33: 139-155.
80. Rosario B, Hearst MA. Multi-way relation classification: application to protein-protein interactions; 2005. Association for Computational Linguistics Morristown, NJ, USA. pp. 732-739.
81. Craven M, Kumlien J. Constructing Biological Knowledge Bases by Extracting Information from Text Sources; 1999. AAAI Press. pp. 77-86.
82. Rindflesch TC, Tanabe L, Weinstein JN, Hunter L. EDGAR: extraction of drugs, genes and relations from the biomedical literature; 2000. pp. 515-524.
83. Chun H-W, Tsuruoka Y, Kim J-D, Shiba R, Nagata N, *et al.* (2006) Extraction of gene-disease relations from Medline using domain dictionaries and machine learning. Proceedings of the Pacific Symposium on Biocomputing: 4-15.

84. Tsai RT-H, Lai P-T, Dai H-J, Huang C-H, Bow Y-Y, *et al.* (2009) HypertenGene: Extracting key hypertension genes from biomedical literature with position and automatically-generated template features. BMC Bioinformatics 10: S9.
85. Feng C, Yamashita F, Hashida M (2007) Automated Extraction of Information from the Literature on Chemical-CYP3A4 Interactions. Journal of Chemical Information and Modeling 47: 2449-2455.
86. Chang DT-H, Ke C-H, Lin J-H, Chiang J-H (2012) AutoBind: automatic extraction of protein-ligand-binding affinity data from biological literature. Bioinformatics 28: 2162-2168.
87. Bui Q-C, Katrenko S, Sloot PMA (2011) A hybrid approach to extract protein-protein interactions. Bioinformatics 27: 259-265.
88. Jang H, Lim J, Lim J-H, Park S-J, Lee K-C, *et al.* (2006) Finding the evidence for protein-protein interactions from PubMed abstracts. Bioinformatics 22: 220-226.
89. Jiao D, Wild DJ (2009) Extraction of CYP Chemical Interactions from Biomedical Literature Using Natural Language Processing Methods. Journal of Chemical Information and Modeling 49: 263-269.
90. Miyao Y, Sagae K, Sætre R, Matsuzaki T, Tsujii Ji (2009) Evaluating Contributions of Natural Language Parsers to Protein-Protein Interaction Extraction. Bioinformatics 25: 394-400.
91. Krallinger M, Leitner F, Rodriguez-Penagos C, Valencia A (2008) Overview of the protein-protein interaction annotation extraction task of BioCreative II. Genome Biology 9: S4.
92. Wong L (2001) PIES, a protein interaction extraction system. Proceedings of Pacific Symposium on Biocomputing 6: 520-531.
93. Castaño J, Zhang J, Pustejovsky J. Anaphora resolution in biomedical literature; 2002.
94. Pustejovsky J, Castano J, Sauri R, Rumshinsky A, Zhang J, *et al.* Medstract: creating large-scale information servers for biomedical libraries; 2002. Association for Computational Linguistics Morristown, NJ, USA. pp. 85-92.
95. Nguyen N, Kim J-D, Tsujii Ji (2008) Challenges in Pronoun Resolution System for Biomedical Text. Proceedings of the Sixth International Language Resources and Evaluation (LREC'08).
96. BA UÖ, S. S, T. F, J. P, B S (2012) Evaluating the state of the art in coreference resolution for electronic medical records. Journal of the American Medical Informatics Association.
97. Dai H-J, Wu C-Y, Tsai RT-H, Pan W-H, Hsu W-L (2013) T-HOD: A Literature-based Candidate Gene Database for Hypertension, Obesity, and Diabetes. Database (Oxford).
98. Pafilis E, O'Donoghue SI, Jensen LJ, Horn H, Kuhn M, *et al.* (2009) Reflect: augmented browsing for the life scientist. Nat Biotech 27: 508-510.

99. Huang HS, Lin YS, Lin KT, Kuo CJ, Chang YM, *et al.* (2007) High-Recall Gene Mention Recognition by Unification of Multiple Backward Parsing Models. Proceedings of the Second BioCreative Challenge Evaluation Workshop: 109?111.
100. Dai H-J, Hung H-C, Tsai RT-H, Hsu W-L (2007) IASL Systems in the Gene Mention Tagging Task and Protein Interaction Article Sub-task. Proceedings of Second BioCreAtIvE Challenge Workshop.
101. Wei C-H, Harris BR, Li D, Berardini TZ, Huala E, *et al.* (2012) Accelerating literature curation with text-mining tools: a case study of using PubTator to curate genes in PubMed abstracts. Database 2012.
102. Tsai RT-H, Dai H-J, Lai P-T, Huang C-H (2009) PubMed-EX: A web browser extension to enhance PubMed search with text mining features. Bioinformatics 25: 3031-3032.
103. Perez-Rey D, Jimenez-Castellanos A, Garcia-Remesal M, Crespo J, Maojo V (2012) CDAPubMed: a browser extension to retrieve EHR-based biomedical literature. BMC Medical Informatics and Decision Making 12: 29.
104. Aronson A (2001) Effective Mapping of Biomedical Text to the UMLS Metathesaurus: The MetaMap Program. JOURNAL OF BIOMEDICAL INFORMATICS 35: 17-21.
105. Neves M, Carazo J-M, Pascual-Montano A (2010) Moara: a Java library for extracting and normalizing gene and protein mentions. BMC Bioinformatics 11: 157.
106. Huang M, Liu J, Zhu X (2011) GeneTUKit: a software for document-level gene normalization. Bioinformatics 27: 1032-1033.
107. Hearst MA, Divoli A, Guturu H, Ksikes A, Nakov P, *et al.* (2007) BioText Search Engine: beyond abstract search. Bioinformatics 23: 2196.
108. Hoffmann R, Valencia A (2005) Implementing the iHOP concept for navigation of biomedical literature. Bioinformatics 21: 252-258.
109. Korkontzelos I, Mu T, Ananiadou S (2012) ASCOT: a text mining-based web-service for efficient search and assisted creation of clinical trials. BMC Medical Informatics and Decision Making 12: S3.
110. Muller HM, Kenny EE, Sternberg PW (2004) Textpresso: an ontology-based information retrieval and extraction system for biological literature. PLoS Biol 2: e309.
111. Lu Z (2011) PubMed and beyond: a survey of web tools for searching biomedical literature. Database: the journal of biological databases and curation.
112. Kano Y, Baumgartner WA, McCrohon L, Ananiadou S, Cohen KB, *et al.* (2009) U-Compare: share and compare text mining tools with UIMA. Bioinformatics 25: 1997-1998.
113. Ferrucci D, Lally A, Gruhl D, Epstein E, Schor M, *et al.* (2006) Towards an interoperability standard for text and multi-modal analytics. IBM Res Rep.
114. Thamrongrattanarit A, Shafir M, Crivaro M, Borukhov B, Meteer M. What can NLP tell us about BioNLP?; 2012; Montreal, Canada. Association for Computational Linguistics. pp. 122-129.

115. Kim J-D, Ohta T, Pyysalo S, Kano Y, Tsujii Ji (2009) Overview of BioNLP'09 Shared Task on Event Extraction. Proceedings of the Workshop on Current Trends in Biomedical Natural Language Processing: Shared Task. Boulder, Colorado: Association for Computational Linguistics pp. 1-9.
116. Oda K, Kim J-D, Ohta T, Okanohara D, Matsuzaki T, *et al.* (2008) New challenges for text mining: mapping between text and manually curated pathways. BMC Bioinformatics 9: S5.
117. Ongenaert M, Van Neste L, De Meyer T, Menschaert G, Bekaert S, *et al.* (2008) PubMeth: a cancer methylation database combining text-mining and expert annotation. Nucl Acids Res 36: D842-846.
118. Rzhetsky A, Seringhaus M, Gerstein M (2008) Seeking a new biology through text mining. Cell 134: 9-13.
119. Jensen LJ, Saric J, Bork P (2006) Literature mining for the biologist: from information retrieval to biological discovery. NATURE REVIEWS GENETICS 7: 119-129.
120. Raychaudhuri S, Plenge RM, Rossin EJ, Ng AC, Purcell SM, *et al.* (2009) Identifying relationships among genomic disease regions: predicting genes at pathogenic SNP associations and rare deletions. PLoS genetics 5: e1000534.
121. Harmston N, Filsell W, Stumpf MP (2010) What the papers say: Text mining for genomics and systems biology. Human genomics 5: 17-29.
122. Von Mering C, Jensen LJ, Snel B, Hooper SD, Krupp M, *et al.* (2005) STRING: known and predicted protein–protein associations, integrated and transferred across organisms. Nucleic Acids Research 33: D433-D437.
123. Altman R, Bergman C, Blake J, Blaschke C, Cohen A, *et al.* (2008) Text mining for biology - the way forward: opinions from leading scientists. Genome Biology 9: S7.
124. Arighi CN, Roberts PM, Agarwal S, Bhattacharya S, Cesareni G, *et al.* (2011) BioCreative III interactive task: an overview. BMC Bioinformatics 12: S4.
125. Zhang Z, Cheung K-H, Townsend JP (2009) Bringing Web 2.0 to bioinformatics. Brief Bioinform 10: 1-10.
126. Mons B, Ashburner M, Chichester C, van Mulligen E, Weeber M, *et al.* (2008) Calling on a million minds for community annotation in WikiProteins. Genome Biology 9: R89.

Chapter 13

Learning to Rank Biomedical Documents with only Positive and Unlabeled Examples: A Case Study

Mingzhu Zhu

Information Systems Department, New Jersey Institute of Technology
University Heights, Newark, NJ 07102, USA
mz59@njit.edu

Yi-Fang Brook Wu

Information Systems Department, New Jersey Institute of Technology
University Heights, Newark, NJ 07102, USA
wu@njit.edu

Meghana Samir Vasavada

Bioinformatics Program, New Jersey Institute of Technology
University Heights, Newark, NJ 07102, USA
mv55@njit.edu

Jason T. L. Wang

Bioinformatics Program and Computer Science Department, New Jersey Institute of Technology
University Heights, Newark, NJ 07102, USA
wangj@njit.edu

In the text mining field, obtaining training data requires human experts' labeling efforts, which is often time consuming and expensive. Supervised learning with only a small number of positive examples and a large amount of unlabeled data, which is easy to get, has attracted booming interests in the field. A recently proposed relabeling method, which assumes unlabeled data as negative data for text classification, has been shown successful in identifying relevant biomedical documents on a specific topic. However, it's not known whether and how feature selection affects the performance of the method. In addition, no extensive research has been conducted to evaluate how the performance of the method changes when the proportion of positive examples in the unlabeled dataset varies. Following the relabeling method, we train Support Vector Machines using positive and unlabeled data to rank incoming documents based on their probability values of being predicted as positive. Using an RNA-protein binding (RNAPB) dataset collected from PubMed, we conduct a series of experiments to evaluate the performance of the proposed text ranking algorithm. Our experimental results show that 1) feature selection is helpful in improving the performance of the algorithm; 2) the increase of the proportion of the positive examples in the unlabeled dataset decreases the performance of the algorithm, but when no reliable negative training data is available, the classifiers, built based on the unlabeled data that contains a small proportion of positive examples, may have comparable performance with the classifiers built based on pure negative data.

1. Introduction

Text mining tools aim to extract information of interest to a user effectively and efficiently. The literature, both on-line and off-line, is increasing at a considerable rate, which makes it almost impossible for a researcher to keep up-to-date with all the relevant literature manually, even on specialized topics [1]. To interpret the large-scale data sets that are being generated from diverse disciplines, it's necessary for researchers to expand their research fields beyond their core realm. Literature mining tools become essential for researchers to get access to the large amount of documents in multidisciplinary fields.

Take bioinformatics as an example. For the average user, the most frequently used literature mining or information retrieval (IR) tools are keyword-based search engines like PubMed. However, it is hard to use IR tools to get all the relevant literature on a specific topic, which is referred to as topic search here. For instance, a junior researcher or a student may hope to get all the published articles that are related to single-nucleotide polymorphism (SNP), which is a very important topic in genetics. It might seem that a keyword search using "single-nucleotide polymorphism" can meet the requirement of this task, but keyword-search is tedious and time consuming in that we need to try different keywords many times to identify the documents that we are interested in. For a layman in biology, sometimes it's not easy to find the appropriate keywords for a specific topic. Moreover, the returned documents from IR tools may be irrelevant even the keywords are constructed by experts. As a result, it is important to develop novel tools for users to solve the information search problems.

Machine learning methods, such as supervised learning, have been widely used in information retrieval to help users locate the information they are interested in [1, 2]. In supervised learning, a large set of labeled training data is required to make sure the trained model has good performance. However, it is often the case that there is not enough reliable labeled data. With the development of the World Wide Web, the volume of data is increasing dramatically, which makes it even more difficult to create labeled data manually.

Recently, supervised learning with positive and unlabeled data (known as PU learning) [2] has attracted substantial interests from researchers in data mining and machine learning fields. It deals with text classification problems where only the positive examples and the unlabeled examples are available. Since it is easy to get unlabeled data, it is important to learn how to take advantage of these unlabeled data. In fact, many researches advocating PU learning have shown that it is beneficial to use unlabeled data with positive data rather than use the positive data only [2]. Since it is easy for a researcher to collect a small number of documents of interest, and get a large number of unlabeled documents, PU learning seems a good alternative to solve the problem of topic search. In this chapter, we explore how to take advantage of PU learning to get as many relevant documents as possible that are related to a specific topic, and rank these documents according to their relevance to the topic.

There are two machine learning paradigms about learning with unlabeled data [3][4][5][6][7]. The first one is called learning from labeled and unlabeled examples (LU learning). The second one is called learning with only positive and unlabeled data (PU learning). In both paradigms, unlabeled data has been shown useful for boosting learning accuracy. The main difference between LU learning and PU learning is that the latter has more restrictions. In LU learning, there is a small set of labeled data, which contain both positive and negative instances, and the unlabeled data are used to augment the available labeled training examples. However, in PU learning, there is no negative training data. The positive training data are used to separate and identify positive and negative examples in the unlabeled data, and then a model is learned based on the identified positive and negative examples [6][8].

Many techniques about learning with unlabeled data have been proposed. Nigam et al. [3] use a small set of labeled instances and a large set of unlabeled instances to build a classifier. They show that the classifier built based on the labeled and unlabeled documents has better performance than that built based on a small set of labeled documents alone.

When both positive and unlabeled data are available, the positive training data can be used to estimate the positive class conditional probability, $p(x|+)$, and the unlabeled data can be used to estimate $p(x)$. With the prior $p(+)$, which is known or can be estimated using other sources, the negative class conditional probability can be obtained as follows:

$$p(x|-) = \frac{p(x) - p(+)\,p(x|+)}{1 - p(+)} \tag{1}$$

In [4], Denis et al. adopt the conditional probability $p(x|-)$ to perform text classification with Naïve Bayes method.

Another commonly used method in PU learning is called self-training [5]. The basic idea is that a classifier is first trained with a small set of labeled data. Then the classifier is used to classify the unlabeled data. The most confident unlabeled instances and their predicted labels will be used for iterative training.

In [6], Liu et al. adopt an EM algorithm and naïve Bayesian classification method to separate positive and negative examples. They first put some positive examples, called "spies", in the unlabeled data set. After completing the EM algorithm, they use the probabilistic labels of the spies to determine the likelihood that a document is negative. The final classifier is built based on the reliable negative documents identified from the unlabeled data set.

In [9], Yu et al. propose a mapping-convergence algorithm for PU learning. There are two stages in their algorithm: mapping stage and convergence stage. In the mapping stage, they perform initial approximation of highly negative examples. In the convergence stage, they iteratively run an internal classifier that maximizes margins to progressively achieve the true boundary of the positive class in the feature space.

In [10][11], Elkan et al. describe an iterative relabeling algorithm, which assumes that the unlabeled training examples are negative to learn a classifier in each of the iterations of the algorithm. The authors show that their algorithm works well in identifying biomedical documents related to proteins [10], but it is not known whether such a method is applicable to other biomedical document data sets.

Following the approach described by Elkan et al., we propose here a new method for ranking documents without negative training data. Unlike Elkan et al.'s approach, which focuses on binary classification of biomedical documents, the proposed method is mainly concerned with ranking these documents, and hence our method can be easily incorporated into a search engine. In addition, we want to investigate the role of feature selection in document ranking, which was not considered in Elkan et al.'s work. Since feature selection is of great importance in dealing with high dimensional data, we conduct experiments to see whether the performance of the proposed method changes when the number of features varies.

If we assume the unlabeled data (U) as negative examples for classifier training, then the positive documents in U, denoted as PU, become noises. The higher proportion of PU in U, the noisier the assumed negative examples are. So we hypothesized that with the increase of the proportion of PU in U, the performance of the trained classifier tends to degrade in terms of ranking testing documents. We conduct extensive experiments to test this hypothesis using the RNAPB data set collected from PubMed, which contains articles related to RNA-protein binding.

Our experimental results on the RNAPB data set show that when a small number of features are selected using the Chi-square statistic (χ^2-statistic) method [12], the performance of the proposed method can be as good as or even better than when no feature selection is used. Another finding from this research is that the performance of the proposed method tends to degrade with more positive examples being included in the unlabeled training set U. However, the classifier that is built with positive examples and unlabeled data that contains a small proportion of positive examples, can achieve comparable performance with the classifiers that are built with positive and pure negative examples.

2. Background

Our work is closely related to two fields, namely information retrieval and supervised learning. We review some basic concepts in these two fields below.

2.1. *Information Retrieval*

Information retrieval is concerned with obtaining information resources meeting a user's need from a data collection. The user's need or request is usually represented as queries and the data collection may contain structured data or unstructured text documents. Automated information retrieval systems such as search engines are the most widely used tools for people to solve information overload problems [13].

The core of information retrieval is to model how people compare texts and design computer algorithms to accurately perform this comparison. With the development of the World Wide Web, information retrieval involves several tasks and applications, which include text search, multimedia search and other media search. A usual search scenario is that a user submits a query to a search engine, which will return a list of documents ranked based on some criterion. Often, the documents are ranked based on the extent to which they are relevant to the query. Thus, relevance is a fundamental concept in information retrieval. Simply speaking, a relevant document contains the information that the user who submits the query to a search engine is looking for. Because the same concept can be expressed in different words, simply comparing the text of a query with the text of documents to conduct exact match retrieval usually produces very poor results. To address this issue, many information retrieval models have been proposed and tested to see how well they work. An information retrieval model defines the process of how to match a query with a document, which forms the basis of the ranking algorithms used by today's search engines for ranking search results.

State-of-the-art information retrieval models are usually based on the statistical properties of text rather than the linguistic structure of the text. For example, modern ranking algorithms are typically designed by considering the occurrence frequency of a word rather than whether the word is a noun or a verb. Although some models do incorporate linguistic features, they are proven less effective.

Performance evaluation of search engines is an important subject in information retrieval, as it is necessary to gauge the effectiveness of a search engine. Widely used performance measures include precision and recall. Precision is the proportion of retrieved documents that are relevant to the query, and recall is the proportion of relevant documents that are retrieved. Since most of the information retrieval models produce a ranked output, to summarize the effectiveness of a ranking algorithm, precision values are often calculated and combined into the Mean Average Precision (MAP) measure, whose definition will be given later in this chapter.

2.2. *Supervised Learning*

Supervised learning, also known as classification, is the task of automatically assigning labels to data, such as web pages, articles, or images. It finds many applications including spam detection, sentiment analysis and information retrieval, to name a few. It is analogous to human gaining new knowledge by learning from the past experiences [2]. For a machine, the "past experiences" are encoded in a set of data records.

A data record is also called an example, an instance, a case or a vector. It is described by a set of attributes or features $A = \{A_1, A_2, \ldots, A_n\}$. In addition, each data record has a special target attribute C, which is called the class attribute. The class attribute C has a set of discrete values, i.e. $C = \{C_1, C_2, \ldots, C_m\}$, where $m \geq 2$. A class value is also called a class label. For example, to classify food as "healthy" or "not healthy", there are two class values. When there are two class values or labels (i.e., $m = 2$), we can refer to one of them as the positive label and the other as the negative label, and the classification is referred to as binary classification. Data with positive labels are called positive data while data with negative labels are called negative data.

In general, supervised learning is a two-step process. In the first step, a function is built from a data set D, which is called the labeled training data set, to relate values of attributes in A to class values in C. The function is also called a classification model, a predictive model or simply a classifier. It can be in any form, e.g., a decision tree, a set of rules, a Bayesian model or a hyperplane. In the second step, the function is used for classification. Here, a testing data set is used to assess the predictive accuracy of the function. The testing data set is randomly selected from a general data set, and is different from the training data set, which means the testing data set is not used to build the function or classifier.

The accuracy of a classifier on a testing data set is defined as the proportion of testing records that are correctly classified by the classifier. For each data record in the testing set, its predicted class value is compared with its true class label. If the accuracy of a classification model is considered acceptable, the model can be used to classify future data whose class values or labels are unknown.

3. Methods

In this chapter, we present a new method for ranking text documents without negative training data by following the iterative relabeling approach proposed by Elkan et al. [10][11]. Our focus here is not about iterative relabeling, but to see whether feature selection matters in each run of model training and prediction. Since feature selection is of great importance in dealing with high dimensional data, we conduct experiments to see whether the performance of our method changes when the number of features varies. We also explore how the number of positive examples in the unlabeled training data affects the performance of the proposed method.

Specifically, we adopt the relabeling process proposed by Liu and Yu [6][9] to learn a classification model to rank the documents in a given testing set. An SVM model is trained based on some positive examples and unlabeled examples, where the unlabeled examples are assumed as negative data. This model is then run on separate testing data to predict the probability that a document in the testing set is positive. The documents in the testing set are ranked by the probability value returned by the SVM. The performance of our method is evaluated using the Mean Average Precision (MAP) measure, which is a popular measure in the information retrieval field.

We use the Libsvm toolkit [14] to conduct this research. Documents are transformed into vectors after carrying out stop words removal and stemming. The weight of each feature is calculated using the *tf-idf* method [13]. The χ^2-statistic (CHI) method [12][15] is adopted as the feature selection method to select the top *M* features that have the highest Chi-square scores. We evaluate the performance of our method with varying *M* values. We use the features that result in the best performance of our method to conduct experiments to study how the performance of the method changes with the increase of the proportion of the positive examples in the unlabeled training set.

3.1. *Feature Selection*

Feature selection is an important step in developing machine learning based systems. It refers to the process in which a subset of the features in the training set is selected and used for classification. In text mining and ranking, which is the main subject of this chapter, the terms occurring in documents are considered as features. Feature selection here serves two main purposes. First, it mitigates the problem of dimension curse by decreasing the size of the effective vocabulary. Second, feature selection solves overfitting problems by removing noisy features, which may result in the increase of the classification error on testing data. Feature selection is based on an algorithm in which a utility measure for each of the terms to a class is computed and the *M* terms that have the largest values of this measure will be selected. Other terms that have smaller values of the measure will not be used in text ranking.

In a comparative study of feature selection methods in statistical learning for text categorization, Yang and Pedersen [12] evaluated five feature selection methods including document frequency (DF), information gain (IG), mutual information (MI), χ^2-statistic (CHI) and term strength (TS). The authors found that IG and CHI are the best methods. In this chapter, we use the CHI method for feature selection. We want to identify a subset of features through which a model can be learned to optimize the performance of the proposed text ranking algorithm.

Let A be the number of times a term t and a class c co-occur. Let B be the number of times t occurs without c. Let C be the number of times c occurs without t. Let D be the number of times neither c nor t occurs. N is the total number of documents. We define the term-goodness measure to be:

$$\chi 2(t,c) = \frac{N \times (AD - CB)^2}{(A+C)(B+D)(A+B)(C+D)} \quad (2)$$

The larger the term-goodness measure, the more relevant t to the class c is. In our case, c represents the positive class, and we will select the top M terms, t, that have the largest $\chi^2(t, c)$ values and use these terms for text ranking.

3.2. *Feature Weight Calculation*

In text mining and information retrieval fields, a document is usually represented as a vector, where the weight of each of the features (terms) in the vector is determined using the *tf-idf* method. Here, the term frequency $tf(t, d)$ is defined as the number of times the term t occurs in the document d. The higher value of $tf(t, d)$, the more important t is in d. On the other hand, the inverse document frequency $idf(t)$ indicates how important a term t is in distinguishing the documents in a collection of documents. The inverse document frequency is calculated using the following formula:

$$idf(t) = \log (N/df(t)) \quad (3)$$

where N is the total number of documents in the collection, and $df(t)$ is the document frequency of t, which is defined as the number of documents containing *t*. The *tf-idf* value of the term *t* is defined as the product of its *tf* and *idf* values, i.e.

$$tf\text{-}idf(t,d) = tf(t,d) \times idf(t). \tag{4}$$

3.3. *Text Ranking Algorithm*

We use Support Vector Machines (SVMs) [14][16] to predict the likelihood that a document in the testing set is positive. The larger the likelihood is, the higher the testing document is ranked. A two-class, or binary, SVM classifier assigns labels to a testing document based on the sign of the decision function

$$f(\bar{x}) = \sum_i \alpha_i y_i K(\bar{x}_i, \bar{x}) + b \tag{5}$$

where $\bar{x}$ is the testing document to be classified or ranked, $\bar{x}_i$ are the training documents, $y_i \in \{-1,1\}$ are the class labels, positive or negative, for $\bar{x}_i$, α_i are weights assigned to the training documents during training, K is the kernel function, and b is a bias term. There are several kernel functions available. In this work, we use the radial basis function (RBF) kernel, which is defined as

$$K(\bar{u}, \bar{v}) = e^{-\gamma\|\bar{u} - \bar{v}\|^2} \tag{6}$$

where γ is a user-determined parameter.

In ranking the testing documents, we train a binary SVM using positive and unlabeled data where the unlabeled data are treated as negative data. The trained model is applied to the testing data set to predict the likelihood that a testing document is positive. The larger the likelihood value, the higher rank the testing document receives. We change the proportion of positive documents in the unlabeled training set to see how the proportion affects the performance of the proposed text ranking algorithm.

4. Experiments and Results

4.1. *Data Sets*

We conducted a series of experiments using the RNAPB data set collected from PubMed, which contains articles related to RNA-protein binding. There are 1160 positive documents and 3929 negative documents in the data set. The positive data set consists of biomedical articles that are about RNA-protein binding while the negative data set is comprised of biomedical articles that are irrelevant to RNA-protein binding. All these documents are collected manually through the PubMed search engine, and verified by our collaborators and domain experts [17][18][19][20][21]. Each document here only contains the title and abstract of a PubMed article. Other information of a published article, such as author and publication date, is not included. Table 1 shows the five most frequently occurring terms in the positive and negative data set, respectively.

Table 1. The five most frequently occurring terms in the data

Positive set	protein	binding	complex	interaction	RNA-protein
Negative set	gene	tree	sequence	phylogenetic	species

4.2. *Performance Measure*

Each document in the testing set is ranked based on its probability that it is predicted as being positive. A good ranking means all the relevant (i.e. positive) results are in the top ranked positions. We adopt the MAP measure [22][23] widely used in the information retrieval (IR) field to evaluate the performance of our approach. In ranking the results of a query, MAP represents the mean of the average precision scores of the results. Formally, let L be a ranked list of retrieved documents, and R be the set of relevant documents being retrieved. The MAP value of L is calculated using the following formula:

$$\text{MAP}(L) = \frac{1}{|R|}\sum_{k=1}^{n}\left(p(k)\times rel(k)\right) \tag{7}$$

where $|R|$ is the number of retrieved relevant documents, k is the rank in the list L of retrieved documents, n is the total number of retrieved documents, $p(k)$ is the precision at cut-off k in the list, $rel(k)$ is an indicator function equaling 1 if the document at rank k is a relevant (i.e. positive) document, zero otherwise [24]. The precision at cut-off k, $p(k)$, equals the number of retrieved relevant documents in the top k ranked documents divided by the number of retrieved documents in the top k ranked documents (hence the denominator is k). The final MAP value is the mean of the MAP value of each query. In our experiments, the results ranked by a trained SVM classification model are equivalent to the search results returned by a query. Positive examples are considered as retrieved relevant documents while negative examples are considered as retrieved irrelevant documents. Experimental results on the RNAPB data set show that our approach combining support vector machines and feature selection performs well.

4.3. *Experimental Design*

Let P denote the set of positive examples in the training set, U denote the set of unlabeled examples in the training set, PU denote the set of positive examples in U, and NU denote the set of negative examples in U. Let $|P|$, $|U|$, $|PU|$ and $|NU|$ denote the size of P, U, PU and NU respectively. Thus, $|U| = |PU| + |NU|$.

In each run of our experiments, we randomly sample a subset of documents from the positive data set and negative data set respectively to form our testing data set. These testing documents are removed from the positive and negative data sets. From the remaining positive data set, we randomly sample a subset of $|P|$ examples to form the positive training data, and $|PU|$ examples for the unlabeled training data, with the constraint that there is no overlap between P and PU. We also randomly sample a subset of $|NU|$ examples from the remaining negative data set for the unlabeled training data. Next, an SVM model is learned from the positive training dataset P and unlabeled training dataset U, where the unlabeled training data in U is treated as negative training data. The trained model is then applied to the testing data to predict the probability that a document in the testing set is positive. The documents in the testing set are ranked based on the probability values. An MAP value is calculated for the document ranking. We carry out 10 runs of the experiments, and the final MAP value is calculated by averaging over the MAP values from the 10 runs.

In the experiments, each document is preprocessed and transformed into a vector. We use Stanford CoreNLP [15], available at http://nlp.stanford.edu/software/corenlp.shtml, for stop words removal, stemming and lemmatization. The CHI method (χ^2-statistic) [12] is adopted for feature selection. The top M features or terms that have the largest CHI scores are selected as features to create the document vectors. We then use the *tf-idf* method [13] to calculate the feature weights in each document vector.

Libsvm [14] is adopted as the supervised learner to perform document ranking. We employ the default parameters and the RBF kernel in Libsvm. Since Libsvm enables one to predict the probability that a document is positive, it is appropriate for our study, in which we rank each document in the testing set based on its probability of being predicted as positive. The larger the probability is, the higher the testing document is ranked.

4.4. *Experimental Results and Analysis*

Our first experiment is to evaluate the impact of the number of features, M, used in the text ranking algorithm on the performance of the algorithm. We fix $|P|$, $|PU|$ and $|NU|$ at 30. The testing set contains 30 positive examples and 30 negative examples.

Fig. 1 shows the experimental results. When M is less than 5, the more features are used, the better performance the algorithm achieves. The algorithm achieves the best performance when M is 5. When M is greater than 5, some noise features are included, and hence the performance of the algorithm degrades. When the number of features exceeds 500, the effect of increasing features and the effect of noise features counter each other. As a result, the performance of the algorithm remains stable, though the computational time increases with a larger M. In subsequent experiments, we fix M at 5, and use the top 5 features to

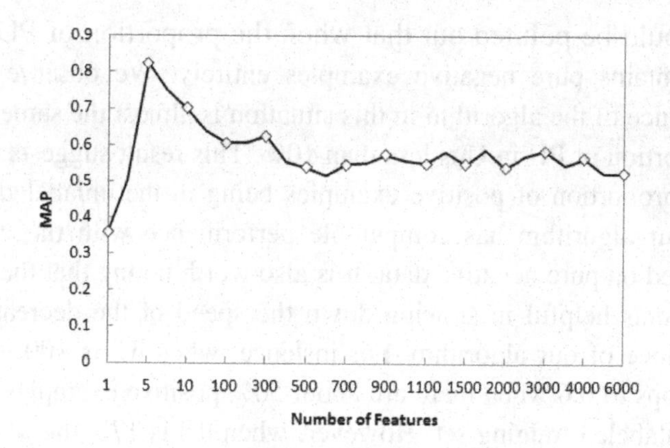

Fig. 1. Impact of the number of selected features on the performance of the text ranking algorithm.

evaluate how the proportion of the positive examples in the unlabeled training set affects the performance of our algorithm.

In our second experiment, we fix |P| and |U| values, and change |PU| and |NU| to make the proportion of PU in U increase from 0 to 100 percent. Fig. 2 summarizes the results, where four observations are made:

1. As expected, with a fixed |U|, the higher proportion of positive examples in U, the worse performance the algorithm achieves.

2. When the proportion of PU in U is less than 10%, the change in MAP values is less than 0.05, suggesting that when no reliable negative data is available, the unlabeled training data with a small proportion of positive examples can be used instead.

3. When the proportion of PU in U is less than 40%, the performance of the algorithm decreases about 15% with the increase of the proportion. On the other hand, when the proportion is between 40% and 80%, the performance of the algorithm decreases more than 20% with the increase of the proportion.

4. When the proportion of PU in U is less than 40%, the change of |U| has little effect on the performance of the algorithm. However, when the proportion is larger than 40%, the increase of |U| (e.g. from 100 to 170) seems helpful in improving the performance of the algorithm. This suggests that with a larger |U|, the impact of the proportion of PU in U is less significant.

It should be pointed out that when the proportion of PU in U is 0, U contains pure negative examples entirely. We observe that the performance of the algorithm in this situation is almost the same as when the proportion of PU in U is less than 10%. This result suggests that with a small proportion of positive examples being in the unlabeled training set U, our algorithm has comparable performance with the classifiers built based on pure negative data. It is also worth noting that the increase of |U| seems helpful in slowing down the speed of the decrease of the performance of our algorithm. For instance, when |U| is 100, the MAP value drops to 0.6 when there are about 50% positive examples included in the unlabeled training set. However, when |U| is 170, the MAP value doesn't drop to 0.6 until there are about 90% positive examples included in the unlabeled training set.

5. Conclusions and Outlook

In this chapter, we present a framework of adopting PU learning for ranking biomedical documents. The core of the framework is a text ranking algorithm that combines methods of feature selection, feature weighting and support vector machines. Using the RNAPB data set, we show experimentally that feature selection is helpful in improving the performance of the text ranking algorithm. A small number of features not only reduce the size of document vectors thus saving computational resources, but also lead to as good as or even better performance of the algorithm than when a larger number of features are used. Our experimental results on the set of biomedical documents also indicate that the performance of the text ranking algorithm tends to decrease when increasingly more positive examples are included in the unlabeled training set. However, when the proportion of the positive examples in the unlabeled training set is less than 10%, the increase of the proportion only changes the performance of the algorithm slightly, suggesting that unlabeled training data with a small proportion of positive examples can be used when no reliable negative examples are available. We have tested our approach on other data sets, and the qualitative conclusion obtained from those experiments remains the same.

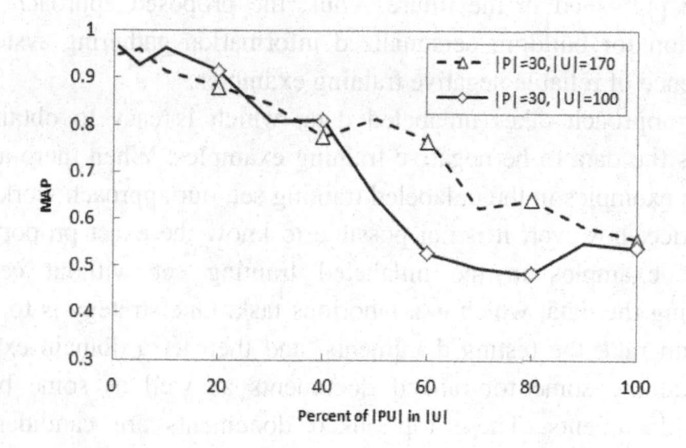

Fig. 2. Impact of |PU|/|U| on the performance of the text ranking algorithm.

In this research, we only adopt the CHI method (χ^2-statistic) to perform feature selection, without using domain-specific knowledge. It is known that the effectiveness of a feature selection process can be significantly enhanced when incorporating domain expertise into the process [17]. However, our goal here is to provide a general framework, which different researchers can use to rank different types of biomedical documents. Domain knowledge related to different types of biomedical documents would not be the same. For example, domain-specific features related to RNA-protein binding would be different from those related to single-nucleotide polymorphism (SNP). Incorporating domain-specific features into our framework would limit the application of this framework.

Our algorithm can be used to build personalized information gathering systems, which aim to locate and rank documents of interest to a particular person. The documents that a user is interested in can be regarded as positive examples, and the results returned by a search engine can be regarded as unlabeled examples. For example, a biologist working on RNA-protein binding collects biomedical documents related to this subject, which are positive examples. Results returned by the PubMed search engine are unlabeled examples. Our algorithm is trained by these positive and unlabeled examples, and the trained algorithm can then be used to rank and gather articles concerning RNA-protein binding that are published in the future. Thus, the proposed approach lays a foundation for building personalized information gathering systems in the absence of reliable negative training examples.

Our approach takes unlabeled data, which is easy to obtain, and assumes the data to be negative training examples. When there are few positive examples in the unlabeled training set, our approach works well. In practice, however, it is not possible to know the exact proportion of positive examples in the unlabeled training set without carefully examining the data, which is a laborious task. One strategy is to let our algorithm rank the testing documents, and then let a domain expert or user examine some top-ranked documents as well as some bottom-ranked documents. These top-ranked documents are candidates for positive examples and the bottom-ranked documents are candidates for negative examples. Through human verification, we obtain more positive examples and negative examples, and can use these validated data to train the proposed text ranking algorithm again to obtain a better model. This iterative procedure will continue refining the labels of the data at hand, yielding a better ranking list of documents, and hence a better personalized information gathering system.

In general, in supervised learning, both positive and negative examples are needed, but it is often the case that no reliable negative training data is available. Learning a model from positive and unlabeled data is then useful in this case. Moreover, with PU learning, the unlabeled data can be used to create more labeled examples, as described above. We foresee the hybrid approach combining information retrieval, supervised learning, document ranking and human-computer interaction will be the key technology for building future personalized information gathering systems.

References

1. L. J. Jensen, J. Saric, and P. Bork, *Literature mining for the biologist: from information retrieval to biological discovery*, *Nature Reviews Genetics,* 7(2), 119 (2006).
2. B. Liu, *Web Data Mining: Exploring Hyperlinks, Contents and Usage Data.* Springer (2006).

3. K. Nigam, A. McCallum, S. Thrun, and T. Mitchell, *Learning to classify text from labeled and unlabeled documents,* in *Proceedings of the 15th National Conference on Artificial Intelligence*, (1998), p. 792.
4. F. Denis, R. Gilleron, and M. Tommasi, *Text classification from positive and unlabeled examples,* in *Proceedings of the 9th International Conference on Information Processing and Management of Uncertainty in Knowledge-Based Systems*, (2002), p. 1927.
5. G. Haffari and A. Sarkar, *Analysis of semi-supervised learning with the Yarowsky algorithm,* in *Proceedings of the 23rd Conference on Uncertainty in Artificial Intelligence*, (2007), p. 159.
6. B. Liu, W. S. Lee, P. S. Yu, and X. Li, *Partially supervised classification of text documents,* in *Proceedings of the 19th International Conference on Machine Learning*, (2002), p. 387.
7. W. S. Lee and B. Liu, *Learning with positive and unlabeled examples using weighted logistic regression,* in *Proceedings of the 20th International Conference on Machine Learning*, (2003), p. 448.
8. X. Li and B. Liu, *Learning to classify texts using positive and unlabeled data,* in *Proceedings of the 18th International Joint Conference on Artificial Intelligence*, (2003), p. 587.
9. H. Yu, J. Han, and K. C. Chang, *PEBL: web page classification without negative examples, IEEE Transactions on Knowledge and Data Engineering*, 16(1), 70 (2004).
10. K. Noto, M. H. Saier Jr., and C. Elkan, *Learning to find relevant biological articles without negative training examples,* in *Proceedings of the 21st Australasian Joint Conference on Artificial Intelligence*, (2008), p. 202.
11. C. Elkan and K. Noto, *Learning classifiers from only positive and unlabeled data,* in *Proceedings of the 14th International Conference on Knowledge Discovery and Data Mining*, (2008), p. 213.
12. Y. Yang and J. O. Pedersen, *A comparative study of feature selection in text categorization,* in *Proceedings of the 14th International Conference on Machine Learning*, (1997), p. 412.
13. G. Salton and M. J. McGill, *Introduction to Modern Information Retrieval.* McGraw-Hill, Inc. (1986).
14. C.-C. Chang and C.-J. Lin, *LIBSVM: a library for support vector machines, ACM Transactions on Intelligent Systems and Technology*, 2(3), 1 (2011).
15. Stanford Core NLP Software. http://nlp.stanford.edu/software/corenlp.shtml (2012).
16. T. Joachims, *Text categorization with support vector machines: learning with many relevant features,* in *Proceedings of the 10th European Conference on Machine Learning*, (1998), p. 137.
17. C. Laing, D. Wen, J. T. L. Wang, and T. Schlick, *Predicting coaxial helical stacking in RNA junctions, Nucleic Acids Research*, 40(2), 487 (2012).

18. R. S. Bindra, J. T. L. Wang, and P. S. Bagga, *Bioinformatics methods for studying microRNA and ARE-mediated regulation of post-transcriptional gene expression, International Journal of Knowledge Discovery in Bioinformatics*, 1(3), 97 (2010).
19. M. Khaladkar, J. Liu, D. Wen, J. T. L. Wang, and B. Tian, *Mining small RNA structure elements in untranslated regions of human and mouse mRNAs using structure-based alignment, BMC Genomics*, 9, 189 (2008).
20. J. T. L. Wang, H. Shan, D. Shasha, and W. H. Piel, *Fast structural search in phylogenetic databases, Evolutionary Bioinformatics*, 1, 37 (2005).
21. J. T. L. Wang, T. G. Marr, D. Shasha, B. A. Shapiro, G.-W. Chirn, and T. Y. Lee, *Complementary classification approaches for protein sequences, Protein Engineering*, 9(5), 381 (1996).
22. E. Agichtein, E. Brill, and S. Dumais, *Improving web search ranking by incorporating user behavior information,* in *Proceedings of the 29th Annual International ACM SIGIR Conference on Research and Development in Information Retrieval*, (2006), p. 19.
23. J. Xu and H. Li, *Adarank: a boosting algorithm for information retrieval,* in *Proceedings of the 30th Annual International ACM SIGIR Conference on Research and Development in Information Retrieval*, (2007), p. 391.
24. A. Turpin and F. Scholer, *User performance versus precision measures for simple search tasks,* in *Proceedings of the 29th Annual International ACM SIGIR Conference on Research and Development in Information Retrieval*, (2006), p. 11.

Chapter 14

Automated Mining of Disease-Specific Protein Interaction Networks Based on Biomedical Literature

Rajesh Chowdhary[1], Boris R Jankovic[2], Rachel V. Stankowski[1], John A.C. Archer[2], Xiangliang Zhang[2], Xin Gao[2] and Vladimir B. Bajic[2]

[1]Marshfield Clinic-Marshfield Center, MCRF, 1000 North Oak Avenue, Marshfield, WI 54449, USA.
chowdhary.rajesh@mcrf.mfldclin.edu

[2]King Abdullah University of Science and Technology (KAUST), Computational Bioscience Research Center, Computer, Electrical and Mathematical Sciences and Engineering Division, Thuwal 23955-6900, Kingdom of Saudi Arabia.
vladimir.bajic@kaust.edu.sa

Elucidation of protein interaction networks (PINs) is essential for understanding disease-related biological processes and mechanisms. While significant information regarding PINs is available in the literature as free-text, such information is difficult to retrieve and synthesize. The following chapter describes machine learning techniques and web-based tools for literature-based extraction of protein interactions and their networks. Two case studies are provided to illustrate the efficient use of automated text-mining based applications in biomedical research using the tools PIMiner (http://www.biotextminer.com/PPI/index.html) and CPNM (http://www.biotextminer.com/CPNM/).

1. Introduction

Cellular biological processes rely on protein interaction (PI) networks (PINs). Comprehensive understanding of PINs is essential to elucidating the cellular processes that are activated as a response to external stimuli in either normal or diseased conditions [46]. Although the biomedical literature contains a wealth of information on PIs, such information can be difficult to retrieve. Recent efforts have focused on building databases to store information on PIs that has been manually curated from the literature. Such resources include, among others, HPRD [27], MINT [8], BioGRID [53], MIPS [41], PDZBase [5], IntAct [2], and STITCH [31]. These databases are useful, but are limited by low overall coverage and a tendency to become rapidly outdated as curation lags behind rapid expansion of the literature. Automated text-mining methods for PI extraction offer a complementary approach. Such methods have evolved over time as demonstrated by the consistent improvements reported by the BioCreative Event, which relies on a competitive yet collaborative approach to the development of text-mining modules [30,37,51]. Computational methods for literature-based PI/PIN extraction and related sub-problems vary in their objectives, approaches, and design. For example, methods have been developed for recognition and normalization of protein names [15–17,21,36,39,40,44,52,56,57,59,62], detection of PIs [1,3,4,6,7,9,10,12–14,22,23,26,28,29,33–35,47,48,55,60,61,64], and detection of PINs [11,20,54]. These tools offer various degree of user-friendliness or expertise required for use [24,25].

Two recently developed tools that can be used for extraction of disease-specific PIs/PINs are PIMiner [10] and Context-Specific Protein Network Miner (CPNM) [11]. These tools and their methodologies are described below. Case studies are also provided to illustrate the efficient use of PIMiner and CPNM for a class of practical medical research problems allowing the reader to see the advantages and simplicity of text-mining based applications.

2. Machine Learning Techniques for Protein Extraction

Automated PI extraction techniques use rules, patterns, or templates to detect PIs in text [6,7,9,20,22,47,48,54,55]. The two major approaches to such detection include manually specifying rules or using machine learning techniques to computationally infer rules from manually annotated text. Initial techniques for PI extraction [20,54] used simple term co-occurrence based methods, which resulted in a high rate of false positives. Efforts to manually specify a larger set of rules generally reduced false positives, but resulted in low overall coverage. Compared to other methods, machine learning techniques have better performance resulting in a decrease in false positives rate and increased coverage [6,7,9,22,47,48,55]. The PIMiner and CPNM systems described below use a machine learning approach to PI extraction that relies on Bayesian Network modeling [9,38].

Machine learning models are developed using labeled datasets to train and test the model. Training data are used to train the model to correctly predict the labels of the testing set. The testing set is unobserved during the training process and the labels of the testing set are unobserved during the testing process. In practice, the training set is often partitioned into two subsets. One subset is used for training and the other for testing the model. However, since the partition can be biased, the training set is often partitioned randomly into k subsets. In each iteration of model performance evaluation, one of the subsets is selected and designated as the test set, while the remaining subsets are considered training sets. The procedure is repeated k times until each subset has been selected as the test set once. The final performance of the model is estimated as the average performance of the model over these k runs. This approach is referred to as k-fold cross-validation. In some cases, k is set to the total number of data points in the training set, with each subset containing a single data point. This is called leave-one-out cross-validation.

The machine learning model described here uses Bayesian Networks. A Bayesian Network is a statistical model that graphically encodes the relationships between variables of interest, which in this case includes text features that describe PIs. Assuming that a PI triplet is described by a feature vector X, the class label (C = true or false) prediction is based on the Bayes theorem:

$$P(C|X) = \frac{P(X \mid C)P(C)}{P(X)} \tag{1}$$

where $P(X|C)$ is the probability of observing PI triplet X given that its class is C, while $P(X)$ and $P(C)$ are the prior probability of X and C, respectively. When posterior probability $P(C = \text{true}|X)$ is greater than $P(C = \text{false}|X)$, the PI triplet X is classified as true. Otherwise, it is classified as false. Bayesian Network training is performed to learn dependencies of features in each class (true or false) of PI triplets, which is used to calculate $P(X|C)$ for a given X. To perform the model training process in the PIMiner and CPNM systems, manually classified true and false PI triplets were obtained from the BioGRID database [53]. The well-trained Bayesian Network captures language rules related to PIs based on triplet features. To evaluate the predictability of the trained Bayesian Network model, it was tested on a large set of sentences containing true and false PIs randomly selected from PubMed and manually curated. Accurate prediction results revealed that the method was capable of extracting large numbers of new PIs from text [9].

Bayesian Network models were built by manually selecting 12 features related to the language rules for PI interactions in text. Each of the 12 features captured information associated with language or grammar rules that typically characterize PI relationships, such as distance in number of words between elements of a PI triplet. Triplets with shorter distances are more likely to be true. Word order was also an important feature to consider as humans often, when talking in a given context, use words in a certain prevalent order. For example, an interaction between two proteins is indicated only when the word 'interacts' is located between the two protein names, not before or after two protein names. On the other hand, 'interaction' can be before, between, or after the two protein names in different contexts. Features of interest were then modeled in a Bayesian Network. Several other features were explored in addition to the 12 used to train the model, but were filtered out as they were found to be insignificant by Chi-square test (p-value > 0.05). Chi-square is a statistical test that can be used to determine the discriminating power of a categorical feature in classifying data. Given a feature, the observed frequency distribution of different classes is calculated by applying the feature to a given dataset. The observed distribution is then compared to the expected frequency distribution of the classes in the dataset. The value of the Chi-square test statistic is calculated by the equation:

$$\chi^2 = \sum_{i=1}^{n} \frac{(O_i - E_i)^2}{E_i} \tag{2}$$

where E_i is the expected frequency for class i, O_i is the observed frequency of class i for a given feature, and n is the number of classes.

The p-value of the calculated χ^2 value is then estimated by looking up a χ^2 distribution table. A p-value < 0.05 typically implies that the feature is an important one [9].

3. PIMiner: A Web-based Tool for Protein Interactions Extraction from Biomedical Literature

The online PIMiner tool is flexible, user-friendly, and easy to test. PIMiner is based on the PI extraction method described above to identify PI triplets using Bayesian inference [10,38]. PIMiner is easy to use for routine PI extraction tasks and benefits from a comprehensive protein name dictionary, extensive inclusion of protein relationship concepts (*e.g.*, phosphorylation, methylation), the capacity to tag multiple-word protein names, and the use of different syntactical forms of interaction words based on word endings (*e.g.*, regulate, regulates, regulated, regulating, *etc*). PIMiner also provides an optional filter function to allow extraction of only specific interaction types as specified by the user. PIMiner is a useful tool for biological curators and annotators or biologists performing large-scale extraction of PIs from the biomedical literature (*e.g.*, for the building of databases such as BioGRID) [10].

The PIMiner system consists of two modules, A and B, that function separately to extract PI information from raw text and to evaluate PIMiner performance with labeled training and test data, respectively. Figure 1 shows the web-interface for the two modules. In Module A, PubMed abstracts are converted into individual sentences and then tagged for protein names and interactions. The dictionary for protein names includes over 8 million names of proteins and their variants based on data extracted from several sources. The tagger can detect a certain amount of variation in protein names within a sentence by examining words surrounding the protein name. For example, 'X receptor' can be detected by the tagger within a given sentence even for those cases when 'X receptor' protein is not contained in the dictionary, but the protein 'X' is. Over 2000 unique terms that describe the potential type/nature of the interactions of two proteins are present in the list of interaction words.

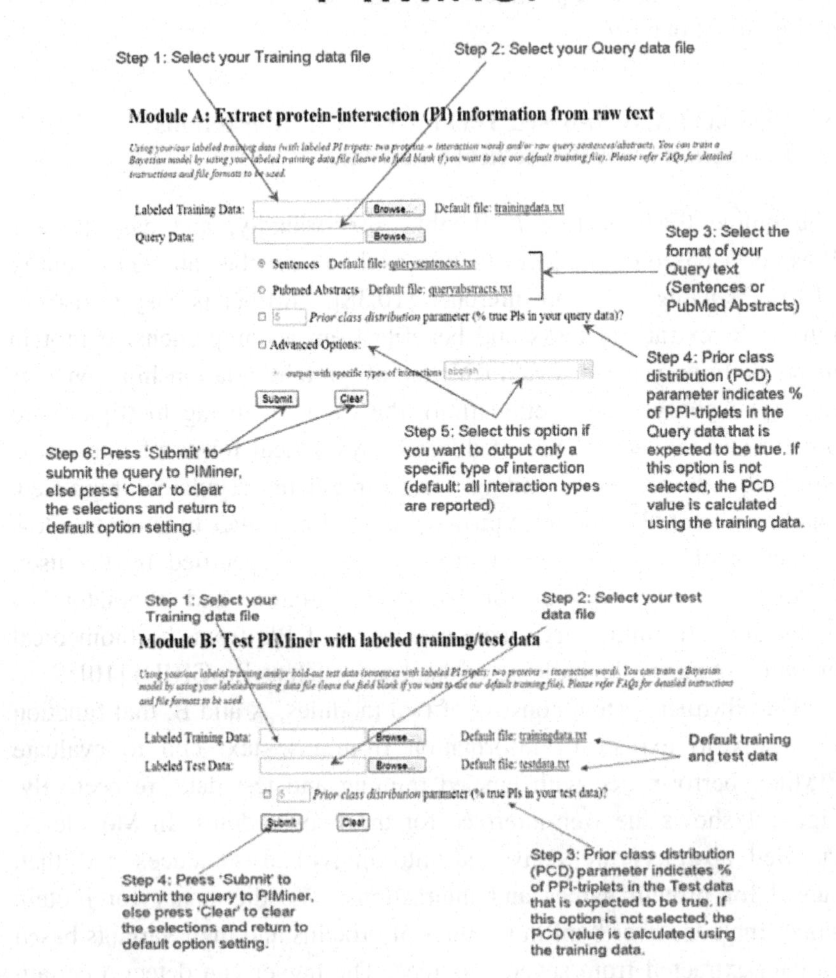

Fig. 1. **PIMiner web-interface.** PIMiner consists of two separate modules, A and B, for extracting PI information and testing the PIMiner system, respectively. Module A allows the user to input his or her own training data or to use the training data provided and query data in either sentence or abstract format. Advanced options include the option to filter results to focus on a specific type of interaction. Module B allows the user to test the performance of PIMiner by entering pre-labeled training and test data in sentence format. Adapted from Chowdhary *et al* [10].

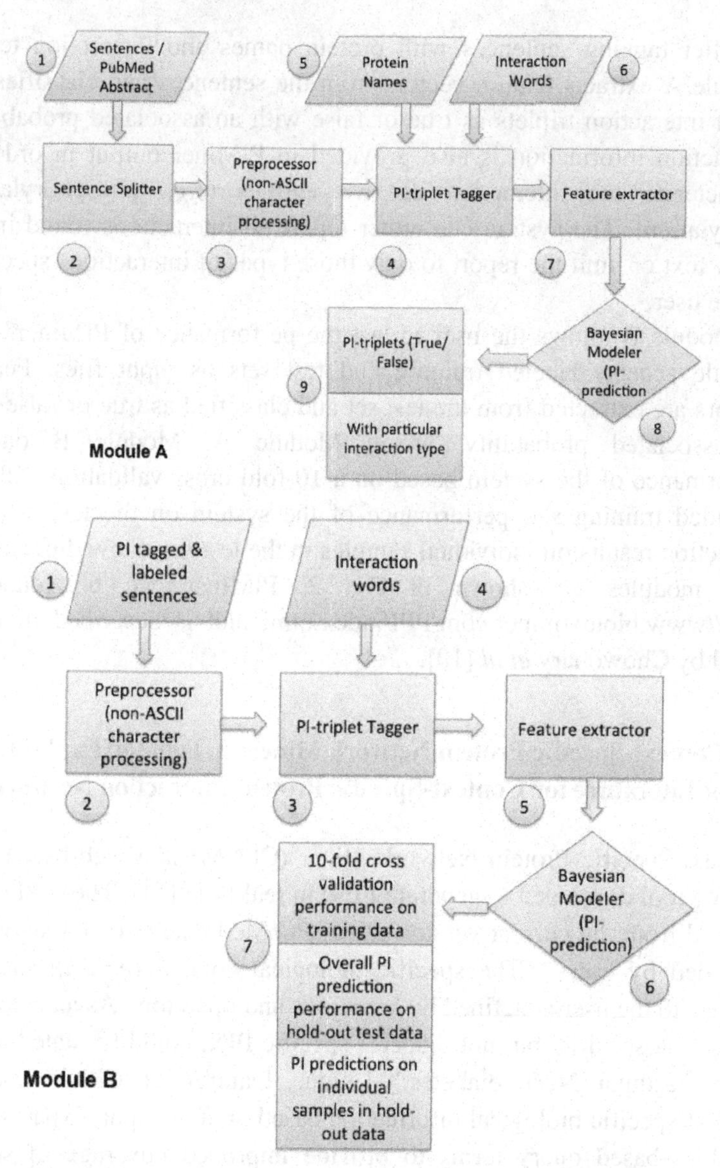

Fig. 2. **Flow diagrams of PIMiner modules.** In Module A, the sentences or PubMed abstracts provided as query data are split into sentences and pre-processing is performed. The protein name dictionary and interaction words are used to tag PI-triplets and extract features, which are then used to predict PIs using Bayesian inference modeling. Output includes PI-triplets with probability of being true or false and the particular interaction type. Module B functions similarly except that input sentences are already tagged with protein names and output evaluates PIMiner performance. Adapted from Chowdhary *et al* [10].

After tagging sentences with protein names and interaction terms, Module A extracts feature vectors from the sentences and classifies the target interaction triplets as true or false with an associated probability. Interaction information is also provided in PIMiner output in order to characterize the interaction of two entities (*e.g.*, phosphorylation, methylation). The system can either report all interactions found in the query text or limit the report to only those types of interactions specified by the user.

Module B allows the user to test the performance of PIMiner. This module accepts labeled training and test sets as input files. Feature vectors are extracted from the test set and classified as true or false with an associated probability, as in Module A. Module B outputs performance of the system based on a 10-fold cross validation with the provided training set, performance of the system on the test set, and prediction results on individual samples in the test set. Flow diagrams of both modules are shown in Fig. 2. PIMiner can be found at http://www.biotextminer.com/PPI/index.html and is described in more detail by Chowdhary *et al* [10].

4. Context-Specific Protein Network Miner: A Tool for Exploration of Literature for Context-Specific Protein Interaction Networks

Context-Specific Protein Network Miner (CPNM) is a web-based text-mining tool developed to generate PINs in real-time [11]. These PINs are derived from the current version of the PubMed database and keywords provided by users. The specific biological context (*e.g.*, disease) of interest to the user is defined by keywords and operators. As an example, an asthma-specific, but not diabetes-specific PIN, could be stated as the query 'asthma NOT diabetes.' Unique features of CPNM include context-specific biological information based on user input, expansion of ontology-based query terms to provide improved coverage of search results, online-processing of PubMed abstracts for up-to-date search results in a consistent manner, as well as meaningful output including PI interaction type and directionality that includes resulting interaction network summary statistics, as described in detail below [11].

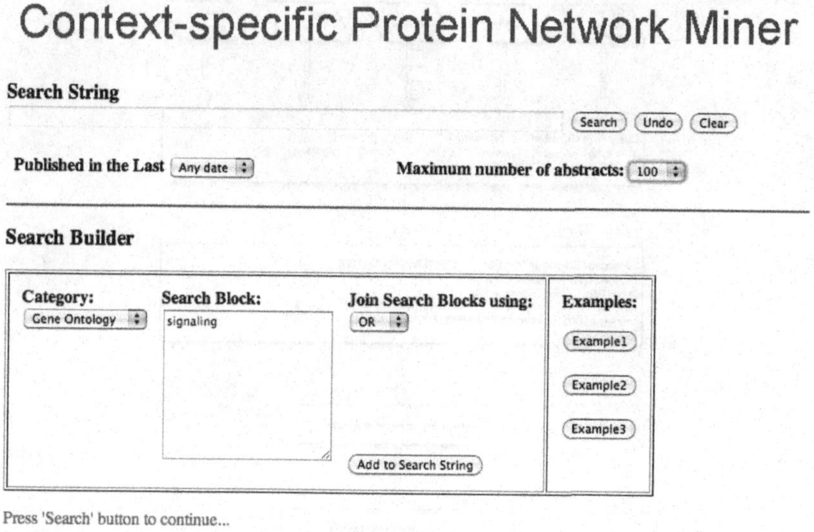

Fig. 3. **CPNM web-interface.** In CPNM, the user inputs a search string built using the search builder to provide biological context for PIN generation. Categories include Gene ID/Name, GO-terms, species, disease, and tissue. The user can also enter additional terms as free-text. Adapted from Chowdhary *et al* [11].

The web interface of CPNM allows the user to build a query with keywords in the categories of gene ID/name, GO-terms, species, disease, and tissue (Fig. 3). The user may also provide keywords as free-text. The query builder allows separation of keywords in each category using the operators AND/OR/NOT. Following keyword input by the user, CPNM expands the query by including all synonyms as well as other terms related to the keyword that are below the query keyword node on the ontology tree network provided by the Open Biological and Biomedical Ontologies (OBO) foundry [50]. The expanded query is used to search PubMed and retrieve abstracts of interest. An example query expansion is shown in Fig. 4. PIMiner system functionalities are then used to split and process sentences, tag protein and interaction words, detect and extract PI relationships, and predict PI direction. Following PI identification, CPNM features a module for normalization of the protein name to official symbols in the Entrez Gene database through a series of matching steps. PINs are reported as a table and displayed in graphical format with additional functionality that allows the user to navigate, view, and explore the results. Figure 5 shows a flow diagram of the CPNM system [11].

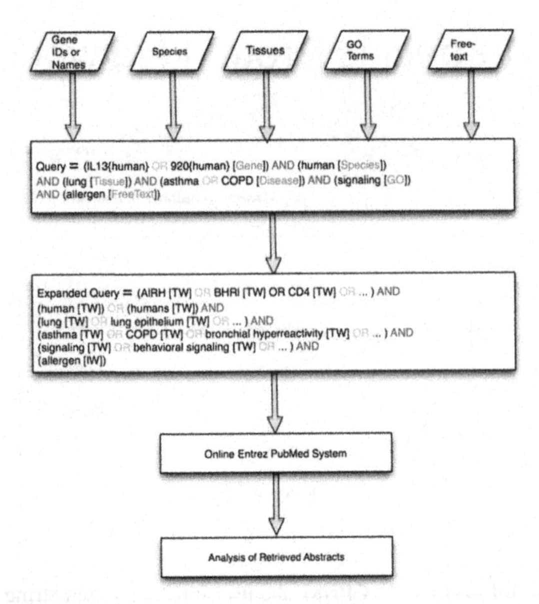

Fig. 4. **Sample CPNM query expansion.** After the user inputs keywords and a query is built, CPNM expands the query by retrieving all synonyms and other related terms below the query keyword node on the ontology tree network provided by the Open Biological and Biomedical Ontologies (OBO) foundry [50]. The expanded query is used to search the online Entrez PubMed database and retrieved abstracts are used for analysis. Adapted from Chowdhary *et al* [11].

As with PIMiner, CPNM output for individual PIs contains the interaction of interest together with the interaction type, a probability score, and if applicable, direction of interaction. In table format, results are sortable by column, and the evidence sentence is presented with the PI triplet highlighted. Keywords are highlighted in the sentence as well as in the corresponding abstract. CPNM also provides links from protein names to the associated Entrez Gene IDs for more direct access to specific details about the protein of interest. A PIN diagram is generated and summary statistics rank proteins on the basis of the number of interacting proteins, directly connected pairs, and inbound and outbound directed edges. The CPNM filter function can be used to control date of retrieved abstracts, types of interactions, and probability threshold. CPNM can be found at http://www.biotextminer.com/CPNM/ and is described in more detail by Chowdhary *et al* [11].

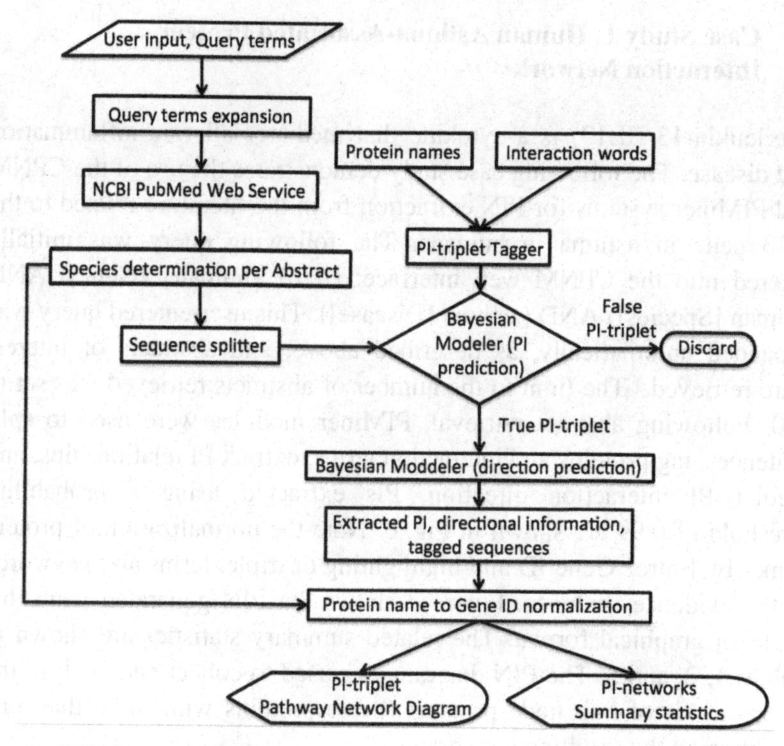

Fig. 5. **CPNM system flow diagram.** The user builds a query using the CPNM web-interface, which is then expanded to include synonyms and related terms. The expanded query is used to search the online PubMed database and species is determined if specified by the user. Using PIMiner functionalities, the abstracts are split into sentences and tagged with protein names and interaction words. PI-triplets are passed through a Bayesian modeling system to predict true or false interactions and false interactions are discarded. True interactions are passed through the Bayesian modeling system again to predict the direction of the interaction and the extracted PIs, directional information, and tagged sentences go through a process to normalize protein names. These interactions are then used to build a context-specific PIN. PINs are displayed as a PI-triplet pathway network diagram and summary statistics are given. Adapted from Chowdhary *et al* [11].

It should be noted that while CPNM takes advantage of several features of the PIMiner system, the two applications serve fundamentally different purposes. PIMiner uses raw text as input and predicts PI-triplets, and thus may be most suitable for biocuration-type work. CPNM, on the other hand, uses context-specific keywords as input to predict PINs, and is more suitable for biomedical researchers interested in exploring PINs specific to a certain biological condition.

5. Case Study 1: Human Asthma-Associated Protein Interaction Network

Interleukin-13 (IL13) is a cytokine that mediates allergic inflammation and disease. The following case study demonstrates the use of the CPNM and PIMiner systems for PIN extraction from the literature related to the IL13 gene in asthma in humans. The following query was initially entered into the CPNM web interface: (IL13 {human} [Gene]) AND (human [Species]) AND (asthma [Disease]). This user-entered query was expanded automatically, as described above, and abstracts of interest were retrieved. The limit to the number of abstracts retrieved was set to 500. Following abstract retrieval, PIMiner modules were used to split sentences, tag proteins and interaction terms, extract PI relationships, and predict PI interaction direction. PIs extracted using a probability threshold of 0.99 are shown in Fig. 6. Note the normalization of protein names by Entrez Gene ID and highlighting of triplet terms and keywords in the evidence sentence. Figure 7 shows the PIN generated from this query in graphical format. The related summary statistics are shown in Tables 1, 2, and 3. The PIN diagram was used to collect and analyze the complete set of hub node proteins (those proteins with more than one neighbor) in the resulting network.

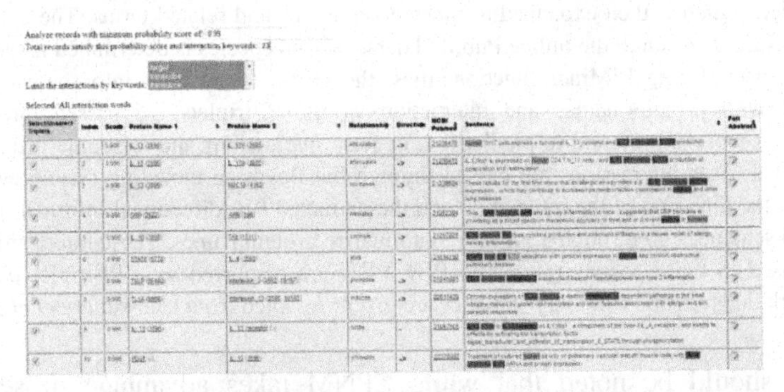

Fig. 6. **Case study 1: extracted protein interactions.** PIs extracted related to IL13 and asthma with a minimum probability of 0.99 are shown. The names and gene IDs of both interacting proteins are shown and the interaction word and direction for each relationship are given. Links to the evidence in PubMed are provided and the evidence sentence is highlighted to emphasize keywords, interacting proteins, and interaction words in different colors not visible on this figure. Clicking on the link to the full abstract provides a similarly highlighted version of the abstract. Adapted from Chowdhary *et al* [11].

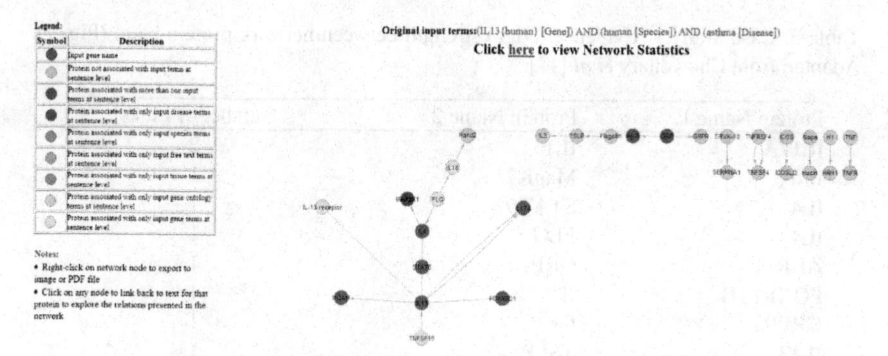

Fig. 7. **Case study 1: PIN diagram.** The PIN diagram demonstrates protein interactions related to IL13 and asthma in a graphical manner. Color coding (not visible) indicates types of association, and arrows describe directionality of the interaction. Adapted from Chowdhary *et al* [11].

Table 1. Case Study 1: Hub node protein statistics in network diagram (Fig. 7). Adapted from Chowdhary *et al* [11].

Protein	Neighbors	Percent Coverage (# neighbors/# total network nodes)
IL13	6	20%
IL4	3	10%
FLG	2	6.7%
GRP	2	6.7%
IL10	2	6.7%
STAT6	2	6.7%
TSLP	2	6.7%

Table 2. Case Study 1: Outbound, inbound, and undirected edge connectivity by node (Fig. 7). Adapted from Chowdhary *et al* [11].

Protein	Outward	Inward	Undirected
IL13	4	1	2
IL4	1	1	1
FLG	2	0	0
GRP	2	0	0
IL10	1	1	0
TSLP	1	1	0
STAT6	0	0	2

Table 3. Case Study 1: Evidence (edge) strength between network protein pairs (Fig. 7). Adapted from Chowdhary *et al* [11].

Protein Name 1	Protein Name 2	Number of Links[1]
IL17A	IL13	2
IL4	MapK21	1
IL4	STAT6	1
IL4	FLG	1
AHR	GRP	1
FOXRED1	IL13	1
GRPR	GRP	1
IL13	TSLP	1
IL13	STAT6	1

[1]More links between two nodes indicates greater literature support.

Hub node proteins interact with multiple partners in the network and have the potential to contain important information about the biological context of interest. For that reason, we further investigated those proteins satisfying hub-protein criteria including IL13, IL4, GRP, FLG, STAT6, IL10 and TSLP. This set of proteins was then queried against the pathway database hiPathDB, which itself integrates a number of other pathway databases [63]. Table 4 summarizes the results of this analysis. Manual verification of the pathways identified by hiPathDB found that several of the identified pathways were previously associated with asthma, which is the context disease term of interest in this example. Therefore, we were able to use CPNM, which calls upon the PIMiner system, to connect the context of interest (IL13 and asthma in humans) with pathway information via PIN generation. Furthermore, we identified additional pathways with no association with asthma documented in the literature, providing novel candidate associations for hypothesis generation and further exploration via other experimental methods.

All hub-proteins, with the exception of FLG (Filaggrin), were associated with known pathways in hiPathDB (Table 4). We searched PubMed for associations between asthma and FLG and consequently found that the *FLG* gene was previously related to increased risk of asthma in a number of studies [42,43,49,58], though some evidence suggests otherwise [45]. We also found that in the case of asthma, the hub-protein GRP appeared to be investigated in mice as an anti-inflammatory therapeutic agent [65]. Real time operation with the current version of PubMed allows CPNM to capture such current information.

Table 4. Case Study 1: Pathway involvement of hub node proteins. Adapted from Chowdhary *et al* [11].

Pathway Name	Total Interactions	Source [63]	Participating Network Proteins	Asthma Association
IL4-mediated signaling events	62	Nci-Nature	IL4, IL10, STAT6	None documented
Jak-STAT signaling pathway	9	KEGG	STAT6	Part of KEGG asthma pathway
Calcineurin-regulated NFAT-dependent transcription in lymphocytes	8	Nci-Nature	IL4	PMID: 12452838
gata3 participate in activating the th2 cytokine genes expression	7	BioCarta	IL4, IL13	Association in pathway annotation
cytokine-cytokine receptor interaction	5	KEGG	IL4, IL13, IL10, TSLP	Part of KEGG asthma pathway
glucocorticoid receptor regulatory network	5	Nci-Nature	IL4, IL13	None documented
Leishmaniasis	4	KEGG	IL4, IL10	None documented
IL12-mediated signaling events	3	Nci-Nature	STAT6, IL4	None documented
Chagas disease	3	KEGG	IL10	None documented
il-10 anti-inflammatory signaling pathway	2	BioCarta	IL10	None documented
CD40/CD40L signaling	2	Nci-Nature	IL4	None documented
IL12 signaling mediated by STAT4	2	Nci-Nature	IL4, STAT6	None documented
Regulation of nuclear SMAD2/3 signaling	1	Nci-Nature	IL10	None documented
IL2 signaling events mediated by STAT5	1	Nci-Nature	IL4	None documented
Calcium signaling in the CD4+ TCR pathway	1	Nci-Nature	IL4	None documented
Downstream signal transduction	1	Reactome	STAT6	None documented
Peptide ligand-binding receptors	1	Reactome	GRP	None documented

Table 5. Case Study 2: Differentially expressed genes. Adapted from Chowdhary *et al* [11].

Gene IDs	Regulation
A2M, LAMP1, MYBL2, HLA-DQA1, MMP12, LIPA, HG1723-HT1729, GSTM4, CDA, HG4068-HT4339, SPP1	Up-regulated > 3-fold
RPE65, SLC14A1, CXCL6, LAMB1, SNAH14, CNTF, D14822, M64936, IFI27, PFDN4, COL4A5, PDE3A, HG3934-HT4204, HTN1, BAMBI, MAP2, HG2260-HT2349	Down-regulated > 3-fold

6. Case Study 2: Protein Interaction Network Associated with Differential Gene Expression

Gene expression experiments typically result in large amounts of data, but exploration of the association between PINs and gene expression data presents a challenge for researchers. The following case study demonstrates the use of the CPNM system to generate a context-specific PIN using gene expression experiment results.

Gene expression data were selected from an in-house set of gene expression datasets related to common respiratory diseases (www.respiratorygenomics.com/GeneExpression/). The selected experiment compared gene expression in alveolar macrophages from non-smokers and smokers with chronic obstructive pulmonary disease (COPD) [18]. Genes that were up- or down-regulated more than three-fold were selected for analysis. Table 5 shows the genes of interest. A CPNM query was formulated to use context-specific information including disease name (COPD), smoking status, and the 28 genes that were differentially expressed. The final query entered into the CPNM web interface was: (gene names separated by OR {human} [Gene]) AND (COPD [Disease]) AND (smokers OR non-smokers OR nonsmokers [FreeText]). The PIN diagram generated using a probability threshold of 0.58 and sub-networks displaying hub nodes are presented in Figs. 8 and 9, respectively. Two hub node proteins ITGAM and SERPINE2 were identified. These two hub node proteins were not in the input set of genes. This analysis demonstrates the utility of CPNM to elucidate PINs/hub proteins that are potentially associated with differential gene expression. The resulting information is based on the entire current body of biomedical literature, providing a broad picture that is not limited to the genes examined in a single experiment. Application of CPNM in this way may be particularly valuable to those researchers who perform gene expression experiments to examine biological pathways associated with drugs or diseases.

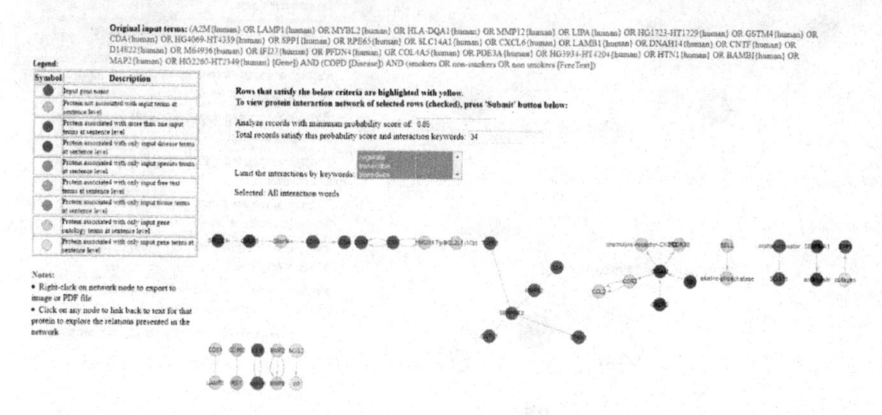

Fig. 8. **Case study 2: PIN diagram.** The PIN diagram demonstrates protein interactions related to differentially expressed genes, COPD, and smoking in a graphical manner. Color coding (not visible) indicates types of association, and arrows describe directionality of the interaction. Adapted from Chowdhary *et al* [11].

7. Conclusion and Future Directions

The text-mining capabilities of CPNM in conjunction with PIMiner are useful for curating biological data, validating known hypotheses, and generating novel hypotheses that are related to certain biological contexts, such as genes and diseases, in order to provide further insight into associated biomedical problems. Text mining allows for comprehensive examination of the biomedical literature and allows researchers to take advantage of the vast body of published literature when developing and testing hypotheses. The applications described here are easy to use and provide rapid and customizable results for biomedical researchers without requiring specialized knowledge of the computational and machine learning techniques upon which the application is based. In the future, integration of other third party tools, such as pathway databases, may further improve the functionality and utility of these automated systems.

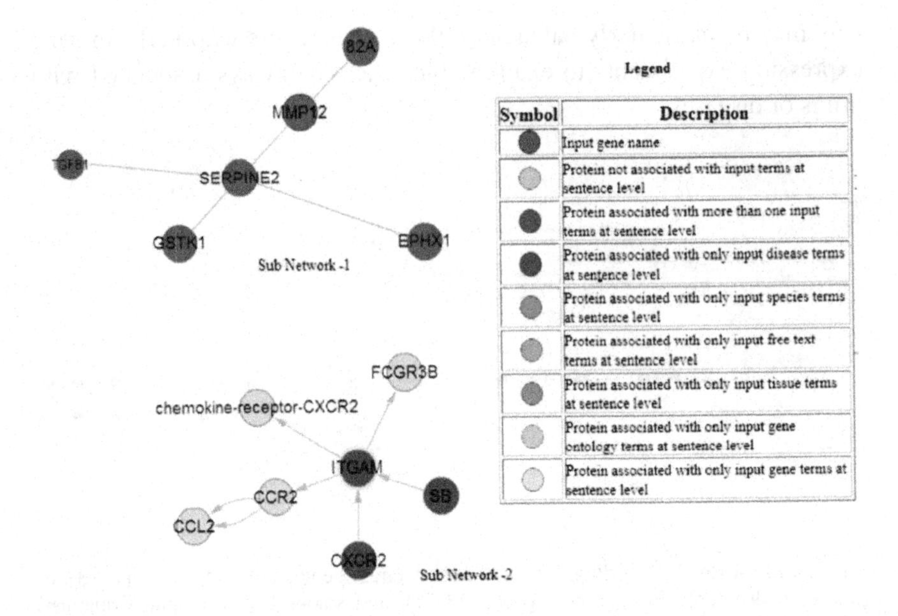

Hub nodes in PIN generated by CPNM for Case Study II

Fig. 9. **Case study 2: sub-networks of hub nodes.** Enlarged sub-networks of hub nodes surrounding the proteins SERPINE2 and ITGAM are shown. Color coding (not visible) indicates types of association, and arrows describe directionality of the interaction. Adapted from Chowdhary *et al* [11].

References

1. Agarwal, S., Liu, F., Li, Z., and Yu, H. (2010) Machine learning based approaches for Biocreative III tasks. *Proc. BioCreative III Workshop*, pp. 46–51.
2. Aranda, B., Achuthan, P., Alam, F. Y., Armean, I., Bridge, A., Derow, C., Feuermann, M., Ghanbarian, A. T., Kerrien, S., Khadake, J., Kerssemakers, J., Leroy, C., Menden, M., Michaut, M., Montecchi-Palazzi, L., Neuhauser, S. N., Orchard, S., Perreau, V., Roechert, B., van Eijk, K., and Hermjakob, H. (2010) The IntAct molecular interaction database in 2010. Nucleic Acids Res., 38, pp. 525–531.
3. Barbosa-Silva, A., Fontaine, J. F., Donnard, E. R., Stussi, F., Ortega, J. M. and Andrade-Navarro, M. A. (2011) PESCADOR, a web-based tool to assist text-mining of biointeractions extracted from PubMed queries. BMC Bioinformatics, 12, p. 435.
4. Barbosa-Silva, A., Soldatos, T. G., Magalhães, I. L., Pavlopoulos, G. A., Fontaine, J. F., Andrade-Navarro, M. A., Schneider, R. and Ortega, J. M. (2010) LAITOR–Literature Assistant for Identification of Terms co-Occurrences and Relationships. BMC Bioinformatics, 11, p. 70.

5. Beuming, T., Skrabanek, L., Niv, M. Y., Mukherjee, P. and Weinstein, H. (2005) PDZBase: a protein-protein interaction database for PDZ-domains. Bioinformatics, 21, pp. 827–828.
6. Björne, J., Ginter, F., Pyysalo, S., Tsujii, J. and Salakoski, T. (2010) Complex event extraction at PubMed scale. Bioinformatics, 26, pp. i382–i390.
7. Bui, Q. C., Katrenko, S. and Sloot, P. M. (2011) A hybrid approach to extract protein-protein interactions. Bioinformatics, 27, pp. 259–265.
8. Ceol, A., Chatr, A. A., Licata, L., Peluso, D., Briganti, L., Perfetto, L., Castagnoli, L. and Cesareni, G. (2010) MINT, the molecular interaction database: 2009 update. Nucleic Acids Res., 38, pp. 532–539.
9. Chowdhary, R., Zhang, J. and Liu, J. S. (2009) Bayesian inference of protein-protein interactions from biological literature. Bioinformatics, 25, pp. 1536–1542.
10. Chowdhary, R., Zhang, J., Tan, S. L., Osborne, D., Bajic, V. B. and Liu, J. S. (2013) PIMiner: a web tool for extraction of Protein Interactions from Biomedical Literature. Int. J. Data Min. Bioinform, 7, pp. 450–462.
11. Chowdhary, R., Tan, S. L., Zhang, J., Karnik, S., Bajic, V. B. and Liu, J. S. (2012) Context-specific protein network miner - an online system for exploring context-specific protein interaction networks from the literature. PLoS One, 7, p. e34480.
12. Dogan, R., Yang, Y., Neveol, A., Huang, M. and Lu, Z. (2010) Identifying protein-protein interactions in biomedical text articles. *Proc. BioCreative III Workshop*, pp. 61–66.
13. Fontaine, J. F. and Navarro, M. A. (2010) Fast classification of scientific abstracts related to protein-protein interaction using a Naive Bayesian linear classifier. *Proc. BioCreative III Workshop*, pp. 67–72.
14. Gerner, M., Nenadic, G. and Bergman, C. M. (2010) An Exploration of mining gene expression mentions and their anatomical locations from biomedical text. *Proc. 2010 Workshop on Biomedical Natural Language Processing*, pp. 72–80.
15. Hakenberg, J., Plake, C., Leaman, R., Schroeder, M. and Gonzalez, G. (2008) Inter-species normalization of gene mentions with GNAT. Bioinformatics, 24, pp. 126–132.
16. Hakenberg, J., Gerner, M., Haeussler, M., Solt, I., Plake, C., Schroeder, M., Gonzalez, G., Nenadic, G. and Bergman, C. M. (2011) The GNAT library for local and remote gene mention normalization. Bioinformatics, 27, pp. 2769–2771.
17. Hanisch, D., Fundel, K., Mevissen, H. T., Zimmer, R. and Fluck, J. (2005) ProMiner: rule-based protein and gene entity recognition. BMC Bioinformatics, 6, p. S14.
18. Heguy, A., O'Connor, T., Luettich, K., Worgall, S., Cieciuch, A., Harvey, B. G., Hackett, N. R. and Crystal, R. G. (2006) Gene expression profiling of human alveolar macrophages of phenotypically normal smokers and nonsmokers reveals a previously unrecognized subset of genes modulated by cigarette smoking. J. Mol. Med. (Berl.), 84, pp. 318–328.

19. Hirschman, L., Colosimo, M., Morgan, A. and Yeh, A. (2005) Overview of BioCreAtIvE task 1B: normalized gene lists. BMC Bioinformatics, 6, p. S11.
20. Hoffmann, R. and Valencia, A. (2004) A gene network for navigating the literature. Nat. Genet., 36, p. 664.
21. Huang, M., Liu, J. and Zhu, X. (2011) GeneTUKit: a software for document-level gene normalization. Bioinformatics, 27, pp. 1032–1033.
22. Hunter, L., Lu, Z., Firby, J., Baumgartner, W. A., Johnson, H. L., Ogren, P. V. and Cohen, K. B. (2008) OpenDMAP: An open source, ontology-driven concept analysis engine, with application to capturing knowledge regarding protein transport, protein interactions and cell-type-specific gene expression. BMC Bioinformatics, 9, p. 78.
23. Iossifov, I., Rodriguez, E. R., Mayzus, I., Millen, K. J. and Rzhetsky, A. (2009) Looking at cerebellar malformations through text-mined interactomes of mice and humans. PLoS Comput. Biol., 5, p. e1000559.
24. Jose, H., Vadivukarasi, T. and Devakumar, J. (2007) Extraction of protein interaction data: a comparative analysis of methods in use. EURASIP J. Bioinform. Syst. Biol., p. 53096.
25. Kabiljo, R., Clegg, A. B. and Shepherd, A. J. (2009) A realistic assessment of methods for extracting gene/protein interactions from free text. BMC Bioinformatics, 10, p. 233.
26. Kemper, B., Matsuzaki, T., Matsuoka, Y., Tsuruoka, Y., Kitano, H., Ananiadou, S. and Tsujii, J. (2010) PathText: a text mining integrator for biological pathway visualizations. Bioinformatics, 26, pp. i374-381.
27. Keshava, P., Goel, R., Kandasamy, K., Keerthikumar, S., Kumar, S., Mathivanan, S., Telikicherla, D., Raju, R., Shafreen, B., Venugopal, A., Balarishnan, L., Marimuthu, A., Banerjee, S., Somanathan, D. S., Sebastian, A., Rani, S., Ray, S., Harrys Kishore, C. J., Kanth, S., Ahmed, M., Kashyap, M. K., Mohmood, R. Ramachandra, Y. L., Krishna, V., Rhiman, B. A., Mohan, S., Ranganathan, P., Ramabadran, S., Chaerkady, R. and Pandy, A. (2009) Human Protein Reference Database–2009 update. Nucleic Acids Res., 37, pp. 767–772.
28. Kim, S., Shin, S. Y., Lee, I. H., Kim, S. J., Sriram, R. and Zhang, B. T. (2008) PIE: an online prediction system for protein-protein interactions from text. Nucleic Acids Res., 36, pp. W411–W415.
29. Kim, S. and Wilbur, W. J. (2010) Improving protein-protein interaction article classification performance by utilizing grammatical relations. *Proc. BioCreative III Workshop*, pp. 83–88.
30. Krallinger, M., Vazquez, M., Leitner, F., Salgado, D., Chatr-Aryamontri, A., Winter, A., Perfetto, L., Briganti, L., Licata, L., Iannuccelli, M., Castagnoli, L., Cesareni, G., Tyers, M., Schneider, G., Rinaldi, F., Leaman, R., Gonzalez, G., Matos, S., Kim S., Wilbur, W. J. Rocha, L., Shatkay, H., Tendulkar, A. V., Agrawal, S., Liu, F., Wang X., Rak, R., Noto, K., Elkan, C., Lu, Z., Dogan, R. I., Fontaine, J. D., Andrade-Navarro, M. A. and Valencia, A. (2011) The Protein-Protein Interaction tasks of

BioCreative III: classification/ranking of articles and linking bio-ontology concepts to full text. BMC Bioinformatics, 12, p. S3.

31. Kuhn, M., Mering, C., Campillos, M., Jensen, L. J. and Bork, P. (2008) STITCH: interaction networks of chemicals and proteins. Nucleic Acids Res., 36, pp. 684–688.
32. Leaman, R. and Gonzalez, G. (2008) BANNER: an executable survey of advances in biomedical named entity recognition. Pac. Symp. Biocomput., pp. 652–663.
33. Leaman, R., Sullivan, R. and Gonzalez, G. (2010) A top-down approach for finding interaction detection methods. *Proc. BioCreative III Workshop*, pp. 99–104.
34. Lourenco, A., Conover, M., Wong, A., Pan, F., Abi-Haidar, A., Nematzadeh, A., Shatkay, H. and Rocha, L.. (2010) Testing Extensive Use of NER tools in Article Classification and a Statistical Approach for Method Interaction Extraction in the Protein-Protein Interaction Literature. *Proc. BioCreative III Workshop*, pp. 113–118.
35. Matos, S., Campos, D. and Oliveira, J. L. (2010) Vector-space models and terminologies in gene normalization and document classification. *Proc. BioCreative III Workshop*, pp. 119–124.
36. Mika, S. and Rost, B. (2004) NLProt: extracting protein names and sequences from papers. Nucleic Acids Res., 32, pp. W634–W637.
37. Morgan, A. A., Lu, Z., Wang, X., Cohen, A. M., Fluck, J., Ruch, P., Divoli, A., Fundel, K., Leaman, R., Hakenberg, J., Sun, C., Liu, H. H., Torres, R., Krauthammer, M., Lau, W. W., Liu, H., Hsu, C. N., Schuemie, M., Cohen, K. B. and Hirschman, L. (2008) Overview of BioCreative II gene normalization. Genome Biol., 9, p. S3.
38. Needham, C. J., Bradford, J. R., Bilpitt, A. J. and Westhead, D. R. (2007) A primer on learning in Bayesian networks for computational biology. PLoS Comput. Biol., 3, p. e129.
39. Neves, M. L., Carazo, J. M. and Pascual, M. A. (2010) Moara: a Java library for extracting and normalizing gene and protein mentions. BMC Bioinformatics, 11, p. 157.
40. Okazaki, N., Cho, H. C., Sætre, R., Pyysalo, S., Ohta T. and Tsujii, J. (2010) The gene normalization and interactive systems of the University of Tokyo in the BioCreative III challenge. *Proc. BioCreative III Workshop*, pp. 125–130.
41. Pagel, P., Kovac, S., Oesterheld, M., Brauner, B., Dunger-Kaltenbach, I., Frishman, G., Montrone, C., Mark, P., Stümpflen, V., Mewes, H. W., Ruepp, A. and Frishman, D. (2005) The MIPS mammalian protein-protein interaction database. Bioinformatics, 21, pp. 832–834.
42. Palmer, C. N., Ismail, T., Lee, S. P., Terron-Kwiatkowski, A., Zhao, Y., Zhao, Y., Liao, H., Smith, F. J., McLean, W. H. and Mukhopadhyay, S. (2007) Filaggrin null mutations are associated with increased asthma severity in children and young adults. J. Allergy Clin. Immunol., 120, pp. 64–68.
43. Ponińska, J., Samolinski, B., Tomaszewska, A., Raciborski, F., Samel-Kowalik, P., Walkiewicz, A., Lipiec, A., Piekarski, B., Komorowski, J., Krzych-Falta, E.,

Namysłowski, A., Borowicz, Kostrzewa, G., Majewski, S. and Płoski, R. (2011) Filaggrin gene defects are independent risk factors for atopic asthma in a Polish population: a study in ECAP cohort. PLoS One, 6, p. e16933.
44. Rebholz-Schuhmann, D., Arregui, M., Gaudan, S., Kirsch, H. and Jimeno, A. (2008) Text processing through Web services: calling Whatizit. Bioinformatics, 24, pp. 296–298.
45. Rogers, A. J., Celedón, J. C., Lasky-Su, J. A., Weiss, S. T. and Raby, B. A. (2007) Filaggrin mutations confer susceptibility to atopic dermatitis but not to asthma. J Allergy Clin. Immunol., 120, pp. 1332–1337.
46. Rzhetsky, A., Seringhaus, M. and Gerstein, M. (2008) Seeking a new biology through text mining. Cell, 134, pp. 9–13.
47. Saetre, R., Sagae, K. and Tsujii, J. (2008) Syntactic features for protein-protein interaction extraction. *Proc. 2nd Int. Symposium on Languages in Biology and Medicine*, pp. 6.1–6.14.
48. Saetre, R., Yoshida, K., Miwa, M., Matsuzaki, T., Kano, Y. and Tsujii, J. (2010) Extracting protein interactions from text with the unified AkaneRE event extraction system. IEEE/ACM Trans. Comput. Biol. Bioinform., 7, pp. 442–453.
49. Schuttelaar, M. L., Kerkhof, M., Jonkman, M. F., Koppelman, G. H., Brunekreef, B., de Jongste, J. C., Wijga, A., McLean, W. H. and Postma, D. S. (2009) Filaggrin mutations in the onset of eczema, sensitization, asthma, hay fever and the interaction with cat exposure. Allergy, 64, pp. 1758–1765.
50. Smith, B., Ashburner, M., Rosse, C., Bard, J., Bug, W., Ceusters, W., Goldberg, L. J., Eibeck, K., Ireland, A., Mungall, C. J., OBI Consortium, Leontis, N., Rocca-Serra, P., Ruttenberg, A., Sansone, S. A., Scheuermann, R. H., Shah, N., Whetzel, P. L. and Lewis, S. (2007) The OBO Foundry: coordinated evolution of ontologies to suppor biomedical data integration. Nat. Biotechnol., 25, pp. 1251–1255.
51. Smith, L., Tanabe, L. K., Ando, R. J., Kuo, C. J., Chung, I. F., Hsu, C. N., Lin, Y. S., Klinger, R., Friedrich, C. M., Ganchev, K., Torii, M., Liu, H., Haddow, B., Struble, C. A., Povinelli, R. J., Vlachos, A., Baumgartner, W. A. Jr., Hunter, L., Carpenter, B., Tsai, R. T., Dai, H. J., Liu, F., Chen, Y., Sun, C., Katrenko, S., Adriaans, P., Blascke, C., Torres, R., Neves, M., Nakov, P., Divoli, A., Maña-López, M., Mata, J. and Wilbur, W. J.. (2008) Overview of BioCreative II gene mention recognition. Genome Biol., 9, p. S2.
52. Solt, I., Gerner, M., Thomas, P., Nenadic, G., Bergman, C. M., Leser, U. and Hakenberg, J. (2010) Gene mention normalization in full texts using GNAT and LINNAEUS. *Proc. BioCreative III Workshop*, pp. 143–148.
53. Stark, C., Breitkreutz, B. J., Reguly, T., Boucher, L., Breitkreutz, A. and Tyers, M. (2006) BioGRID: a general repository for interaction datasets. Nucleic Acids Res., 34, pp. 535–539.
54. Szklarczyk, D., Franceschini, A., Kuhn, M., Simonovic, M., Roth, A., Minguez, P., Doerks, T., Stark, M., Muller, J., Bork, P., Jensen, L. J. and von Mering, C. (2011) The STRING database in 2011: functional interaction networks of proteins, globally integrated and scored. Nucleic Acids Res., 39, pp. 561–568.

55. Tikk, D., Thomas, P., Palaga, P., Hakenberg, J. and Leser, U. (2010) A comprehensive benchmark of kernel methods to extract protein-protein interactions from literature. PLoS Comput. Biol., 6, p. e1000837.
56. Tsuruoka, Y., Tateishi, Y., Kim, J. D., Ohta, T., McNaught, J., Ananiadou, S. and Tsujii, J. (2005) Developing a robust part-of-speech tagger for biomedical text. Advances in Informatics, *Proc. 10th Panhellenic Conference on Informatics*, 3746, pp. 382–392.
57. Wei, C. H., Huang, I. C., Hsu, Y. Y. and Kao, H. Y. (2009) Normalizing biomedical name entities by similarity-based inference network and de-ambiguity mining. *Proc. Ninth IEEE International Conference on Bioinformatics and Bioengineering*, pp. 461–466.
58. Weidinger, S., O'Sullivan, M., Illig, T., Baurecht, H., Depner, M., Rodriguez, E., Ruether, A., Klopp, N., Vogelberg, C., Weiland, S. K., McLean, W. H., von Mutius, E., Irvine, A. D. and Kabesch, M. (2008) Filaggrin mutations, atopic eczema, hay fever, and asthma in children. J. Allergy Clin. Immunol., 121, pp. 1203–1209.
59. Wermter, J., Tomanek, K. and Hahn, U. (2009) High-performance gene name normalization with GENO. Bioinformatics, 25, pp. 815–821.
60. Wong, L. and Liu, G. (2010) Protein interactome analysis for countering pathogen drug resistance. J. Comp. Sci. Tech., 25, pp. 124–130.
61. Wren, J. D., Bekeredjian, R., Stewart, J. A., Shohet, R. V. and Garner, H. R. (2004) Knowledge discovery by automated identification and ranking of implicit relationships. Bioinformatics, 20, pp. 389–398.
62. Xu, H., Fan, J. W., Hripcsak, G., Mendonça, E. A., Markatou, M. and Friedman, C. (2007) Gene symbol disambiguation using knowledge-based profiles. Bioinformatics, 23, pp. 1015–1022.
63. Yu, N., Seo, J., Rho, K., Jang, Y., Park, J., Kim, W. K. and Lee, S. (2012) hiPathDB: a human-integrated pathway database with facile visualization. Nucleic Acids Res., 40, D797–D802.
64. Zhang, L., Berleant, D., Ding, J., Cao, T. and Syrkin Wurtele, E. (2009) PathBinder—text empirics and automatic extraction of biomolecular interactions. BMC Bioinformatics, 10, p. S18.
65. Zhou, S., Potts, E. N., Cuttitta, F., Foster, W. M. and Sunday, M. E. (2011) Gastrin-releasing peptide blockade as a broad-spectrum anti-inflammatory therapy for asthma. Proc. Natl. Acad. Sci. U. S. A., 108, pp. 2100–2105.

Index